I0576705

Henry Scherren

Popular history of animals for young people

With 13 coloured plates and numerous illustrations in the text

Henry Scherren

Popular history of animals for young people
With 13 coloured plates and numerous illustrations in the text

ISBN/EAN: 9783337228781

Printed in Europe, USA, Canada, Australia, Japan

Cover: Foto ©berggeist007 / pixelio.de

More available books at **www.hansebooks.com**

POPULAR

HISTORY OF ANIMALS

FOR

YOUNG PEOPLE

BY

HENRY SCHERREN, F.Z.S.

AUTHOR OF "PONDS AND ROCK POOLS"

*WITH 13 COLOURED PLATES AND NUMEROUS ILLUSTRATIONS
IN THE TEXT*

PHILADELPHIA: J. B. LIPPINCOTT COMPANY
LONDON: CASSELL AND COMPANY, LIMITED
1895

PREFACE.

THE object of this book is to give a short account of the Animal Kingdom in clear and simple language. The book being intended chiefly for young people, no formal classification has been given, and popular names have been used throughout. But the main divisions of the Animal Kingdom have been plainly indicated ; and modern classification has been practically followed. The Author's aim has been to write in such fashion that the book may serve to waken, or quicken, interest in the observation of the habits of the lower animals, and as an introduction to the study of their relations to us and to each other.

H. S.

CONTENTS.

LIST OF COLOURED PLATES.

For the numerous Illustrations in the Text, *see* INDEX at the end.

POPULAR HISTORY OF ANIMALS

FOR

YOUNG PEOPLE.

———•••———

CHAPTER I.

HOW ANIMALS ARE CLASSIFIED.

IN dealing with Natural History, the first requisite is a classification or arrangement of some kind. This need not be
elaborate ; but, to be of real service, it must be based on
right principles.

Resemblance in external form and similarity in habits
were formerly taken as a guide, and led to many wrong conclusions—
such as classing the hyrax among the Rodents and the kinkajou among
the Lemurs. Nor was the element in which an animal lived a safer
means of judging, for it led Pliny to put the whales, which Jeremiah*
knew to be Mammals, with the Fishes ; and even down to the middle
of the seventeenth century naturalists classed the Bats with Birds,
till John Ray taught them better.

The principle now adopted is that of relationship, which teaches
that all forms of life at present existing have arisen from simpler
forms—these, in their turn, being derived from other still more simple ;
and so backwards, till the period when the only life on this planet
was represented by forms as lowly as the Amœba.

Hence, schemes of classification now set forth our knowledge, so
far as it goes, of the relationship of animals to each other, and in
many cases take the actual shape of a genealogical tree, in which the
principal groups are represented by branches, giving off smaller ones,
from which branchlets and twigs arise, representing the divisions of
the principal groups.

It was formerly the practice to divide all animals into two groups—
Vertebrates and Invertebrates—according as they did, or did not, possess
a backbone. And for a long time this division seemed to work very well,
till increase of knowledge made it clear that among the Invertebrates there
were some which showed more or less definite traces of a backbone,

---

* "Even the sea monsters draw out the breast, they give suck to their young ones."—
Lamentations iv. 3.

B

or something very much like one; and that in some of the creatures which had a backbone it was not divided into vertebræ, or joints. In others, again, as the Shark and the Skate, the spinal column is gristly in substance, and not bony.

This gives us three groups, instead of two, to deal with :—

Animals with a backbone (Man, Monkeys, Lions, Elephants, Whales, Birds, Reptiles, Frogs and Newts, and Fishes).

Animals with traces of a backbone (the Lancelets, Tunicates, or Sea-squirts, and Acorn-worms).

Animals without a backbone or traces of one (Cuttle-fish, Spiders, Insects, Crabs and Lobsters, Starfish and Sea-urchins, "Worms," Sponges, Stinging Animals, and Primitive Animals, or Animalcules).

Smaller groups of animals which, though they differ in many respects from each other, possess some common character not found in the rest, constitute *Classes*. Thus, the Beasts of Prey, the Whales, and the Bats agree in this, that the young are suckled by their mothers. These animals, and all others whose young are nourished with their mothers' milk, constitute the class of Mammals. Eagles, Ostriches, and Pigeons agree in that they are clothed with feathers; and, with all other animals similarly clad, make up the class of Birds.

Then among the Mammals it is easy to distinguish several main groups. The Cats, Dogs, and Bears subsist chiefly by preying on animals of the same Class; Horses, Oxen, and Deer have the toes encased in hoofs; Beavers, Rabbits, and Mice have the front teeth adapted for gnawing, and so on. This division gives us the *Orders*—Flesh-eaters or Beasts of Prey, Hoofed Mammals, Gnawers or Rodents, etc.

If we consider the structure and habits of the Cats (Lion, Tiger, Leopard, etc.), and the Dogs and Wolves, points of agreement will be found common to the Tiger and Leopard, and the Domestic Cat, which do not exist in the Dogs and Wolves. Hence the former are placed in the Cat *family*, and the latter in the Dog *family*. Then further, in the Cat family is one form — the Cheetah, or Hunting Leopard—whose relationship to the rest is more distant than that subsisting between the other members. We may say that Lions, Tigers, Leopards, Pumas and Jaguars, and the smaller Cats, are brothers, while the Cheetah is only a cousin. This difference of relationship is expressed by putting the True Cats into the Cat *genus*, and the Cheetah into another *genus*, and these two constitute the Cat family.

The final unit of classification is the *Species*. This term is difficult to define, but it may be taken to denote "a number of animals so closely resembling each other that they might be supposed to be the offspring of the same parents, and in turn giving birth to animals like themselves."

VERTEBRATES.

Having thus mapped out the ground over which we are to travel, our next step will be to gain some idea of the plan of a Vertebrate, or backboned animal.

Most children possess a strange fancy for covering slates and the covers of copy-books with "drawings." Many of these are "animal" subjects, treated in what artists call a conventional manner : that is, it is generally understood that such or such a figure represents such or such an animal- for example, a horse or a lion. A few strokes on the top of the head give it horns, and make it into a cow or a buffalo. Sometimes a legend is put beneath in clear print hand, "THIS IS A COW," to prevent the possibility of mistake.

PLAN OF A VERTEBRATE SKELETON.

Some of the earliest attempts bear close resemblance to the above— a diagram rather than a picture. Nevertheless, it will serve our purpose quite as well as, or even better than, the most elaborate anatomical drawing : for while that, by reason of its correctness, would only serve for an individual, our diagram, with a little imaginative modification, will do duty for the skeleton of a Lamprey, a Fish, an Amphibian, a Reptile, a Bird, or a Mammal.

The horizontal line represents the backbone, or vertebral column, which forms the chief internal support of the body, and which is called the axial skeleton. It is made up of a number of separate bones, as we may see for ourselves when a hare or rabbit, fowl, or fish, is sent up to table. These bones are fitted together, with a gristly pad between them, so as to allow of free motion ; and this arrangement enables us to bend our backs and turn our heads. Most of these bones are perforated by a hole or canal, through which runs the spinal cord, terminating at the front end or top in a big mass—the brain, enclosed for protection in the brain-box, or skull.

The four strokes, forming two angles, may well stand for the limbs, which are never more than four in number in any backboned animal ; though they may be reduced to two, as in some lizards and in the whales ;

B 2

they may be altogether absent, as in the slow-worm, snake, and viper of our copses and plantations; or of the two pairs, one pair may be but partially developed, as in the so-called wingless birds of New Zealand.

The outstanding portion of the figure at the end opposite the head is, it is hardly necessary to say, the tail, the bony framework for which exists at some period of life, though this appendage is not possessed by Man, some monkeys, Manx cats, and guinea-pigs. The dog uses his tail to show that he is pleased; horses and cows use the tail as a fly-flapper; birds and fishes as a rudder, and it also serves the last-mentioned animals as a natural screw-propeller; while to some monkeys it is almost as useful as an additional hand would be.

We shall be able to test the truth of these statements by examining our own bodies, or by handling the family cat, or a pet rabbit. It will be sufficient to run the hand gently down the back from the neck to the tail, to assure ourselves of the continuity of the backbone; and in the same way we may feel that the limbs and skull are connected with it. We must, however, notice that the limbs are turned away from the main nervous system—running through the backbone, and remember that this arrangement is universal in Vertebrate Animals. But though we shall generally find four limbs in a Vertebrate or Back-boned Animal, they are not always of the same shape, nor are they always used for the same purpose. The arms of a man correspond to the forelegs of a horse, a lion, an elephant, a lizard, or a frog; to the wings of a bird or a bat; and to the pair of fins that are called *pectoral*, and generally situated near the head, in fishes, just as the *ventral* pair correspond to his legs. The limbs differ also in their use in the lower Vertebrates: generally they serve for locomotion, to carry their owner from place to place; but sometimes one pair and sometimes another are modified into grasping organs and fulfil the purpose of hands. Everyone has seen a squirrel sit up and nibble a piece of biscuit which he holds in his forepaws, while a parrot will use its foot to convey a dainty morsel to its mouth.

In some cases where, as in the Boas, there are no external limbs, there are internal traces of one pair—foreshadowings of what was to come; or these traces, as in the Whales and Dolphins, may represent limbs lost through disuse. This question of loss through disuse is very important; for it shows that, while movement upward is the general law of Nature, there may also be degeneration, or movement backwards and downwards. On this point it is well to read what Kingsley says in his "Water-Babies" of the Doasyoulikes, who left the country of Hardwork for the land of Readymade, at the foot of the Happy-go-Lucky Mountains, where flapdoodle grows wild. Of course, "Water-Babies" is only a fairy tale : the author said so, and he ought to know. But it is a fairy tale with a good deal of truth in it, and some excellent natural history into the bargain.

MAMMALS.

The importance of Mammals to Man is greater than that of any other group of animals, and chief among the class, in this respect, stand the Ungulates, or Hoofed Mammals, containing the Horses, Oxen, Sheep, and Goats. Some of these serve as beasts of draught and burden, others for food, and when dead their skin, hair, wool, hoofs, etc., are all turned to good account.

We shall get a good idea of the bony framework of a Mammal from the figure herewith, which represents the skeleton of a camel. The general plan should be compared with the rough diagram on page 3, and the bones with the human skeleton and its parts on pages 8 to 11.

All Mammals have warm red blood, and breathe by lungs ; gills are never developed. Except in the Duck Mole and Spiny Anteaters, the young are brought forth alive, and during their growth they do not undergo any change or metamorphosis, like that of frogs and newts: in other words, there is no larval stage. The young of the Pouched Mammals are not fully developed when they come into the world, and most of them pass some time in the pouch of the mother. But the new-born young of all are nourished with milk, secreted by the mother, and from this circumstance the name of the class is derived.

Another characteristic of this class is the hairy covering of the skin. This is complete, or nearly so, in most Mammals, but extremely scanty in the Whales, being limited to a few bristles round the mouth, and even these disappear when the animals become full grown. Hair is often of two kinds—one long and stiff, that appears on the surface ; the other short,

SKELETON OF CAMEL.

a, Skull ; 1, shoulder-blade (*scapula*) ; 2, arm (*humerus*) ; 3, fore-arm (*ulna*, or *cubitus*) ; 4, wrist (*carpus*) : 5, *metacarpus*, corresponding to the human palm ; 6, digits, corresponding to the fingers ; 7, thigh-bone (*femur*) ; 8, leg (*tibia*) ; 9, ankle (*tarsus*), the bone that stands out behind is the heel ; 10, *metatarsus*, corresponding to the sole of the human foot ; 11, digits, corresponding to the human toes : 12, cervical vertebræ ; 13, dorsal vertebræ ; 14, lumbar vertebræ ; 15, sacral vertebræ, 16, caudal vertebræ ; 17, ribs ; 18, pelvis.

soft, and downy, and called fur. A very good example of the two kinds of hair is seen in the fur seal, so many of which are killed that their skins may be made into jackets and mantles for the ladies of Europe and America. In Pigs, the hairs form stiff bristles, and in Hedgehogs and

HAIR OF FUR SEAL.
s, Skin; f, fur; h, hairs.

Porcupines, they are so thick as to form spines, those of the latter animals being popularly known as "quills." In the Scaly Ant-eaters, the body is covered with scales, and in the Armadillos with bony plates; but between these scales and plates true hairs grow more or less thickly. Hair, like that of the Sheep, which "felts," or forms a compact mass, owing to its surface being covered with minute scales, is called wool. In Man, the hairy covering, except on the head, is generally scanty; but the natural clothing of the Ainu of Yezo is so thick that they are generally spoken of as the "hairy" Ainu. The object of this covering is, of course, warmth; and its loss in the Whales is made up by a thick layer of fat, called "blubber," immediately beneath the skin. In Mammals inhabiting cold regions, the coat generally becomes much thicker in winter, falling off again in summer, and there is in many cases a change in the colour at this season. Thus, the Ermine, which yields such valuable fur, is white in winter (at which time it is hunted for its skin), and brown in summer.

Concerning the remaining classes of Vertebrates, we have already seen (page 3) that a common plan of structure runs through them all, from the highest to the lowest. What has been said about Mammals, and comparison of the skeleton of the camel (page 5) with the human skeleton (page 8), with the short descriptions prefixed to the other Classes and their Orders, will enable us to discover wherein Birds, Reptiles, Amphibians, Fishes, and Lampreys agree with or differ from the Mammals. And this also holds good with respect to the dwellers in the Borderland.

INVERTEBRATES.

With the Invertebrates it is quite different. This immense group has no common type, but comprises a collection of sub-kingdoms offering a strange diversity of plan. Yet, even through these, relationships can be traced more or less clearly, as will be seen later on when dealing with each sub-kingdom.

CHAPTER II.

MAN.

N Natural History it is usual to give to the separate parts of all Vertebrate Animals the same names as are applied to similar parts of the human body. For this reason, we must examine the human skeleton on the next page rather closely.

This bony framework is a very complicated piece of mechanism, and consists of more than 200 separate bones. There are two principal parts—the one corresponding to the straight line in our rough diagram (page 3), and called the *axial* skeleton, because like an axis, or rod, it runs down between the two halves (right and left) of the body; the other, corresponding to the Λ-shaped marks, and called the *appendicular* skeleton, consisting of the appendages of the body—the arms and legs, generally termed the limbs. At the top of the axial skeleton is the brain-box, or skull, forming, with the bones of the face, the skeleton of the head.

If a perpendicular line be drawn through the skull, backbone, and pelvis, the skeleton would be divided into a right half and a left half, the bones on one side corresponding to those of the other. This is called bilateral symmetry, and will be found in all backboned animals.

The limbs are not only in pairs—two arms and two legs—but there is a correspondence in the bones of the pectoral pair (the arms) with those of the pelvic pair (the legs). The arm and the thigh have each a single bone, while in the fore-arm and leg there are two; the bones of the wrist correspond broadly with those of the ankle; and the likeness between our fingers and toes is a matter of common knowledge. These, of course, are five in number on each limb; and this is the case with most Monkeys, many of the Flesh-eaters, etc. The foot is planted flat on the ground; and the great toe (in civilised Man, at least) is in a line with the other toes, and is never used as a thumb.

There is a similarity also in the method by which the pairs of limbs are joined to the trunk: the arms are jointed to shoulder-blades, which are connected with the breast-bone by the collar-bones, and with the trunk by powerful muscles; the thighs are received in sockets on each side of the pelvis, the bones of which are fused together and united with the sacral vertebræ (*see* Spinal Column, page 9). These means of attachment are called the shoulder and hip girdles.

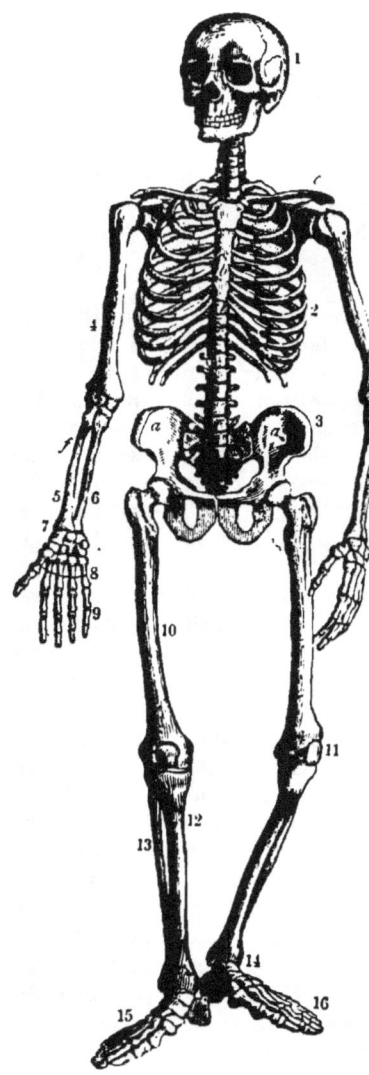

HUMAN SKELETON.

1, Skull (*cranium*); 2, thorax; 3, breast-bone ; c, collar bone (*clavicle*); 3, pelvis a, a, haunch bone ; 4, arm (*humerus*), 5 and 6, bones of fore-arm (*ulna* and *radius*); 7, wrist (*carpus*) ; 8, bones of palm (*metacarpus*) ; 9, bones of fingers (*phalanges*); 1 , thigh-bone (*femur*); 11, knee-pan (*patella*); 12 and 13, bones of leg (*tibia* and *fibula*); 14 ankle (*tarsus*); 15, bones of foot (*metatarsus*); 16, bones of toes (*phalanges*).

The backbone, or spine, consists of a series of bones placed one above another, and called vertebræ, or turning-bones (from the Latin *verto*=to turn), because each can turn a little, as when we bend the body from side to side. In early life there are thirty-three distinct bones, each made up of separate pieces, which become united in the adult.

The spine is marked off into five separate parts or regions (page 9). The region of the neck has seven (cervical) vertebræ ; the region of the back (dorsal) twelve ; and the region of the loins (lumbar) seven. These twenty-four bones can all be separated from each other, and on that account are called *true* vertebræ. Those of the sacral and coccygeal regions (both of which are fused into distinct bony masses in the adult) are called *false* vertebræ. Thus, in the grown man the spine consists of but twenty-six separate bones. In the infant this bony support is straight, but as we grow older it is bent into a series of curves, corresponding to the regions into which it is divided.

These curves are of great use : (1) By their means the spine can bear a much greater weight on the head and shoulders than it otherwise could, the proportion of the strength of this curved column to one perfectly straight being as 9 to 1. (2) They render the movements of the body, especially when running, much more easy. (3) In the movement of the column this arrangement protects the spinal cord, which joins the brain above, and is continued below, giving off nerves on each side in its course, as far as the lower part of the first vertebra of the loins, below which it dwindles into a bundle of white thread-like nerves. The Man-like Apes approach Man most nearly in the curves of the spine; but even in them the curve of the loins is not so strongly marked, and this is due to Man's erect position.

The regions marked in the figure correspond to those of Mammals generally; and in the neck region, whether it be long as in the giraffes, or short as in the whale or the porpoise, the number of vertebræ is always seven, excepting in the three-toed Sloths, which have nine, the Scaly Ant-eaters, which sometimes have eight, and of a two-toed Sloth and the Manatee, which have six. The number of vertebræ in the other regions differs greatly in the lower Mammals.

The last region in the human spine is often spoken of as the " coccygeal " region; the term *caudal* (or tail) region being reserved for Mammals with a free tail. This is perhaps due to the desire to mark off Man from the rest of the class to which, as an animal, he belongs; or perhaps because "tailed" men are not of frequent occurrence. A scientific journal, however, has said that there are probably always a few men

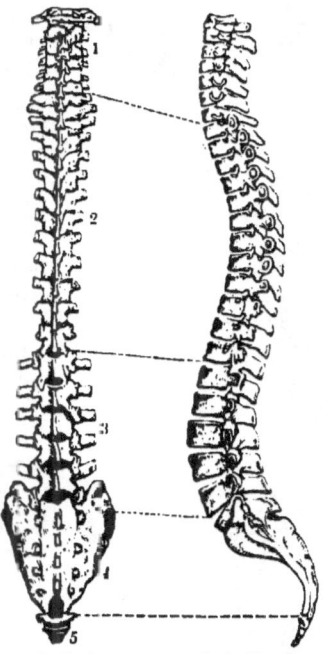

Back view.　　　Left side view.

SPINAL COLUMN.

1, Region of the neck (*cervical*); 2, region of the back (*dorsal*); 3, region of the loins (*lumbar*); 4, region of the haunches (*sacral*); 5, region of the tail (*coccygeal* or *caudal*).

living in whom a free tail has been developed; and in his last book Dr. Oliver Wendell Holmes tells us that during the session of the Medical Congress at Washington, a distinguished London physician showed him the photograph of a small boy, some three or four years old, "who had a very respectable little tail, which would have passed muster on a pig, and would have made a frog or a toad ashamed of himself."

The top, or highest vertebra is called the atlas, because it supports the head, just as the mythic leader of the Titans was supposed to support the heavens. The bony groove on each side receives a corresponding projection or pivot on the occipital bone (Human Skull, *o*).

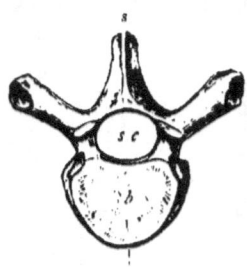

SIXTH VERTEBRA
OF THE BACK FROM ABOVE.
b, Body ; *s c,* spinal canal ;
s, spinous process.

The vertebra just below is called the axis, because in front is a strong bony peg, which fits into the atlas, and serves to turn the head, for when we look round, not only is the head turned, but the atlas is turned with it. In Birds and Reptiles there is but a single pivot in the skull, while the Amphibians (Frogs and Newts) have two like the Mammals. The rest of the lower vertebræ do not differ greatly from each other, except in point of size, growing larger and stronger until the last of the twenty-four rests on the solid mass of the vertebræ of the haunches, which forms the keystone of the pelvic arch. Each of these vertebræ consists of a front portion, called the *body,* and behind of a ring of bone, enclosing the spinal canal, through which runs the spinal cord. The ring is furnished with various projections of bone, which form joints, so to speak, with the other vertebræ, and also serve for the attachments of muscles. These spinous processes may be easily felt, if one puts his hand between his shoulders and draws it downwards along the spine. The prominent bone at the back of the neck is the spinous process of the seventh (and last) vertebra of that region, which ends in a rounded knob.

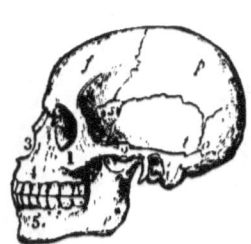

HUMAN SKULL.

f, Frontal bone ; *p,* parietal bone (one on each side) ; *t,* temporal bone (one on each side) ; *o,* occipital bone ; *s,* sphenoid bone (which forms part of the floor of the skull, and also projects on the right side) ; 1, left cheek-bone ; 2, zygomatic arch ; 3, bone of the nose ; 4, left half of upper jaw ; 5, lower jaw.

The head has twenty-two bones, and its skeleton is called the skull. Of these, eight form the brain-box, or skull proper, and the remaining fourteen make up the bones of the face. Of all these only one is movable—that of the lower jaw, which is employed in masticating food, and sometimes less usefully in talking.

From the figure we shall be able to make out the principal bones of the skull and face (the latter are marked with figures). The ethmoid, or sieve-like bone, below the frontal bone, and at the back of the face, is hidden. The bones of the skull are fitted together by uneven edges, somewhat like the teeth of a saw, and the joinings, some of which may be traced on the figure, are called sutures. The bony arch binds the face bones to those of the skull, and serves for the attachment of muscles. In many Mammals the bones of the upper jaw are four instead of two, the parts carrying the cutting-teeth being distinct from the parts carrying the other teeth. In Man and some Monkeys these bones form a single one on each side. The two branches of the lower jaw in man are united to form a single horseshoe-shaped bone, but in some Mammals they are bound together by ligaments. On the lower jaw on each side is a projecting part, or pivot, which fits loosely into a hole in the temporal bone, thus giving great freedom of motion. The human jaw can be moved directly up and down as in the Flesh-eating Mammals, backwards and forwards as in the Gnawing Animals, or Rodents, and it has also the circular motion so noticeable in sheep and cattle when they are chewing the cud.

Man has two sets of teeth. The first, called the milk set, consists of twenty teeth, and is replaced in later life by the permanent set, in which there are thirty-two. These teeth are of three kinds—cutting, or incisor teeth; tearing, or canine teeth—so called from their being of great size in the Dogs: and the grinding teeth, or molars. In early childhood we have

DIAGRAM SHOWING THE TEETH IN THE UPPER JAW; A, IN CHILDHOOD; B, IN MANHOOD.

I, Incisor teeth; c, canine teeth; P, premolars; M, molars.

but two grinders, replaced by five in later life. The two nearest the canine teeth are called premolars, and the three behind them molars, or "true" molars. The figures on the teeth in the diagram denote the age (in years above and in months below) at which the teeth generally appear. With the teeth of Man those of most other Mammals may be compared, and the same names are used for the teeth of the lower animals. Gnawing Animals, or Rodents, have no canines; and the sharp-pointed molars of the Flesh-eaters differ greatly in shape from those of the Hoofed Mammals, which live on vegetable food.

The chest, or thorax, is formed behind by the vertebræ of the back, and in front by the breast-bone, while the ribs connect these two columns. Of the twelve pairs of ribs, seven, called the *true* ribs, are united to the breast-bone by gristly bands (called the costal cartilages). Of the remaining five pairs, called *false* ribs, three pairs are joined by similar bands to the ribs immediately above, and the two lowest pairs, which are quite free in front, are also called floating ribs. To the top of the breast-bone the clavicles are attached at one end, the other being connected with the shoulder-blade, the whole forming the shoulder girdle, which serves to connect the fore (or pectoral) limbs with the body.

The office of the chest is to contain the heart and lungs, two of the chief centres of life—the former the engine which sends the blood circulating through the body, and the latter the organs of breathing or respiration. In the latter process the chest has important work to do, the front and side walls moving regularly up and down—that is, becoming expanded and contracted as air passes into and out of the nostrils and mouth.

The pelvis (so-called from its basin-like shape) serves to support the trunk, and to give attachment to the lower limbs. The haunch-bones are united in front, and between them, behind, the bony mass of the five sacral vertebræ is wedged in like the keystone of an arch. The name "pelvis" is given not only to the cavity, but to the bones which bound it, the latter being also called the pelvic arch or pelvic girdle. This arch or girdle is universally present, though in different degrees of completeness, in Mammals, and in the Whales and Manatees is represented by two small bones. In the Pouched Mammals, Duck Mole, and Spiny Ant-eaters, two small bones project from the front part of the pelvis, and in the females generally give support to a pouch.

Man belongs to the Order Primates, which also contains the Man-like Apes, the Monkeys, and the Lemurs. The great characteristic of the other members of the order is their more or less close resemblance in shape to Man. There are usually five fingers and five toes, but the thumb may be small or altogether wanting. The great toe generally bears a flat nail, and may very often be used as a thumb, so that the foot becomes a grasping organ. It was formerly the fashion to mark Man off in a separate order, named Bimana, or animals with two hands, as distinct from the Man-like Apes, Monkeys, and Lemurs, which were called Quadrumana, or animals with four hands; but this distinction is now abandoned. The foot of a child is, to some extent, a grasping organ, as it is also among adults of some of the lower races. The Australian savage can pick up his spear with his toes; and Indian workmen use the foot in hand-like fashion to hold their work. Europeans, who have the feet covered, have lost the power of thus employing the foot; but very little practice will enable a boy of not more than average patience to pick up a pencil from the floor with his naked toes. There are three kinds of teeth, except in the Aye-Aye, which has no canines.

Science tells us in pretty plain terms the animal origin of Man, but she is silent as to how and when he made his appearance on this earth. Two theories have been put forward: one, that all men have descended from common ancestors; the other, that each ot the different races of man developed, independent of the others, in its own region. The first is that now generally held. All races, from the most highly civilised to the lowest barbarians, resemble each other, not only in general form and bodily structure, but in the working of their minds, as is shown to some extent by the existence of similar beliefs and folk-stories among widely different peoples—these being, in many cases, the independent efforts of men in a low stage of civilisation to account for natural phenomena—the rising and setting of the sun, the succession of day and night, thunder and lightning, etc. In addition to this, people of the most dissimilar races intermarry freely, and the

SKETCH OF MAMMOTH ON A PIECE OF MAMMOTH IVORY.
(*From the Cave of La Madeleine, France.*)

fact that offspring result from such marriages is another indication of descent from a common ancestor.

We know what Man of the nineteenth century is like; but early Man we know only by his flint weapons, the remains of his refuse heaps, some artistic scratchings like the above, and a few bones. The early home of Man is unknown, but the oldest remains known have been found in Europe; though this may be due to the fact that the other continents have not been searched so thoroughly. A French author makes Asia the birthplace of the human race; but Dr. Brinton, when lecturing on "The Earliest Men," before the American Association for the Advancement of Science, in 1893, placed the first home of Man in Southern Europe or Northern Africa, or on the continuation of these latitudes in Western or Central Southern Asia. He does not think that the upward course was gradual, but that Man was suddenly evolved from the highest Man-like animal in the glacial, or possibly just before the glacial epoch, giving an antiquity of 50,000 to 100,000 years. In his opinion the earliest men walked erect, had full foreheads, red hair, and blue or grey eyes, were about of the same size and general appearance as now, perhaps were not even hairy, were kind to each other, social and artistic, had some sort of language, and knew how to make fire.

Man, as an animal, is chiefly distinguished from the family next below him by possessing a larger brain and a larger brain-case, or skull, as compared with the bones of the face; by the fact that his body is fitted for an upright position; by the small size of the canine teeth, and by the absence of a space in the opposite jaw for their reception. Occasionally one meets people in whom these teeth are large and prominent. This is probably a throwing back to the condition

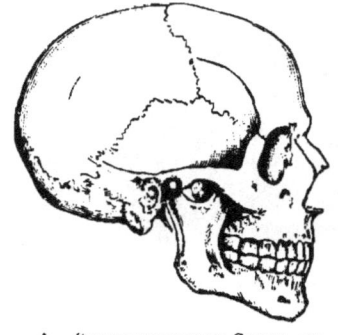 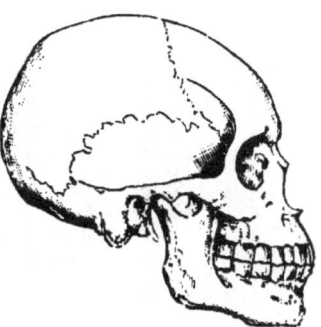

| A.—STRAIGHT-JAWED SKULL OF EUROPEAN. | B.—FORWARD-JAWED SKULL OF AUSTRALIAN. |

when our ancestors had these teeth largely developed, for use as weapons, as they still are in the Man-like Apes.

Man is generally considered to consist of a single species, the different races being taken as so many varieties; and from time to time many systems have been adopted for classifying them. Perhaps the best is that put forward by Sir William Flower, who groups all existing races round—

   I. The White, or Caucasian type.

   II. The Yellow, or Mongolian type.

   III. The Black, or African type.

In these types there is considerable difference in the shape of the skull (and consequently of the head), and in the position of the jaws with respect to the bones of the nose. In looking at Fig. A, it is evident that an upright line might be drawn to touch both the bones of the nose and the chin. In Fig. B, a vertical line touching the chin will be at some distance from the nasal bones. The first skull is said to be straight-jawed, or orthognathous, while the second is forward-jawed, or prognathous—a term which, Professor Huxley says, "has been rendered with more force than elegance by the Saxon equivalent—snouty."

I. The people of the Caucasian type with whom we are best acquainted—the people of the British Isles, Europe, and the United States—form the highest branches of the human genealogical tree, though many of the races ranged under this type hold by no means such a lofty position. Professor Huxley divided the peoples of this type into (1) Fair Whites and (2) Dark Whites; or, as one may say, Blondes and Brunettes. Of the first, a fair-haired Englishman is a good example; of the second, a Frenchman from Marseilles.

(1) The Fair Whites generally have white skin, ruddy complexion, fine flaxen, brown, or auburn hair, and blue or brown eyes. They are above the average height, the majority being from 5 ft. 6 in. to 5 ft. 10 in. and 6 ft.; though, of course, many individuals are shorter than 5 ft. 6 in., and some few taller than 6 ft. In Scotland, Norway and Sweden, and Denmark, and the North of Germany, the Fair Whites predominate. Many live in England—as, indeed, do many Dark Whites; and every intermediate grade may be met with, often during a morning's walk, if one keeps one's eyes open. They occur also in North Africa and Afghanistan, and from the intermarriage of Fair Whites with Mongols have sprung the Finns and Lapps of Europe and some tribes of Asiatic Russia. It is to this branch of the Caucasian type that America, Africa, and Australia owe the greater part of their white population.

(2) In the Dark Whites the skin may be white, olive, or shades of brown, sometimes so dark as to be scarcely distinguishable from that of the Negro. The hair is generally brown or black, usually straight, but sometimes curly, and the eyes are black and sparkling. In height, they are below the Fair Whites, ranging from little more than 5 ft. to about 5 ft. 6 in., a man above that stature being considered tall. To this branch belong the people of Southern Europe generally, of South West Asia and the North of Africa (in Plate I., No. 1 represents a Caucasian from Georgia, No. 2 an Arab, and No. 3 a native of the Soudan). The intermarriages of this branch have left their mark on some of the Indo-Chinese tribes; and to marriages with some of the lower Dark Whites the Australian natives probably owe the peculiar character of their hair. From a mixture of Dark White with Negro blood spring the Copts and fellaheen of Egypt, some tribes on the west of the Red Sea, and some other tribes farther south. Among the lowest

peoples of the branch are the Todas of the Neilgherry Hills, the Veddahs of Ceylon, and the Ainu of Yezo, recently visited by Mr. A. H. S. Landor, who considers them to be "the farthest behind in the great race of human development." So that in the Caucasian type we have the cultured European and American, ranking highest, and the hairy Ainu, ranking (perhaps) lowest, among the peoples of the world.

So little is known about the Ainu that we give a few particulars from Mr. Landor's book ("Alone With the Hairy Ainu"). The skin is light reddish, and they are very hairy. One he describes as resembling an orang rather than a human being; and the Ainu themselves told him that the country was formerly much colder, and asked, "Why should we be as hairy as bears if it were not to keep out the cold?" "The skin is greasy," Mr. Landor tells us, "the natural result of many years of an unwashed existence; and this gives to the hairy people a peculiar and strong odour, much resembling that of monkeys. Many are familiar with the peculiar odour of an uncleaned monkey's cage, and the same, intensified a thousand times, characterises an Ainu village." When actively angry the Ainu "sneer and snarl at one another, frowning ferociously and showing all their front teeth, but specially uncovering their fangs or dog-teeth." The author tells us that "their toes are supplementary fingers, and they often hold things between the big toe and the next. . . . . Then, again, the toes are often used to pick up small objects out of the reach of the hands, and also to scratch the lower extremities." It is remarkable that the Japanese have a legend that, long ago, the Ainu women suckled young bears, which gradually developed into men.

II. In peoples of the Yellow or Mongolian type, the skin varies in colour from a sallow hue, such as is often seen in our own countrymen, to lighter or darker shades of brown. The hair is black, coarse, and straight, and among the North American Indians, very long. The face is broad and flat, the cheek-bones are prominent, the eyes almond-shaped and set obliquely. The jaws project more than they do in peoples of the Caucasian, but less than in those of the Black type. The upper lip is hairy, but the beard is scanty. In height they range from a little less to a little more than 5 ft. 6 in., but the variation is not great either way. Some of the North American Indians, however, are very tall, while the Tibetans and Bolivians are short and squat in figure.

Of these Yellow People we may count five subdivisions or branches, the first of which is made up of the native inhabitants of Northern and Central Asia, the Chinese (Plate I., No. 4) and Japanese, and the people of Tibet, Burma, and Siam. The central tableland

PLATE I.

1. Georgian.   2. Arab.   3. Nubian.   4. Chinese.   5. Negro.
6. Australian.   7. North American Indian.   8. Malay.
9. Polynesian.

of Asia was the home of the Mongolian races that have again and again moved westward to attack and ravage Europe. And from those who settled in the parts where their arms were victorious, have sprung the Laplanders, the Finns, the Hungarians, and the Turks.

The second branch is made up of the Malays, whose home is in the Malay Peninsula (Plate I., No. 8) and the islands of the Archipelago lying to the south and south-east of Siam. The third branch consists of the natives of Polynesia (Plate I., No. 9), New Zealand being included in this term. In some respects many of the peoples of this branch resemble the Caucasian type; but their fine bodily frame is probably due to intermarriage with the Negroes, and white settlers have no doubt contributed to the same result. To the fourth branch belong the descendants of the races that were native in North and South America (excluding Greenland) before the conquest and colonisation of the New World by Europeans. Here are included tribes differing greatly from each other in stature, customs, mode of life, and in the stages of culture to which they have attained. Best known to us—it may be through Fenimore Cooper and Mayne Reid—are the so-called Red Indians (Plate I., No. 7). The "noble savage," however, is rapidly vanishing from the earth. The Indian wars of the United States have utterly exterminated some tribes, while others have been driven away from their hunting-grounds, which have been appropriated by the white man. On page 19 is represented one of the native Americans of the coast of Brazil. They are seldom more than 4 ft. high, and are at a very low stage of civilisation. The name of the tribe (Botocudos) is derived from the Portuguese *botoque*, which means a bung or a plug, and refers to the pieces of wood worn in the ears and lower lip. To this branch also belong the well-developed Patagonians and the stunted Fuegians. The fifth branch consists of the Eskimo, probably Mongol immigrants, who, having been hemmed in, as it were, by the ice on the north and by the Indian population on the south, have to a great extent preserved their peculiar characters, for between them and the tribes to the south of them there has been but little intermarriage. In the east of Greenland the Danes have brought many of the Eskimo to some extent under the influence of civilisation.

The Chinese and Japanese stand highest among the peoples of this type. The civilisation of the former is of very ancient date, while the latter, who have quite recently adopted European habits and customs, bid fair to equal, if they do not outstrip, their teachers. Japanese names will be found among the list of writers in many of the scientific journals of Europe, and to them are due the solution of

C

several vexed questions. It was a Japanese, trained in a European college, who showed the error of supposing that a hydra, if turned inside out, could make his skin do duty for a stomach (*see* Stinging Animals). One scarcely knows which is the lowest race of this type. Some of them are very low indeed.

III. In the peoples of the Black or Negro type, the cheek-bones are prominent, the eyes large, round, and black, and the cornea—what we call "the white"—has a yellowish tinge. The nose is broad and flat, and the thick lips are turned outwards in a peculiar way. This cannot be imitated by pouting; we must put the fingers at the corners of the mouth and push the lips up and down so as to show the red skin inside. The hair is black and short, generally frizzly, or, to borrow a term from the American Negroes, "woolly." The skin is generally black, cool and soft to the touch, and with a peculiar smell. In stature they are for the most part above the average, ranging from 5 ft. 6 in. to 6 ft. in height, though some do not much exceed 4 ft. The first branch of this type consists of the African Negroes (Plate I., No. 5), some of whom we have all seen in the streets. They live in the central part of Africa, ranging from sea to sea; but many tribes have sprung up from intermarriages with Dark Whites. The Kaffirs of South Africa are somewhat lighter in colour than the tribes to the north of them: their jaws do not project so much, and their teeth are smaller. Next come the Negrillos—among whom are Stanley's "dwarfs"—with yellowish skin, and skull of the round-headed form, who dwell in the forests of Central Western Equatorial Africa, and are the smallest of the human race. the average height being little more than 4 ft. To the third belong the Bushmen of South Africa, with yellow skin, eyes that resemble those of the Chinese, and very short hair, much more frizzly than that of the ordinary Negro, so that it looks as if the head were covered with a number of tiny balls. From a mixture of this race with the true Negroes came the Hottentots. The fourth and last branch—the Melanesians, or Black People of the Islands—includes the Papuans of New Guinea, with hair that grows into enormous mops—and the natives of most of the islands of the South Pacific and (perhaps) of Australia (Plate I., No. 6), though the proper place of these last is not settled. They may be Negroes altered by an infusion of Dark White blood from the South of India, or, as Sir William Flower suggests, "the direct descendants from a very primitive human type, of which the frizzly-haired Negroes may be an offset." Their skins are of a dark coffee-colour; in the form of the skull and the projecting jaws they resemble Negroes; but the nose is wide and not flat, and the curly hair is very different from the "wool" of the true Negro. Here also belonged the extinct natives of Tasmania.

We have now run over the different types of Man, briefly noticing some of the races that compose them. Lest we should be unduly lifted up by our mental and material superiority over what we

BOTOCUDO, WEARING LIP AND EAR ORNAMENTS.

are pleased to call the "lower" races, it may be well to quote some testimony showing that all the advantage is not on our side. In a paper read before the Royal Society of New South Wales in 1882, Mr. Frazer says of the aborigines, "who were regarded as among the most degraded of the races of men":—"They have or had virtues which we might profitably imitate : they are faithful and affectionate to those who treat them kindly." And he quotes the language of a friend who says, "Naturally they are an affectionate, peaceful people, and

C 2

considering that they have never been taught to know right from wrong, their behaviour is wonderful. I leave my house open, with their camp close by, and feel the greatest confidence in them."

Yet one more witness. Dr. Alfred Russel Wallace, in the closing chapter of "The Malay Archipelago," after giving the notions of the best thinkers as to a perfect social state, says: "I have lived with communities of savages in South America and in the East, who have no laws or law courts but the public opinion of the village freely expressed. Each man scrupulously respects the rights of his fellow. In such a community, all are nearly equal. There are none of those wide distinctions, of education and ignorance, wealth and poverty, master and servant, which are the product of our civilisation; there is none of that widespread division of labour, which, while it increases wealth, produces also conflicting interests; there is none of that severe competition and struggle for existence, or for wealth, which the dense population of civilised countries inevitably creates. All incitements to great crimes are thus wanting, and petty ones are repressed, partly by the influence of public opinion, but chiefly by that natural sense of justice and of his neighbour's right which seems to be, in some degree, inherent in every race of man." He goes on to warn us that it is not good to labour for intellectual and material advancement to the neglect of the moral qualities of our nature; and that, if we do so labour "we shall never, as regards the whole community, attain to any real or important superiority over the better class of savages." It is remarkable that much modern legislation is based on similar lines, and is intended to secure a more equal distribution of this world's goods, and so prevent the rich growing richer, while the poor grow poorer and poorer.

A MALAY VILLAGE.

CHAPTER III.

THESE animals, which closely resemble Man in bodily structure, and more or less in outward appearance, are confined to certain tropical regions of the Old World. There is less difference between their bodily structure and that of Man, than there is between them and the lowest monkeys. This resemblance is much more striking when young forms of both families (Man and the Man-like Apes) are compared. The teeth are the same in both.

There are four genera :—(1) The Chimpanzees ; (2) the Gorilla ; (3) the Mias, or Orang-utan ; and (4) the Gibbons. The first two are African, the third and fourth Asiatic.

The Chimpanzee approaches Man in the characters of its skull and teeth, and in the proportional size of the arms. The Gorilla is more Man-like in the proportions of the legs to the body, of the foot to the hand, in the size of the heel, the curvature of the spine, and in the capacity of the skull. The Orangs come nearest to Man in the number of the ribs, and in the form of the hemispheres of the brain ; but they differ from him much more widely in other respects, and especially in the limbs, than do the Chimpanzee and Gorilla. The Gibbons are most remote from Man on the whole, though there is much resemblance in the form of the chest.

The highest of the Man-like Apes are the Chimpanzees. If a line be drawn on the map of Africa from the mouth of the River Gambia as far inland as 28° E. long., and another from the Portuguese town of Benguela to the same meridian, the space enclosed will show the home of these animals. There are probably two species—the Common and the Bald Chimpanzee. Both are now fairly well known, for specimens of each kind have lived a considerable time in confinement, and their habits have been closely watched by skilled observers. The Common Chimpanzee does not exceed five feet in height, and the Bald Chimpanzee is said to be somewhat shorter. But the measurements of young specimens of these two forms, probably about the same age, living in the Zoological Gardens, Regent's Park, in 1889, were nearly equal. In the Common Chimpanzee, the forehead, cheeks, and the whole of the body are covered with long, harsh, black hair ; the upper part of the face, the brows, nose, and muzzle are of a dirty flesh-colour. The hands and feet are of a brownish clay colour, much the same as the general tone of the skin of the Bald Chimpanzee, which animal is thinly covered with dark hair, has a scanty crop of short blackish hair on the top of the head, and large naked ears standing out nearly at right angles.

The Chimpanzees live in the forest, and pass much of their time among the branches of trees, feeding on fruit and tender shoots and buds; but this diet is probably sometimes varied by young birds and small mammals. They live in separate families, or in small groups of families. When upright, the gait is weak; they go generally on all fours, supporting themselves with the back of the closed fingers rather than on the palm of the hand. In their native forests, Chimpanzees seem to romp and play as heartily as they do in confinement; and Dr. Savage, an American missionary, tells how a hollow tree is used as a drum to call the young ones to play, while the old ones sit round in a ring to watch them. Garner confirms the account of the drum-signal, but thinks that the "drum" is a spot of sonorous earth laid upon a soil resembling peat. So human are the ways of the Chimpanzees that the natives believe that they have been degraded from Man's estate; and similar hazy ideas as to the connection between Man and his "poor relations" are current elsewhere. The natives of India have a tradition—versified by Rudyard Kipling—that the ancestors of the monkeys came down to the cornland to teach the farmers to play. But the farmers requited good with evil. They set their visitors to work, and cut off their tails; and the wild monkeys of the forest were afraid to speak to the unfortunate prisoners, lest they also should become the captives of the farmers, and be set to plough and sow. English sailors sometimes say that monkeys can talk, but are afraid to do so lest they should be made to work.

It seems that Chimpanzees are fond of human society. Mr. Garner, who went to Africa to study what he calls the "speech" of Monkeys, writes thus in *Maclure's Magazine* (September, 1893):— "It is not at all rare to find tame Chimpanzees on this coast, going about the premises at large and quite as much at home as any resident. With this short preface, I desire to introduce my own young friend, who lives with me in my forest home. I call him Moses, because he was taken out of a papyrus swamp of the Ogowe. He is devoted to me, and cries after me like a spoiled baby, and follows me like a pet dog. . . . When I leave my cage I usually take him with me, and when he sees me take my rifle he begins to fret, until I let him mount my back, which he does with great skill, and hangs on to me like the ivy to a church wall. A few days since we were returning from a short tour, I saw a young Chimpanzee crossing the path about thirty yards from us, and I tried to induce Moses to call his little cousin; but he declined to do so, and I accused him of being proud because he was mounted, and the other was afoot, and hence he would not speak to him. I am trying to teach Moses to speak English, but up to this time he has not succeeded. However, he has only been in school a very short term, and I think he will learn by-and-by. . . . When he sees or hears anything strange, he always tells me in a low tone, unless it comes too near, and then he announces it with a yell. At times I refuse to pay any attention to him, and he will fall down, scream and sulk like a very naughty child. He is extremely jealous, and does not want any one to come near me. I have made him a neat little house, with hammock and mosquito bar, and at night I tuck him up, when he sleeps quietly until late in the morning. Then he crawls out, rubbing his eyes, and wants his break-fast. He wants to try everything he sees me eat." Unfortunately, Moses died before Mr. Garner left Africa.

YOUNG CHIMPANZEE.

Many animals of this species have been brought to Europe, and have lived for some time in confinement; but the change of climate has generally caused disorders of the chest, and in many cases they have fallen victims to consumption. One that died from this complaint in the Dresden Zoological Gardens, in her last moments put her arms round the director's neck when he came to visit her, looked at him placidly, kissed him three times, stretched out her hand to him, and expired. So that, even in death, there seems something strangely human about these creatures. A somewhat similar story is told on good authority of the death of " Joey," in the Zoological Gardens, Regent's Park.

" Sally," the female Chimpanzee which lived in the Regent's Park

Gardens for over eight years, was quite a celebrity. Crowds of visitors thronged the Sloth's House, where she lived, to see her " performances," and learned papers have been written about her mental faculties. She was not only friendly with her keepers, but recognised acquaintances who visited her from time to time. She fed herself with a spoon from a tin cup, a feat which has been taught to other Man-like Apes. The late Professor Romanes describes her as somewhat capricious in disposition, though on the whole good-humoured, fond of her keepers, and apparently never tired of a kind of bantering play, which off and on they kept up with her continually. They used to invite her to play by an imitation of her own note, and then "she shoots out her lips into a kind of tube, while at the same time she sings a strange, howling note, interrupted at regular intervals; these, however, rapidly become shorter and shorter, while her utterances become louder and louder, winding up to a climax of shrieks and yells, often accompanied with a drumming of the feet, and a vigorous shaking of the network that forms the front of her cage. The whole performances ended with a few grunts."

He was of opinion that "Sally" understood spoken language in a higher degree than that shown by any other brute. She tried, but not very successfully, to reply to what was said to her, for her "language" consisted of three peculiar grunting noises—one that evidently meant " yes "; another (very closely resembling the first) that meant " no "; and a third (quite different from the other two) that meant "thank you."

But the great achievement was teaching "Sally" to count, though the experiment would probably have been more successful could the animal have been kept as a domestic pet, for the constant coming and going of visitors distracted her attention, just as visitors to a schoolroom will distract the attention of pupils.

Professor Romanes arranged that the keepers should ask "Sally" repeatedly for one straw, two straws, or three straws. These she was to pick up and hand out from among the litter in her cage. No constant order was observed in these requests, but when she gave a number not asked for, her offer was refused; while if she gave the right number, she was rewarded with a piece of fruit. When she had learnt to count, without mistake, as far as three, her education was extended from three to four, and from four to five, with favourable results. At this point Professor Romanes allowed the matter to drop; but one of the keepers then went on with the work of teaching on his own account, and tried to carry "Sally's" powers of counting up to ten. The result was not a success and to the end of her days she can only be said to have had knowledge of numbers up to six, or perhaps seven, with some vague perception beyond. She knew, however, that the words "seven," "eight," "nine," "ten," stood for numbers above six. This was shown by the fact that, when asked for any number above six, she handed out more than six, and less than ten, straws.

An attempt was made to teach her the names of colours by means of white, black, red, green, and blue straws; but though she quickly learnt to distinguish between white straws and those of any other colour, she could go no farther. From these experiments Professor Romanes concluded that her failure to distinguish between black, red, green, and blue was not from want of intelligence, but because she was, in some sort, colour-blind.

Next below the Chimpanzee stands the Gorilla (Pl. II., No. 2), the largest and fiercest of the group, though there is every reason to be-

" SALLY."

lieve that the stories formerly told of its ferocity were exaggerated. Gorillas live, in families, in Western Equatorial Africa. The height of an adult male is about 5 ft. 6 in.—something less than the stature of an average Englishman; but they are much more strongly built than Man is. Their legs are short, and their arms disproportionately long, for, when half-erect, they can lay the palms flat on the surface of the ground, though Mr. Garner says they do not do so in walking, but use the back of the fingers from the second joints as a support for the fore part of the body. The skin is black, and covered with long hair, varying in hue from a dusky red to dull black.

From classic times down to about the middle of this century strange stories were told of large man-like apes that dwelt in the forests of Western Africa, and were able to vanquish elephants; but very little was known of these creatures. Now we know that these large man-like apes do exist, for young specimens have been brought alive to Europe and to London, and specimens of full-grown ones—male and female— are to be seen in the Natural History Museum at South Kensington; but we also know that some of the strange stories told of them were greatly exaggerated. These apes live more on the ground than the Chimpanzees, but at night they climb up among the branches of trees to sleep. It is generally said that the old male sleeps on the ground, leaning against the trunk of the tree, so as to protect his family from danger. Mr. Garner doubts this, and remarks that the old male, or "king," as the natives call him, "looks after his own comfort and safety first, and lets his family do as they can." The same writer says that "every instinct of the Gorilla seems to be averse from human society; he delights in a life of seclusion in the most remote and desolate parts of the jungle; and I have never heard of but one gorilla that became even tolerant to man, much less attached to him; and this one was a mere infant. I have seen a few in captivity, but all of them are vicious, and devoid of any sense of gratitude whatever."

The stories of gorillas attacking man are to be doubted. When unmolested they seem to avoid the encounter; but if attacked, their great strength makes them terrible foes. Koppenfels says that when scared by man, the gorilla "sends forth a kind of howl or furious yelp, stands up like an enraged bear, and advances with clumsy gait in this position to attack his enemy. At the same time the hair on his head and the nape of his neck stands erect, his teeth are displayed, and his eyes flash with savage fury. If no further provocation is given, and his opponent retreats, the animal does not return to the attack. In other cases he parries the blows directed against him with the skill of a practised fighter; he grasps his opponent by the arm and crunches it, or else throws him down, and rends him with his terrible teeth."

One that lived in the Berlin Aquarium from July, 1876, to November, 1877, is thus described by Dr. Falkenstein, who brought it to Europe, and in whose charge it had lived for some two years before: "In the course of a few weeks he became so accustomed to his surroundings, and to the people whom he knew, that he was allowed to run about at

THE GORILLA.

liberty without fear that he would make any attempt to escape. He was never chained, nor confined to a cage, and was watched only in the way that little children are watched when they are at play. He was so conscious of his own helplessness that he clung to human companionship, and displayed in this manner a wonderful dependence and trustfulness. He showed no trace of mischievous, malicious, or savage qualities, but was sometimes self-willed. He expressed the ideas which occurred to him by different sounds, one of which was the characteristic tone of importunate petition, whilst others expressed fright or horror, and in rare instances a sullen and defiant growl might be heard."

We are told that when he was anxious to obtain anything, no child could express its wishes in a more urgent or caressing manner. If in spite of this he did not obtain what he wanted, he had recourse to cunning, and looked anxiously about to see if he was watched. If, for example, he was kept prisoner in a room, he would, after several unsuccessful attempts to get his own way, apparently submit to his fate, and lie down near the door with assumed indifference. But he soon raised his head to ascertain if luck favoured him, edging himself gradually nearer and nearer, and then, looking carefully round, he twisted himself about until he reached the threshold; then he got up, peered cautiously round, and with one bound galloped off so quickly that it was difficult to follow him.

Dr. Hartmann gives us the results of his own observations on the animal at Berlin: "The creature generally slept in the bed of his keeper, and ate at the man's table, of plain but nourishing food cooked by the keeper's wife. He sometimes ate fruit, and bananas were occasionally provided for him. . . . He was generally good-tempered, fond of play, but rather mischievous: but he would snatch roughly, and occasionally try the sharpness of his teeth. Sometimes he tried to seize from visitors things which attracted his curiosity, such as the trimmings of ladies' bonnets, lace veils, and the like. But on the whole he behaved with propriety, playfulness, and good temper, and there was much that resembled man in his look and bearing." This gorilla died in 1877 of a galloping consumption.

Since then another specimen was obtained for the Berlin Aquarium, and in 1887 a young specimen, the first acquired by the Society, was exhibited in the Zoological Gardens, Regent's Park.

It is curious that the first gorilla brought alive to England was exhibited as a chimpanzee in Wombwell's Menagerie. It lived but a few months: and when it died, in 1860, Mr. Waterton, who bought the body, discovered what it really was. A likeness of this animal was published in the Proceedings of the Zoological Society for 1877, and the original drawing now hangs in the Society's meeting room in Hanover Square.

The Orangs (Plate II., No. 1) are large red-haired apes, from the islands of Borneo and Sumatra. There is probably but one species, though the orang of Sumatra was formerly considered distinct. There is, however, a dark race, which the natives call *Mias pappan*, and a light race, which they call *Mias rambi*. The males of the dark race have the skin of the face broadened out into folds or ridges on each side, while those of the light race are without these outgrowths of the skin. A smaller variety, also with a fairly smooth face, is called *Mias kassir*. Mias is the native name for all these apes; and the term *orang utan*, by which they are known to us, appears to belong to a savage people dwelling in the woods. A young specimen of this last variety, captured near Saráwak, was presented to the Zoological Society in 1891. Unfortunately, it lived but a short time in the Gardens at Regent's Park. "George" was on excellent terms with his keepers, and enjoyed a mild game of play quite as much as did the more celebrated "Sally."

Wallace, who had good opportunities for studying the habits of this great ape, and who brought home more skins and skeletons than any other collector, tells us that these animals are chiefly confined to the low, swampy forests of Borneo, and he believes that a large stretch of unbroken virgin forest is necessary to their existence. They roam among the tree tops with as much ease as the Indian on the prairie, or the Arab on the desert, and without being obliged to descend to the ground. They live principally on fruit: and the small mountains which rise like islands out of the swamps serve as plantations, where grow the trees yielding the fruit on which the Mias feeds.

With regard to the way in which Orangs travel through this region, he says:—" It is a singular and very interesting sight to watch a Mias making his way leisurely through the forest. He walks deliberately along some of the larger branches in the semi-erect attitude which the great length of his arms and the shortness of his legs cause him naturally to assume; and the disproportion between these limbs is increased by his walking on his knuckles, not on the palm of the hand, as we should do. He seems always to choose those branches which intermingle with an adjoining tree, on approaching which he stretches out his long arms, and seizing the opposing boughs, grasps them together with both hands, seems to try their strength, and then deliberately swings himself across to the next branch, on which he walks along as before. He never jumps or springs, or even appears to hurry himself, and yet manages to get along almost as quickly as a person can run through the forest beneath. The long and powerful arms are of the greatest use to the animal, enabling it to climb easily up the loftiest trees, to seize fruits and young leaves from slender boughs which will not bear its weight, and to gather leaves and branches with which to form its nest."

This nest is simply a lot of small green boughs and twigs broken off by the animal, and piled loosely in the fork of a tree. The mass is about 3 ft. across, and on it the orang lies on his back and sleeps. It seems a very rude affair for such a man-like creature to make. But Professor Hartmann reminds us that several of the lower races of men, in the construction of their huts, do not show much advance beyond the man-like apes. The former, however, build some kind of "shelter"; the latter seem only to make a "resting-place"; and it is doubtful if there is any truth in the stories that the Orang shelters itself from rain with palm leaves and large ferns.

Hornaday, the author of "Two Years in the Jungle," thus describes an old male which he shot. "His back was as broad, and his chest as deep, as a prize fighter's, while his huge hands and feet seemed made with but one end in view—to grasp and hold on. His arms were remarkably long and sinewy, but his legs were disproportionately short and thick. His body was large and heavy, with a chest both broad and full; his eyes were villainously small, and his canine teeth were as large as those of a small bear. His arms and legs were covered with long, coarse brick-red hair, which grew also on his abdomen and sides, but the skin which covered his breast hung in a loose, baggy fold. The face was bare, except for a thin growth of hair on the jaws and chin, which in pictures is usually magnified to a luxuriant beard. His skin was of a shiny brownish-black colour, darkest on the face and throat."

Wallace doubts the existence of Orangs more than 4 ft. 2 in. in height. Sir William Flower, writing since Wallace, puts the greatest height at 4 ft. 4 in., but Mr. Hornaday, who was collecting for an American Museum, claims to have shot one an inch and a half taller than that.

Hornaday notices the difference in the disposition of these animals. Of a young one, about six months old, or eight at the most, he says that it had the temper of a tiger, and made such persistent efforts to pull his hands up to its mouth to bite them, that he tied its elbows behind its back, fastened its feet together, and then bound the creature to the side of the boat. Even then the orang managed to roll over, and bit his captor severely in the calf of the leg. "I gave him," writes Mr. Hornaday, "a sounding slap on the side of the head, which caused him to let me go; but for many days after I carried a large black-and-blue mark in memory of him." He had another specimen which was not only savage, but sullenly refused food; while a third was quite peaccable,

THE ORANG AND ITS NEST.

"not even once attempting to bite, but whined softly when I approached him, and rolled up his big brown eyes appealingly. His petition was not to be refused. I cut the bark that bound his hands and feet, and placed a pile of soft straw in the verandah for him, into the middle of which he immediately crawled and curled himself up. And thus began a great friendship between ape and man."

In the foregoing paragraph there are two remarkable statements. The first is that there existed a great difference in the dispositions of the two orangs kept by Mr. Hornaday. People generally are far too apt to forget that animals·are individuals. No two men were ever alike in all points; and the same may be said of what we are pleased to call the lower animals. The boy who keeps rabbits or white mice will soon learn this by experience. The second statement to be borne in mind is that about one orang's habit, when angry, of seizing its master's hand, and trying to pull it up to its mouth so as to bite. One is reminded of some lines of Lucretius, which may be rendered thus—

> "At first men's weapons were their fists and nails
> And teeth : then stones, and branches torn from trees."

Sometimes one sees an angry child act in a fashion somewhat like that of Mr. Hornaday's orang. Why should it so act? The answer to this question may be read in the lines quoted above.

Hornaday closes his account of the Orang by advising any one who doubted the close relationship between Man and the higher Apes to go to Borneo. "Let him there watch from day to day this strangely human form in all its various phases of existence. Let him see the orang walk, build its nest, eat, drink, and fight like a human rough. Let him see the female suckle her young, and carry it astride her hip, precisely as do the coolie women of Hindostan. Let him witness the human-like emotions of affection, satisfaction, pain and rage, and he will feel how much more powerful is this lesson than all he has read on the subject."

The Gibbons, or Long-armed Apes, are somewhat like dwarfish old men, of slender build, and their arms are so long that some of them can touch the ground as they walk. The head shows none of the coarseness which is so marked in the Gorilla and Orang, but though they are so man-like in appearance and walk upright, or nearly so, when on the ground, their relationship to the lower monkeys is shown by the fact that they possess seat-pads—thickened patches of hairless skin, which seem to act as natural cushions when these creatures sit down. These seat-pads are absent in the Chimpanzee, the Gorilla, and the Orang.

The Gibbons are natives of South-eastern Asia, and are most abundant in the islands of the Malay Archipelago. They live chiefly among trees, swinging themselves from bough to bough by means of their long arms, and feeding on fruit, young shoots and buds, insects, birds' eggs, and probably any birds that are luckless enough to come within reach

SIAMANG.

of their long arms. They all have a powerful voice, and the name of more than one species is taken from its cry.

The largest species, the Siamang, is a native of Sumatra. Its height is about 3 ft., and its extended arms measure nearly twice as much. Its hair is jet-black, and it is distinguished from all other Gibbons by having the second and third toes united by skin as far as the first joint. Mr. H. O. Forbes often met with troops of them. "some of them hanging by one arm to a dead branch of a high-fruiting tree, with eighty unobstructed feet between them and the ground, making

D

the woods resound with their loud barking howls." Its singular cry is produced by inflating a large sac below the skin of the throat, and extending to the lips and cheeks, and then suddenly expelling the air, so as to produce the modulations of the voice. In his "Naturalist's Wanderings," he tells us how his hunter once shot a Siamang, and when the ape fell to the ground a young one was discovered clasped in its embrace. The hunter brought both to Mr. Forbes's hut, when the latter found that the young one was only stunned.

"In a very short time," he says, "it tamed down, and became a most delightful companion. Its expression of countenance was most intelligent, and at times almost human; but in captivity it often wore a sad and dejected aspect, which quite disappeared in its excited moods. With what elegance and gentleness it used to take what was offered with its delicate taper fingers, which, like its head, are more man-like (except for their hairiness) than any other ape's. It would never put its lips to a vessel to drink, but invariably lifted the water to its mouth by dipping in its half-closed hand, and awkwardly licking the drops from its knuckles. The gentle and caressing way in which it would clasp me round the neck with its long arms, laying its head on my chest, uttering a satisfied crooning sound, was most engaging. Every evening it used to make with me a tour round the village square, with its hand on my arm, apparently enjoying the walk as much as I did. It was a most curious and ludicrous sight to see it erect on its somewhat bandy legs, hurrying along in the most frantic haste, as if to keep its head from outrunning its feet, with its long free arm see-sawing in a most odd way over its head to balance itself."

Mr. Forbes doubts if these Gibbons can clear the great distances they are said to do at a jump. He saw a colony of Siamangs, when a forest was being cleared, scampering up and down a tree in abject terror; even when the tree was falling they did not attempt to jump to the ground, but came down with it and perished among its branches.

The other Gibbons are subject to great variations, individuals often differing greatly from each other in their colouring.

The White-handed Gibbon is found throughout the Malay Peninsula, ranging as far north as Tenasserim, where it frequents the wooded hills up to a height of from 3,000 ft. to 3,500 ft. above sea-level. Adult males are about 30 in. high, and the females are a little less. The general colour of the fur may vary from black to yellowish-white, and the back is often variegated. But the hands and feet are always of a pale tint, generally white or yellowish-white above. The naked skin of the face is black, and across the forehead is a white band which sometimes comes down on both sides and meets on the chin, so as to form whiskers and a beard.

When this Gibbon drinks it scoops up the water in its hand. In passing from bough to bough, the feet are seldom, if ever, used, but are left free to pick up any plunder met with by the way ; and a troop of them has been seen making off, with their feet loaded with fruit stolen from the gardens of the natives.

HOOLOCKS.

The Hoolock, or White-browed Gibbon, is said to take its native name from its cry. It is found in North-east India and Bur-mah, and is said to range as far to the south as Upper Tenasserim. The average height is about 32 in., and the colour is gener-ally black, with a white or grey band across the eyebrows ; the females are sometimes of a lighter colour than the males. This Gibbon is good tempered and easily tamed. It is probably no exception to the general rule that when an animal is kept in confinement and does not become tame, part of the fault lies with the keeper or owner.

Mr. Sterndale, the author of " The Mammalia of India," says of his pet Hoolock : " Nothing contented him so much as being allowed to sit by my side, with his arm linked through mine, and he would resist any attempt I made to go away." The pet fell sick—for Gibbons are of delicate constitution—and he was carefully attended by the author's brother, " who had a bed made for him, and the doctor came daily to see the little patient, who gratefully acknowledged their attentions; but to their disappointment he died." Mr. Sterndale says there is but one objection to these monkeys as pets, and that is " their power of whooping a piercing whoop-poo ! whoop-poo ! whoop-poo ! for several minutes till fairly exhausted."

Closely allied to the Hoolock is the Hainan Gibbon, which is a native of the island of Hainan, between the Gulf of Tonking and the China Sea. Only one species has been brought alive to Europe, and that was exhibited for a short time in the Zoological Gardens, Regent's Park. " Jemmy," who was very tame, was quite black in colour, and nearly full-grown. Before his journey to this country he lived for four years in China, and must have been a greater curiosity to the Chinese than he was to us—if one can judge by their accounts of this species, for a magistrate in Hainan told Consul Swinhoe that this Gibbon "had the power of drawing its long arm bones into its body, and that when it drew in one it pushed out the other to such an extraordinary length that he believed they were united."

The Agile Gibbon, which runs into a number of varieties, some of which have been described as distinct species, is found in Borneo, Sumatra, and the islands of the Sulu Archipelago, with a range on the mainland from Cochin-China to Siam. The general colour of the best-known form is a dark brown, and the face is surrounded by a fringe of whitish hair. Martin (" Man and Monkeys ") says of one exhibited in London in 1840, that when a " live bird was let loose in her apartment, she marked its flight, made a long swing to a distant branch, caught the bird with one hand in her passage, and attained the branch with the other hand, her aim both at the bird and the branch being as successful as if one object only had engaged her attention. It may be added, that she instantly bit off its head, picked its feathers, and then threw it down, without attempting to eat it."

The Tufted Gibbon, with a white patch on the crown, the Variegated Gibbon, and Müller's Gibbon are probably varieties of this. Our illustration shows the only specimen of the last-mentioned form yet brought to England. It is not only an excellent portrait, but from it one may see how justly Gibbons are called Long-armed Apes. The long fur was ashen-grey, while the palms and soles were black, as was the face, which was fringed with white.

The Silver Gibbon, like the Hoolock, owes its native name, Wow-wow, to its cry. It is found in Java and, according to some authorities, in Sumatra. The coat is thick, long, and woolly, and of a general dun colour. The upper part of the head is black, and white (or lightish) hair fringes the blackish face. The under-surface is lighter, and the palms and soles are black. Of one that lived for a short time in the menagerie of the Zoological Society in 1828, Martin says that it was usually gentle, but rather uncertain in temper, and would occasionally attempt to bite a stranger. Forbes says, " On first hearing their cries one can scarcely believe that they do not proceed from a band of uproarious and shouting children. Their ' Woo-oo-ŭ:——woo-ut——woo-oo-ŭt——wut-wut-wut——wut-wut-wut,' always more wailing on a dull heavy morning, previous to rain, was just such as one might expect from the sorrowful countenance of this group. They have a wonderfully human look in their eyes ; and it was with great distress that I witnessed the death of the only one I ever shot. Falling on its back with a thud on the ground, it raised itself on its elbows, passed its long taper fingers over the wound, gave a woful look at them, and fell back at full length dead—' saperti orang ' (just like a man), as my boy remarked." Forbes kept one in captivity for a short time, and " it became one of the most gentle and engaging creatures possible." In habits the Wow-wow resembles the Siamang.

MÜLLER'S GIBBON. (*Photographed from Life.*)

CHAPTER IV

THE MONKEYS OF THE OLD WORLD.

IT was formerly the fashion to divide these into three groups— those with long tails being called Monkeys, those with short tails Baboons, while tailless forms were known as Apes. This distinction cannot hold good, for among the Macaques are Monkeys with long tails, Monkeys with short tails, and Monkeys with no tails at all. It is better to keep the word "Ape" for the Man-like forms, "Baboons" for the Dog-headed Monkeys and one or two others closely related to them, using "Monkeys" for all the rest. In the Monkeys of the Old World the space between the nostrils is narrow, and the tail is never prehensile—that is, it cannot be used as a grasping organ. All the species have seat-pads, and in some there are cheek-pouches—that is, the skin of the cheeks is loose, so as to form a kind of natural cupboard on each side, into which food can be put for future use.

Most of the Slender Monkeys are natives of Asia. They have no cheek-pouches, but to make up for this on each side of the stomach are little bags, or pouches, in which the leaves and shoots, which form their chief food, can be stowed and digested at leisure. In all of them the thumb is well developed, and we shall see for ourselves how important this is if we try to pick any small object from the table or the ground with the fingers only. The Indian species are called *langúrs*, and the best known is the Common Langúr, or Sacred Monkey. The average length of the body is a little over 2 ft, while the tail will probably measure half as much again. The general hue of the fur is greyish-brown, but the face, ears, soles, and palms are black. There is no crest, but the hair of the crown spreads out in all directions from a point on the forehead. Owing to the fact that this monkey is looked upon by the Hindus as sacred, and has been protected for a very long period, it has no fear of man, and is found in troops—males, females, and their young—in groves quite close to villages, and even on trees within the village itself; and they swarm in the sacred city of Benares. "They frequently pilfer food from the grain-dealers' shops, whilst the damage they inflict on gardens and fields renders them so great a nuisance that the inhabitants of the country, though they will not, as a rule, kill the monkeys themselves, sometimes beg Europeans to kill the intruders." But it would be scarcely prudent to comply with this request.

Mr. J. L. Kipling ("Beast and Man in India") relates several instances of the dire offence given to natives by Europeans shooting monkeys, and tells an amusing story of a magistrate who, having shot a monkey by accident, stole out by night with a lantern to bury the body, feeling as guilty as if he had been a murderer trying to hide the evidence of his crime.

The Wild Langúrs frequent high trees and rocky hills, never far from water. Jerdon says, "They leap with surprising agility and precision from branch to branch, and when pressed take most astonishing jumps. I have seen them cross from tree to tree, a space of 20 to 30 ft. wide, with perhaps 40 to 50 ft. in descent.

GROUP OF SLENDER MONKEYS.

They can run on all fours with considerable rapidity." But a man well mounted can easily run down a Langúr, and, according to Blanford, it is their power of bounding, and the remarkable appearance they present whilst leaping, with their long tails turned over their backs, that convey the idea of speed rather than the actual rapidity of their motions.

The Himalayan Langúr is a little larger, and lives much farther to the north. It has been seen near Simla sporting amongst the fir-trees when covered with snow.

The Purple-faced Monkeys are natives of Ceylon. The general colour is brown. Hornaday says of them :—" They literally lined the road for seven miles, sometimes in the trees, and sometimes on the ground. One troop of very large old fellows we found playing in the road like schoolboys, galloping up and down, or chasing each other about, with their long tails held up at an angle of forty-five degrees. Their favourite gait is a gallop unless the branches are too thick to permit it, and they can run almost as fast through the tree-tops as over the bare ground. When hotly pursued, it is marvellous to see them run. They head straight away, and gallop madly along the larger branches without a second's pause or hesitation, without a fall or even a false step, spring boldly from one tree-top to the next, and, unless the ground below is very open, they are gone from the hunter's gaze like a flash."

The Douc, from Cochin China, is more stoutly built than the True Langúrs, and more gaily clad. The general hue of the fur is a dingy grey, darker on the upper surface of the body than beneath ; the tail, and a large triangular patch near its root on each side, are white ; the upper parts of the limbs and the hands and feet are black ; the legs are a rich red, and the arms are white. Nothing is known of its habits, and it has not yet been brought alive to this country.

The Tibetan Langúr comes from the Highlands of Tibet, ranging into China, where it has been known from a very early date. The limbs are shorter and stouter than in the Douc, but it is not so brilliantly coloured as that monkey, the general hue being olive-brown above, yellowish beneath and on the inner sides of the limbs, while the face is bluish-grey. The nose is turned up so much that its tip is nearly on a level with the eyes.

Other species are the Madras Langúr, the Malabar Langúr, the Banded Leaf Monkey, the Negro Monkey (the jet-black fur of which is valued for saddlery and military decorations), the Crested Lutong, the Nilgiri Langúr, and some few others.

The Kahaú, or Proboscis Monkey, is a native of Borneo. These monkeys are usually found near water, are swift climbers, and extremely shy. The cry is "*honk*," and occasionally "*kee-honk*," long drawn and resonant, quite like the note of a bass viol. Of the prominent nose, which gives the creature its name, Hornaday says :—" It hangs from the face—well, totally unlike anything else in the world—coming down below the chin, and shaped like a pear, except for a furrow down the middle ; and the division between the nostrils being contracted, causes the organ to terminate in two points. An adult male is about as large as a pointer ; the face is cinnamon-brown, and the body conspicuously marked with reddish-brown and white, the tails of old specimens being white as snow."

The Thumbless Monkeys are African. They are closely allied to the Langúrs, but the thumb is either absent or so small as to be useless. One of the best known is the Guereza, a native of Abyssinia, ranging southward into Somaliland. The fur is long and black, with a mantle of long white hairs hanging down on each side, and the tail is tufted with white; there are long white whiskers on each side of the face. They are said to live in small troops, and to feed on fruit, seeds, and insects. They are hunted for the sake of their skins, which are used by the Abyssinian troops to cover their shields. A variety, in which the mantle and tail-tuft are much more developed, is found at an elevation of 3,000 ft. on Kilima - Njaro. Ten other species are

PROBOSCIS MONKEYS.

known, two of which—the Bay Colobus, from Gambia and the Gold Coast, and the King Monkey, from Sierra Leone—have been exhibited in London.

For the Monkeys of the next genus there is no English collective name, but the French name "Guenons" (which means "Grimacers") is generally employed for them. They are common in Africa, and especially so on the western side of the continent. Very many of them are to be seen in the cages of Zoological Gardens and travelling menageries, and all of them, when young, are lively and amusing, and they rarely, if ever, even when old, exhibit the savageness shown by some of the Macaques, and, to a greater degree, by the Baboons. They are mischievous and destructive, and will often snatch at some article of dress or ornament, and, if successful in the attempt, will promptly pick or tear it to pieces on the spot. And they can bite sharply too, if teased or irritated, and will pinch and scratch anyone who has offended them.

The Diana Monkey has greyish fur, with a long, pointed beard and a crescent on the forehead of pure white. The name was given by Linnæus, from the fancied resemblance of the white on the forehead to the crescent worn by the Roman goddess. One of these monkeys that lived for some time in the Zoological Gardens, Regent's Park, was exceedingly friendly to her acquaintances. On their approaching her cage and whistling, she would manifest her delight by a series of jumps up and down on the floor of her dwelling, and would finish her performance by turning two or three somersaults, one after another. Then she would quietly sit down, holding the netting with both hands, and open her mouth for any dainties her friends might have brought her—raisins, grapes, bananas, Di would take them all. Sometimes she would begin her somersault again of her own accord, as if to show she wanted more ; but if she did not, a wave of the hand and " Now, Di," were quite sufficient to start her, and she seemed to enjoy the fun quite as much as the spectators. She seemed to have some faint notion of " hide-and-seek," for occasionally, at the end of one of her acrobatic exhibitions, she would swing herself upon the branch that crossed her cage and dive into her little sleeping-box at the top, so as to be lost to sight for a minute or two ; then she would come down with a wild leap, and the " show " would begin again.

The Mona Monkey, like the Diana, comes from West Africa. The upper surface is dark, the under surface white, and near the root of the tail on each side is an oval white spot. One that lived in the Jardin des Plantes, at Paris, was allowed its liberty. Its cunning and activity were very great, and its adroitness in performing any little theft was remarkable. It could turn a key, and untie knots and search pockets with a delicacy of touch so little felt that it was not remembered till the theft was discovered. It was gentle and playful, and when caressed uttered a low cry, seemingly an expression of pleasure.

The Vervet has greyish-green fur, with black face, hands, and feet. It is a native of South Africa, and is said to feed on the gum that flows from acacias. These monkeys are often seen in confinement, and one in Regent's Park was as adroit a pickpocket as the Mona Monkey mentioned above. It was particularly pleased to put its hand and arm through the netting and pluck out the hairs from the back of the hand of its acquaintances. But it could be fierce enough when it liked, and it had a command of uncomplimentary language that was astonishing.

The Grivet, from the north-east of Africa, has the hair on the upper surface ringed with black and yellow, which gives the animal a greenish appearance, while the under parts are white. Mr. Blanford met with them in small droves in Abyssinia, and says that their habits appeared to differ little from those of the Macaques, except that they were quicker and less mischievous than the Indian monkeys.

The Green Monkey ranges from Senegal as far south as the Niger.

DIANA MONKEY.

The fur is of a dark green hue above and yellowish below. It appears to be voiceless, for it utters no sound in confinement, and seems to be equally silent in a wild condition, for a French naturalist shot twenty-three out of a large troop, "and yet not one screamed, although they often assembled together, knitting their brows and grinding their teeth, as if they intended to attack me." A celebrated English naturalist adds, "I wish they had, with all my heart."

Stairs' Monkey was discovered quite recently. Dr. Molony brought home a young female in 1892 from the delta of the Zambesi River. It was a gentle, playful little creature, but did not live long in confinement. It was distinguished from other species by a chestnut-red patch in front of each ear. Strange to say, this was not the first specimen brought to England. An old male that had been kept for some years in the open air in a garden in the North of London was presented to the Zoological Society in 1893, but, unfortunately, only lived in the Monkey House for about eight months. A writer in *The Sketch* says :—" He is described by one of the keepers as a ' nice ' monkey—that is, a well-behaved creature, that gives little trouble. As the sole occupant of his dwelling, he cannot quarrel ; he is not given to mischievous tricks, such as snatching off the glasses of any short-sighted person who may come too near the cage ; still less would he behave like his neighbour, the Barbary Ape, who lives opposite, and viciously scratch the hand that offers him some toothsome morsel. But he does not gambol ; his playing days are over. Age sobers monkeys as well as men, and he generally sits sedately at the bottom of his cage, from time to time mounting the traverse bar to take the offerings of visitors, or to put his paw through the wires to be caressed by those on friendly terms with him. But as he takes no liberties, so he suffers none ; and those who wish to see what terrible weapons an old monkey has in his canine teeth should offer him a large nut—a walnut for choice—and, as he pushes it back between the last molars to get the better leverage for cracking it, there will stand out prominently four gleaming ' ivories ' that would not discredit a flesh-eating animal."

The Mangabeys, or White Eyelid Monkeys, also African, are sometimes placed in a separate genus, on account of some difference in the last grinding-tooth. The Sooty Mangabey is the species most often seen in confinement. Its colour is indicated in its name, and it has, like the other species, white eyelids, which show up strongly against the dark coloration of the body. They are larger than most of the Guenons, and in confinement they are well-behaved and gentle. One in Wombwell's Menagerie was very lively and active, and very fond of putting herself into extraordinary attitudes, so as to attract notice. Some that are now in the Gardens at Regent's Park are equally gentle and well-behaved, and offer a decided contrast to the Macaques in a cage close by, who, when teased, grin and show their teeth in a fashion that would bode ill for the teasers if the monkeys were at liberty.

The Macaques are Indian, with the exception of one species, the Barbary Macaque, from North Africa, with a colony, so to speak, at Gibraltar. They have shorter limbs and are more stoutly built than the Guenons, and the muzzle is more dog-like, though less so than in the Baboons. The males are larger than the females, and have the canine teeth more fully developed. Mr. Blanford says that "the species resemble each other in habits. They are found in flocks, often of considerable size, composed of old and young of both sexes. They are active animals, though less rapid in their movements, whether among trees or on the ground, than the Guenons. Most of them, if not all, eat insects, as well as seeds, fruit, etc., and one feeds partly on crabs. They have occasionally been known to devour lizards, and, it is said, frogs also. All have the habit of cramming food into their cheek-pouches, to eat at leisure—a practice that must be familiar to anyone who has fed monkeys in confinement."

Colonel Tickell says that anger is shown by silence, or expressed by a low hoarse "*heu*," not so guttural as a growl; weariness or desire for company by a whining "*hom;*" invitation, deprecation, and entreaty by a smacking of the lips and a display of the front teeth into a broad grin, with a subdued chuckle, highly expressive, but not to be rendered on paper. Fear and alarm are shown by a loud harsh shriek "*Kra*," or "*Kraouk*," which serves also as a warning to others who may be heedless.

Now that so much is written about the "speech" of monkeys, it is interesting to recall the fact that Captain Burton worked at the subject. Lady Burton, in her "Life" of her husband, says that "he collected forty monkeys of all kinds of ages, races, species, and he lived with them, and used to call them by different names. . . . He used regularly to talk to them, and pronounce their sounds afterwards, till he and the monkeys at last got quite to understand each other. He obtained as many as sixty words, I think twenty more than Mr. Garner—that is, leading words—and he wrote them down and formed a vocabulary, meaning to pursue his studies at some future time."

The Bonnet Monkey is a native of Southern India. An adult male is about 3 ft. 6 in. long, of which the tail counts for a little more than half, the fur is brown above and whitish beneath, with the ears and face flesh-coloured. The hair on the crown is long, and spreads out on the top of the head, and this gives the animal its popular name. This is the common monkey of Southern India; it is found wild in the jungles, and particularly tame in the towns. The native shops are open to the streets, and this affords these animals a good opportunity for plundering, the chief sufferers being those who sell fruit and grain. It has been described as the most inquisitive and mischievous of its tribe, with powers of mimicry that cannot be excelled; but it is doubtful if it can surpass the Bhunder for curiosity and mischief, and it is said to be more docile. Closely allied is the Toque Monkey from Ceylon, probably a mere variety. Both these monkeys are trained by Indian showmen to perform a variety of tricks. Mr. J. L. Kipling once saw a travelling showman with a band of performing monkeys. Some wild monkeys which were near the spot where the man began his preparations for the show, took refuge in the neighbouring trees; but when they saw their fellows dance to his piping, and, clothed in strange raiment, ride round and round on a goat, they crept closer with evident surprise and disgust.

The Crab-Eating Macaque is the common Macaque of menageries (Plate II., No. 3). The total length is about 44 in., and the tail is nearly as long as the head and body together. Individuals vary in colour, some being dusky or greyish-brown, whilst in others the brown is tinged with red; the under parts are whitish. It is widely distributed in Burma and Arakan, Siam and the Malay Peninsula. These monkeys live in small groups among the mangrove trees in tidal creeks, feeding principally on seeds, crabs, and insects. They take readily to water, and swim well. A writer in the *Field*, in mentioning the crab-eating habits of this monkey, says that he has good reason to suspect that the True Langúr does the same (near Bombay). Albinos of this species sometimes occur. Some years ago there were a pair in the Zoological Gardens; they were very lively and full of fun, but on bright days they seemed to suffer from the glare, and the male showed his dislike of it by scowling fiercely when the sun was on his face.

The Rhesus Monkey or Bhunder is the common monkey of Northern India, and is found in some places at an elevation of 8,000 ft. The total length is about 32 in., of which the tapering tail is a little more than a third. The general colour is brown, with a tinge of grey, and the under surface is nearly as dark as the fur on the back. It is found in large herds, more generally near the habitations of man than in the jungle, and it varies its vegetarian diet with spiders and insects. Though the Hindus do not regard the Bhunder as sacred, they do not molest it, and in consequence it has grown to be as mischievous as the True Langúr, and is more daring and impudent. The showmen of the North of India train it for exhibition, and it readily learns to perform tricks. These monkeys are extremely mischievous and inquisitive, and get savage as they grow old. In their wild state Blanford says they "are very quarrelsome, perpetually screaming and fighting and teasing each other: in fact, they behave very much like unruly children."

The Pig-tailed Monkey, when full grown, is said to be as big as a good-sized mastiff. It is found in the Malay Peninsula as far north as Tenasserim, and in Sumatra and Borneo. The general coloration is dark, and the shape of the head, especially in old males, approaches that of the Baboons. In Sumatra it is said that these monkeys are trained to climb cocoa-nut trees and gather nuts for their masters, but it is probable that only young animals are so employed.

The Lion-tailed Monkey is often wrongly called the Wanderoo, a name applied by the natives of Ceylon to all monkeys. It is found in troops of from twelve to twenty in the thick hill-forests near the Malabar coast. The general colour is black, and a long ruff of light-coloured

LION-TAILED MONKEY.

hair runs round the face, but does not meet on the forehead. An adult male will measure about 33 in., of which the tail, with a tuft at the tip, counts for little more than a foot. Blanford says they are sulky and savage in captivity. This can scarcely be said of three now in the Zoological Gardens, which share a large cage with two Gibbons and are on friendly terms with all visitors, though they certainly know how to take their own part. It was the writer's custom to feed these monkeys, which generally sat in a row at the bottom of the cage, grasping the netting with their hands, and opening their mouths for grapes or morsels of banana. One morning while the feeding was going on, a Gibbon stole slowly down the wire netting that formed the front of the cage, and putting out its foot seized the fruit intended for the Macaque. The latter jumped up, and chattering loudly, hunted the Gibbon round the cage. The thief swung from projection to projection, uttering cries of terror; but the Macaque gained ground and caught the fugitive, and seizing it by the loins bit it with a vigour that would have been dangerous had not the assailant been so small.

The Japanese Monkey, with long, soft, brownish fur and reddish face, is remarkable for living farther north than any other member of the family. These monkeys live in troops and commit great depredations in gardens and plantations, for they feed on acorns, nuts, oranges, date-plums, and any other fruit that fortune may throw in their way. The Japanese showmen tame this monkey, and train it to walk the tight-rope and to take part in acrobatic performances. Some of the showmen are said to dwarf their monkeys, by arresting their growth; and this is not unlikely, when it is remembered what marvellous results they can bring about in the way of dwarfing trees.

The Tcheli Monkey is found in the mountains to the east of Pekin. The yellowish-brown fur is very thick, and enables these animals to bear the bitter cold of the winters of this part of China, where the thermometer frequently falls to zero—that is, 32 degrees below freezing-point. This species and the former have mere stumps of tails. The Zoological Society possesses specimens of the Japanese and the Tcheli Monkey, but both are kept in cages in the open air, as the warmth of the Monkey House would be injurious to them. But monkeys from much warmer climates can bear cold better than one would think. The old specimen of Stairs' Monkey lived for some years in the open air in the North of London; and in 1893 two Toques just brought home from Ceylon, escaped from the box in which they were confined and remained at liberty for about eight weeks, and during part of the time there was frost on the ground. Frank Buckland wrote of "Jenny," a Barbary Macaque which he had given to a friend :—"She sits nearly all day on the top of a wall, and has only a common dog-kennel for shelter. She is out in frost, snow, and rain, and is none the worse for it. Her fur is magnificent, and she has a beard, that makes her face positively beautiful—for a monkey." And thus "Jenny" lived for sixteen years in the open air.

The Barbary Macaque (a better name than Barbary Ape), or Magot, is found in the north of Africa, and on the Rock of Gibraltar is a small colony of these monkeys. Shortly before the siege in 1779 a party of Spaniards attempted to surprise one of the British outposts. As they were advancing as noiselessly as possible, they came on a troop of Magots, whose cries alarmed the sentinel. The guard turned out, and the Spaniards, seeing that the British were on the alert, hastily retired.

General Elliot never allowed these monkeys to be molested, and now the small colony is under the charge of the signal-sergeant at the Rock. The average length is a little over 2 ft., the general hue of the fur on the upper surface is yellowish-brown, the under parts are whitish, the face and ears flesh-coloured, and the tail is represented by a fold of skin. It is not known how this monkey came to Europe; it is perhaps a descendant of forms now extinct, which formerly lived in many parts of the Continent and even in our own country.

The Black Ape is confined to the Celebes and the island of Batchian, where it was probably introduced by the Malays. The fur is of a deep black, as are the face, hands, and feet, the only exception being the flesh-coloured seat-pads. On the top of the head is a tuft of long hair, spreading out behind and at the sides so as to form a crest. The tail is scarcely an inch long. These monkeys feed on fruit, and live in small troops among the tree-tops—two things that distinguish them from the True Baboons, though the long muzzle shows relationship with that group.

BLACK APE.

The Gelada Baboon, a native of the south of Abyssinia, is the last link joining the Macaques to the true Baboons, from which it differs in that the nostrils are not at the extremity of the snout, but a little higher up, nearer the eyes. The fur is dark coloured; the hair on the top of the head and round the face is very long and turned backwards, flowing over the shoulders so as to form a kind of mane, and the tail is tufted at the end. The Geladas live in troops, which come down from the mountains and rob the fields, for they feed principally on grain. When attacked,

E

they roll down from the heights large stones on their enemies, in much the same manner as men of hill-tribes do. They are occasionally seen in confinement; but "they have no manners, and their customs are beastly."

This brings us to the True Baboons, which are natives of Africa and the country round the northern shores of the Red Sea. They differ from the Monkeys of both hemispheres in being more dog-like, and less human in bodily shape and in disposition. The Man-like Apes and the Old-World Monkeys live chiefly among tree-tops, and feed on fruit; the Baboons live among rocky mountains and in the open country, and supplement the diet of grain with insects, centipedes, and scorpions, occasionally taking lizards and frogs, and in one species, it is said, waging war on flocks of sheep. They go on all-fours, and even on level ground can travel as fast as a horse can trot. The limbs are nearly of the same length, and the seat-pads are sometimes very brightly coloured. They probably do not attack man, unless molested; but if interfered with or roused to anger they are formidable foes, a bite from the large canine teeth being sufficient to kill a dog.

The Arabian or Sacred Baboon was one of the sacred animals of the ancient Egyptians. It was worshipped as the type or symbol of Thoth, the god of letters, who was sometimes represented in the likeness of this creature. But though these animals were held to be sacred, the Egyptians seem to have taught them to do some useful work. A monument still exists in which some of these Baboons are represented gathering fruit, while slaves stand below with baskets to receive it. The adult male is about the size of a large pointer, but of stouter build; the fur is ashen-grey, and the neck and shoulders of the males are covered with a thick mane, making them look like something between a lion and a big French poodle. The tail is fairly long, and tufted at the end; the hands are black, the face and ears flesh-coloured, and the seat-pads bright red. The females are nearly as large as the males, but, like the young, have no manes—those of the older males being probably developed as a defence. Darwin found, from inquiry at the Zoological Gardens, that when baboons fought they tried to bite the back of the neck. These animals are now more common in Abyssinia and the Soudan than in Arabia itself, and in Egypt they are no longer found. Sir Samuel Baker says that they "have a great variety of expressions that may perhaps represent their vocabulary. A few of them I begin to understand, such as the notes of alarm and the cry to direct attention; thus when I am sitting alone beneath the shade of a tree to watch their habits, they are at first not quite certain what kind of a creature I may be, and they utter a peculiar cry to induce me to move and show myself more distinctly."

Fierce as these creatures are when attacked, and resolutely as they defend themselves, they are capable of tender feeling. Professor Romanes when in the Monkey House at the Zoological Gardens, once saw an Anubis Baboon bitten by a neighbour from whom it had attempted to steal a nut. The cries of the victim brought the keeper to its rescue, and by dint of "a good deal of physical persuasion"—that is, the keeper's iron rod—the assailant was induced to let go. "The Anubis Baboon then retired to the middle of his cage, moaning piteously, and holding the injured hand against his chest, while he rubbed it with the other one. The Arabian Baboon now approached him from the top part of the cage, and while making a soothing sound very expressive of sympathy, folded the sufferer in its arms, exactly as a mother would do under similar circumstances. It must be stated, also, that this expression of sympathy had a decidedly quieting effect upon the sufferer, his moans

BABOON.

becoming less piteous so soon as he was enfolded in the arms of his comforter; and the manner in which he laid his cheek upon the bosom of his friend was as expressive as anything could be of sympathy appreciated."

The Yellow Baboon (Plate II., No. 4), so called from the brownish-yellow colour of its fur, has long hair on the crown, and the hands, feet, and face are black. It is a native of Western Africa, ranging across the continent to Kilima-Njaro, where, according to Mr. Johnston, "they were generally found on the outskirts of native plantations, where they almost subsisted on the maize and other food-stuffs stolen from the gardens of their more highly developed fellow-Primates. In the inhabited region generally known as the country of Chaga, baboons were strangely abundant. They went generally in flocks of from fourteen to twenty, of all ages and both sexes. They were so little molested by the natives that they showed small fear of man, and instead of running away would often stop to look at me about twenty yards off, and the old males would show their teeth and

E 2

grunt. I have frequently seen the natives driving them from the plantations as they might a troop of naughty boys, and the Baboons retreating with swollen cheek-pouches, often dragging after them a portion of the spoil. On one occasion, in the river-bed at the foot of Kilima-Njaro, my Indian servant, ordinarily a very plucky boy, met a troop of these animals which, instead of fleeing up the trees, came running towards him in a very menacing manner, and he was so frightened at their aspect that he took to his heels. The Baboons followed, and but that the boy forded the shallow stream and put the water between him and his pursuers, he might have had an awkward contest. I killed a Baboon once in Chaga, one of a troop who were rifling a maize plantation, and its companions, instead of running away, surrounded the corpse and snarled at me. As I had no more ammunition I went back to my settlement to fetch some of my followers, and upon the approach of several men the Baboons ran off."

The Chacma, or Cape Baboon, lives in the mountainous districts of South Africa. An old male is said to be as large as an English mastiff, but it does not appear that any specimen approaching that size has yet been brought to England. The general colour is greyish-black, and the hairs of different length give the fur a shaggy appearance. They live in herds, and generally feed on the bulbs of a lily-like plant, varied with worms, insects, lizards, and birds' eggs ; but it is said that they are, in some cases, becoming flesh-eaters. Mr. Tegetmeier, in the *Field*, says : " It is maintained by some of the farmers that the animal has become carnivororus, leaving its original food and destroying sheep and goats. . . . Should the habit become general it would necessitate very active measures being taken against the Baboons, as their powerful canine teeth, great strength and agility would render them most destructive enemies of sheep, goats, and even calves."

The Anubis Baboon, with olive-green fur, from the west of Africa, lives principally on the stems of the Welwitschia, which it tears open with its large tusks, and nibbles the roots just like a sheep does a turnip. These animals are said to be a great plague to the native cultivators, for they come down in bands and strip whole fields of maize in a single night.

The Guinea Baboon, Sphinx, or Common Baboon, is very often imported for zoological collections. It shows relationship to the Mandrill, but though the cheeks are swollen, they are not brightly coloured. The fur is yellowish-brown, shaded with sandy or light-red tints ; the eyelids are white, the hairless parts black, the tail about half the length of the body and without a tuft. Nothing is known of its habits in a wild state.

The Mandrill is one of the most extraordinary looking animals it is possible to conceive. " Hideous," is an adjective commonly applied to it. Its home is the tropical region of West Africa. The adult male, said to be nearly as large as an Orang, is very stoutly built, with short limbs, and an enormous head without a perceptible forehead. The nostrils are a little behind the lips, and on each side of the face are

prominent swellings, covered with skin, coloured light-blue, scarlet, and purple. The seat-pads are blood-red, and the tail is a mere stump. Mandrills are not often seen in confinement. " Jerry," that was kept in the Surrey Zoological Gardens, was described by Broderip as "a glutton, and ferocious in the extreme. Most kindly he would receive your nuts, and at the same time, if possible, would scratch or pinch your fingers, and then snarl and grunt in senseless anger." He had learnt to drink tea and grog, and to smoke, and he is said to have dined with George IV. at Windsor.

" JERRY."

The Drill, also from West Africa, but spread over a wider range of country, has brownish fur above, and of a much lighter hue beneath. There are roll-like swellings on the face, as in the Mandrill, but the skin covering them is black, as is also the case with the Mandrill. The females and the young have not the repulsive look of the old males.

CHAPTER V.

NEW-WORLD MONKEYS AND MARMOSETS.

THESE animals differ in a marked manner from their cousins of the Old World. All are forest dwellers, frequenting the tree-tops, and most of them have a prehensile tail, which serves in some sort as a fifth hand; but in none are there cheek-pouches for the stowage of food, or those natural cushions which we have described as seat-pads. The partition between the nostrils is very wide; in those monkeys that have a thumb, it cannot be opposed to the fingers so as to pick up or grasp; and in all (excepting the Marmosets) there are four more teeth—that is, 36, against the 32 teeth of the Old-World Monkeys, which are the same in number and kind as our own.

The Capuchin Monkeys (Plate II., No. 5) take their popular name from the fancied resemblance of the long hair on the forehead to the cowl of a friar. In their native forests of Central America they go about in large troops, feeding on the tender shoots and buds of trees, fruit, birds' eggs, and young birds. They are very intelligent, and bear confinement and the climate of England well. They are common in menageries, and itinerant musicians and showmen often train them to perform.

The White-fronted Capuchin, with reddish-brown fur, and white on the face and chest, is plentiful over the level forest-lands of Brazil, and Bates saw large troops on the banks of the Upper Amazon. They spring from tree to tree with marvellous agility, and "grasp, on falling, with hands and tail, right themselves in a moment, and then away they go, along branch and bough to the next tree." Bates kept one as a pet, but it was not a success, as the Capuchin killed an owl-faced monkey, of which its master was very fond. "Upon this," says Bates, "I got rid of him." Belt was more fortunate with "Mickey." His White-fronted Capuchin killed nothing more valuable than ducklings, and a light switch taught him that he must leave off such bad habits. He was a sad thief. "When anyone came near to fondle him he would never miss an opportunity of pocket-picking. . . . One day when he got loose he was detected carrying off the cream-jug from the table, holding it upright with both hands, and trying to move off on his hind limbs. He gave the jug up without spilling a drop, all the time making an apologetic grunting chuckle, which he often used when found out in any mischief, and which always meant 'I know I have done wrong, but don't punish me: in fact, I did not mean to do it—it was accidental.' However, when he saw he was going to be punished, he would change his tone to a shrill threatening note, showing his teeth, and trying to intimidate. He had quite an extensive vocabulary of sounds, varying from a gruff bark to a shrill whistle; and we could tell by them without seeing him when it was he was hungry, eating, frightened, or menacing: doubtless one of his own species would have understood various minor shades of intonation and expression that we, not entering into his feelings and wants, passed over as unintelligible."

SPIDER MONKEYS.

There are many species of Capuchins, or Sapajous, as they are sometimes called. The Brown Capuchin, from Guiana, is often brought to this country, and it was on this species that Miss Romanes made the observations given in Professor Romanes' "Animal Intelligence."

The Woolly Monkeys are Brazilian, and have an under coat of woolly fur beneath the longer hairs. The tail is naked at the tip, on the under side, which gives the animal a securer hold, and the thumb

is well developed. Humboldt's Woolly Monkey is stoutly built, with black
fur and face; its features resemble pretty closely those of an old negro,
whence it is often called the Nigger Monkey. They are good tempered and docile in captivity.   One that lived for a short time in the
Zoological Gardens, Regent's Park, made many friends, and though his
greeting was never demonstrative, it was warm and affectionate.   He
would fondle a hand or a finger, then spring upon the branch, and
hang thereon with his tail, bringing his head close to the netting and
putting out his lips for some fruit as a reward for his good behaviour.
They are among the largest of the American Monkeys. Bates took a
specimen of which the head and body measured 26 in., while the tail
was an inch longer.

The Woolly Spider Monkey, from South-East Brazil, probably links
the Woolly Monkeys to the True Spider Monkeys.   It has the woolly
under-fur of the first, but the thumb is rudimentary.

The Spider Monkeys are found from Mexico in the north to
Uruguay in the south.   Their popular name is due to their long,
slender limbs.   They live principally on fruit, which is often conveyed
to the mouth with the tail, and the stomach is somewhat like that of
the Langurs.   Their activity is very great, as one may often see in the
monkey houses of zoological gardens.   One that shared a very large
cage with some Capuchins in Regent's Park distinguished itself by
chasing them round and round the cage, probably inviting them to a
romp.   But the Capuchins did not enter into the game, and the Spider
swung from rope to bar, and across the cage, without a playfellow.

The Red-faced Spider Monkey, or Coaita, has a wide range in Brazil
and Guiana.   It is of large size, clothed with coarse black fur, and the face
is flesh-coloured.   Bates says that these monkeys are common pets among
the Indians, and he gives them a good character for temper and disposition.   They display some ingenuity in breaking the case in which
what we call Brazil nuts are enclosed, by hammering it on a rock or hard
log.   There seems to be little doubt that they do break off dead branches
with the intention of injuring a supposed foe.   The author of " Canoe and
Camp Life in Guiana " says :—" On seeing us, they used frequently to
hurl down large dead branches, some of which came rather too close to
our heads at times to be comfortable.   The manner in which they performed this was singular : they held on by tail and hind feet to a live
bough in a tree-top, alongside of a dead one, and pushing with their hands
with all their force against the latter, generally succeeding in breaking it
off, when down it came."

The Variegated Spider Monkey, from both sides of the Amazon, is
strikingly coloured.   The long soft fur on the back is black, the checks are

white ; there is a bright reddish-yellow band across the forehead, and the under surface and the sides of limbs and tail are yellow. It is sometimes called Bartlett's Monkey, because Mr. E. Bartlett brought home a specimen in 1866, which was described as a new species. Other species are the Hooded and the Brown Spider Monkeys, from Colombia, and the Black-faced Spider Monkey, from Eastern Peru.

The Owl-faced Monkeys, Night Apes, or Douroucolis (as the natives call them), are more active by night than by day, thus resembling the birds from which they take their popular names. They are of small size, with a large round head, short face, and very large eyes, generally grey or brown; the fur is close, and the long tail can only curl round objects without clinging thereto, and in this respect they differ from the New-World Monkeys mentioned before, and agree with those that follow them, with the exception of the Howlers. These animals are found in the country from Nicaragua to the Amazon and Eastern Peru. There are three or four species, which probably differ little in their habits. Mr. Bates says that they sleep all day long in hollow trees, and come forth to prey on insects and eat fruit only at night. He met with two species—the Three-banded and the Feline. In both the forehead is whitish, and marked with three black stripes, which in the former go back to the crown, and in the latter meet on the forehead. He kept one of the Three-banded Night Apes as a pet. The animal preferred insect food, though it would eat fruit. Bates was told that these monkeys cleared the houses of bats as well as of insect pests. "When approached gently, it allowed itself to be caressed ; but when handled roughly, it always took alarm, biting severely, striking out its little hands, and making a hissing noise like a cat."

The Squirrel Monkeys inhabit the tropical forests from Costa Rica to Brazil and Bolivia. There are three species of these active, graceful little creatures, which in a wild state live in large flocks, and feed in great part on insects. The head is greatly elongated. The common Squirrel Monkey is about 10 in. long in the head and body, with a tail nearly half as long again. The body is olive-grey, the arms and legs bright red, the face white, with a blackish muzzle. Humboldt describes these creatures as having quite child-like faces and being of a very gentle disposition. He had many opportunities of watching their habits, and says that they knew objects when they saw them in pictures, even when they were not coloured ; and when they represented their usual food, such as fruit and insects, they endeavoured to catch hold of them. One may be pardoned for doubting if the "endeavour to catch hold" were prompted by anything more than the spirit of curiosity which is common to all monkeys. If experiments were carried out by showing these monkeys—or, for the matter of that, far higher monkeys—

pictures representing some favourite delicacy, and then others representing, say, a landscape, a battle, or a shipwreck, one would probably be snatched at as eagerly as the other.

The Titi (or Teetee) Monkeys differ from the Squirrel Monkeys in having the head round rather than long, the eyes smaller, and the hair on the tail much longer. They range over South America, from Panama to the southern limits of the great forests of the Amazon. Bates describes them as dull, listless animals, going in small flocks of five or six individuals, running along the main boughs of the trees. He obtained one specimen, which "was caught on a low fruit-tree at the back of our house at sunrise one morning. . . . As the tree was isolated, it must have descended to the ground from the neighbouring forest, and walked some distance to get at it."

The Sakis, or Saki Monkeys, also live in the great forests of South America. Most of them have long hair on the top of the head, which may show a kind of parting down the middle, or may spread from the top so as to form a kind of wig; all have whiskers and a beard, the latter in some cases very long. One of the best-known species is the Hairy Saki, or Humboldt's Saki—for it is called by both names. Bates describes it as a timid, inoffensive creature, with a long bear-like coat of harsh, speckled grey hair. The long fur hangs over the head, half concealing the pleasing, diminutive face, and clothes also the tail to the tip, which member is well developed, being 18 in. in length, or longer than the body. It is a very delicate animal, rarely living many weeks in captivity." Of the American monkeys, he considers this excels the rest in the quality of showing strong personal attachment.

The Uakaris have long hair on the body, but the beard is short, as is the tail, which in some of these monkeys is scarcely more than a stump. They, like the Sakis, dwell in the equatorial forests, rarely descending to the ground. There are several species, probably the best known being the Bald Uakari, which Bates describes as about 18 in. long, clothed from neck to tail with long, straight, and shining whitish hair. The head is nearly bald, and the face glows with the most vivid scarlet-blue; the bushy sandy whiskers meet under the chin, the eyes are reddish yellow. It seems to be confined to the western side of the Japuna, near its principal mouth. It lives in small troops among the crowns of lofty trees, and feeds on fruits of various kinds. It is said to be pretty nimble, but is not much given to leaping, preferring to run up and down the larger boughs in travelling from tree to tree.

The Howlers, or Howling Monkeys, are aptly named. To produce the terrible noises which characterise them, they have on the top of

HOWLER AND YOUNG.

the windpipe a hollow bony structure which intensifies their cries. As in the Spider Monkeys, the tail is prehensile ; but, unlike them, the Howlers have the thumb well developed. Their chief home appears to be in Brazil, but some range into Central America. They live among the tree-tops, and are vegetarian in diet. For size they carry off the palm among the monkeys of the New World, some being nearly 6 ft. in total length, of which the tail counts for half. There are several species differing little in habit, and all of them merit their distinctive name. A traveller in Guiana says :—

"At early morning, at dusk, and through the night, at all our camping places, we were accustomed to hear the Howlers serenading. To my mind the sounds produced by these monkeys more nearly resembled a roar than a howl, and when sufficiently far off are not unpleasant to the ear. When heard from a distance of half a mile or so they seem to begin with low notes, swelling gradually into louder and longer ones, till they merge into a prolonged roar, which dies gradually away with a mournful cadence. When not more than one or two hundred yards away, and consequently plainly heard, they commence with a series of short howls, which break off into grunts, and, at every repetition, become longer and longer till their voices have got fairly in tune, when they give their final roar, which dies as gradually away. Then, after an instant's silence, a few deep grunts are given, as if the remains of the compressed air in their throat drums were being got rid of. Listening carefully to the performance, one can detect a voice at a much higher key than the others, especially in the dying-away portion. The Indians say this is made by a dwarf monkey of the same family which accompanies every troop. I was of the opinion that it was the voice of a female Howler, but the Indians, who are very careful observers, said it was not."

The Red Howler and the Brown Howler have been brought alive to England, but both died soon after their arrival at the Zoological Gardens.

THE MARMOSETS AND TAMARINS.

These tiny creatures have their home in Central and South America. The hands and feet are paw-like, the fingers and toes being armed with claws instead of being furnished with nails; the number of teeth is the same as in the Old-World Monkeys, though there is some difference in their character. The tail is never prehensile, and is often marked with rings of light and dark hairs, and the ears terminate, in many kinds, in a small tuft of hair. They live in small bands among trees, and feed chiefly on insects and fruits. In disposition they are gentle, and they seem to be very affectionate to each other. Accounts differ as to their behaviour in confinement, some authorities asserting that they are easily tamed, others that their confidence is won with difficulty. This may arise from difference of disposition in the animals kept as pets, or it may be the result of the method employed. There are two or three young in a litter, a fact that shows these little creatures to be lower than the Monkeys, which produce but one at a birth.

These little creatures are more like squirrels than monkeys in their habits of climbing, and they confine themselves to the trunks and larger boughs of trees, the long nails enabling them to cling securely to the bark; and Bates saw one species passing rapidly round a tree-trunk. As the tail cannot be used to twist round a branch and the thumb is useless for grasping, the Marmosets are unable to leap from branch to branch, as do some of their relations and neighbours.

The Common Marmoset is a native of Brazil. It has been described as "a little creature resembling a kitten, banded with black and grey all over the body and tail, and having a fringe of long white hairs surrounding the ears." A South American traveller writes of one of these animals which he kept as a pet :—"Nothing pleased him better than to perch on my shoulder, when he would encircle my neck with his long, hairy tail, and accompany me in all my rambles. His tail formed a not very agreeable neck-cloth with the thermometer above one hundred degrees, but he seemed so disappointed when I refused to carry

COMMON MARMOSETS.

him that it was impossible to leave him behind. One reason of our intimacy was that our pursuits were the same, inasmuch as both were entomologists ; but he was a far more indefatigable insect-hunter than myself. He would sit motionless for hours among the branches of a flowering shrub or tree, the resort of bees and butterflies, and suddenly seize them when they little suspected danger."

Some that were kept by a French naturalist would kill small birds that were put into their cage. These they did not eat, though they licked up the blood that fell on the bottom of the cage.

These animals are often seen in confinement in this country, and they have bred several times here and on the Continent. A case was recorded in the *Times* in 1883, and the owner, in announcing the fact, wrongly supposed that it was the first time such an event had happened.

There are several other species, differing somewhat in size and more in coloration from the Common Marmoset, though their habits are pretty much the same. The name "Ouistiti" is often applied to any of them from their shrill, whistling cry.

The Tamarins live in troops in the forests of Panama, Peru, and the Brazils. They are restless little creatures, almost continually in motion, and their food consists of fruit, insects, birds' eggs, etc. There are no hairy tufts to the ears, nor is the tail banded with different colours.

The Negro Tamarin is one of the best-known forms. It is a native of Guiana and the lower part of the Valley of the Amazon. Bates says that "in Pará it is often seen in a tame state in the houses of the inhabitants. When full grown it is about 9 in. long, independently of the tail, which measures 15 in. The fur is thick, and black in colour, with the exception of a reddish-brown streak down the back. When first taken, or when kept tied up, it is very timid and irritable. It will not allow itself to be approached, but keeps retreating backwards, uttering a twittering, complaining noise, its dark, watchful eyes, expressive of distrust, observant of every movement which takes place near it. . . . I once saw one as playful as a kitten, running about the house after the negro children, who fondled it to their hearts' content." The same writer remarks " their knowing expression," which must have struck many other persons. After stating that some anatomists have compared the brain of the Tamarin to that of the Squirrel, he concludes that this is an unsafe character by which to judge of their mental qualities, and adds, " In mobility of expression of countenance, intelligence, and general manners, these small Monkeys resemble the higher Apes far more than they do any rodent animal with which I am acquainted."

The Silky Tamarin, or Lion Marmoset, owes its first name to the character of its fur, and its second "to the long brown mane, which hangs down from the neck and gives it very much the appearance of a miniature lion. Bates says of a tame one : "The first time I went in, it ran across the room straightway to the chair on which I had sat down, and climbed up to my shoulder ; arrived there, it turned round and looked into my face, showing its little teeth and chattering as though it would say, ' Well, and how do *you* do ? '" The colour is bright golden-yellow and the length about 2 ft., of which the tail takes up the half.

CHAPTER VI.

LEMURS AND LEMUR-LIKE ANIMALS.

IN many respects the Lemurs and their allies resemble Monkeys, from which, however, they may be readily distinguished by their sharp, foxy-looking heads, large staring eyes, and the nostrils at the extremity of the snout, in form like those of a dog or a cat. The index-finger in some of them is a mere stump; the second toe always bears a claw, the other digits being furnished with flat nails. The tips of the fingers and toes are flattened into disc-like pads, the skin of which is well supplied with nerves, so that they not only serve as cushions to break the fall of these creatures in the leaps, but as delicate organs of touch. The tail is never prehensile. Most of them are nocturnal—that is, they are more active by night than by day—and live for the most part among the branches of trees, rarely coming down to the ground from choice, and feeding on fruit, insects, reptiles, birds' eggs, and small birds. The teeth vary in number; but the back teeth resemble those of the Insectivores, in that they are furnished with points.

Madagascar is the chief home of Lemurs; some are found in Africa, and others are dotted here and there in the great forests as far east as the Philippines and Celebes. To account for their being thus scattered, it has been suggested that there must have been "a large tract of land in what is now the Indian Ocean, connecting Madagascar on the one hand with Ceylon and with the Malay countries on the other."

The Short-tailed Indris, from the eastern forests of Madagascar, is active in the daytime. The head and body are about 2 ft. long, while the tail is a mere stump. The general hue of the fur is sheeny black, with some white on the back, forearms, and hind-quarters. These animals are held in great veneration by the natives of some villages, though European travellers have not been able to ascertain the reason. A French naturalist has suggested that it may be on account of their sad, wailing cries, not unlike those of a human being in pain.

The Diadem Indris gets its name from a band of white on the forehead, which, as the face is black and fringed with grey, gives the creature a strange appearance. The fur of the upper surface is dark, with some lighter markings. In this animal, and two others closely allied to it, the hinder limbs are longer than the front pair, which, though well suited for a life among the trees, makes walking on all-fours difficult. On the ground they stand half-erect, and move forward by a series of jumps, with the hands raised in the air and the long tail streaming behind. There are two other species—Verreaux's Indris and the Crowned Indris, the former from the west and south, and the latter from the north-west, of Madagascar.

The Woolly Lemur, or Avahi, is generally found alone or with a single companion. It passes the day in sleep among the thick foliage, or in the hollow of a tree, coming out at dusk to feed and gambol. The natives give it a character for stupidity, which is probably not deserved, for the brain is larger in proportion to its body than that of any other of the Lemurs. The fur is distinctly woolly in character, and the general hue is

LEMURS.

reddish, though there is a great difference in individuals. In all these animals the second, third, fourth, and fifth toes are joined by a web up to the first joint. They feed chiefly on fruit.

The True Lemurs (Plate II., No. 6) are found only in the island of Madagascar and the Comoro Islands. They differ from those before described in having the toes free, the limbs more nearly of equal length, and in all the tail is long. Some are diurnal in habit and others nocturnal : but to the fruit diet of the Indris they add eggs, insects, and young birds. Some of them are always to be seen in zoological collections, where they are great favourites with visitors, for while they are active and

sprightly, and indulge in amusing gambols, they show none of the bad temper that is manifested by their higher relatives the Monkeys. If these latter rise above the Lemurs in brain power, they fall below them in conduct.

The White-fronted Lemur, with brownish-red fur and a broad white band across the forehead, is often brought to England. Mr. Broderip kept one as a pet. It was sometimes allowed to wander about the house, and its manifestations of joy when allowed to come into the room with its master are thus described :—

" His bounds were wonderful. From a table he would spring twenty feet and more to the upper angle of an open door, and then back again to the table or his master's shoulder, light as a fairy. In his leaps his tail seemed to act as a kind of balancing-pole, and the elastic cushions at the ends of his fingers enabled him to pitch so lightly that his descent was

hardly felt when he bounded on you. He would come round the back of my neck and rub his tiny head fondly against my face or ear, and, after a succession of fondlings and little gruntings, descend to my instep, as I sat cross-legged before the fire, when he would settle himself down thereon, wrap his tail round him like a lady's fur boa, and go to sleep. When in his cage he generally slept on his perch, rolled up with his head downwards, and his tail comfortably wrapped over all."

The Ring-tailed Lemur is ashy-grey on the upper surface and white beneath, while the tail is banded with black and white. These Lemurs live more on the ground than do any others of the group. They do well in confinement, and are generally gentle in disposition, though one that was presented to the Zoological Society had a nasty trick of making vicious snaps at the hand of anyone who attempted to feed it.

The Mungoose Lemur, with reddish-grey fur; the Ruffed Lemur, generally with black-and-white fur, the colours being disposed in large patches, but sometimes clad in reddish-brown; the Black Lemur, and some others, have lived in the Zoological Gardens, where young Lemurs have been born.

The Gentle or Grey Lemur has a more rounded head than the True Lemurs. It lives in the bamboo forests of Madagascar, sleeping by day and coming out at night to feed on the tender shoots and leaves.

There are two species of Weasel Lemurs, also nocturnal. They are distinguished from the Gentle Lemurs by the fact that full-grown animals have no front teeth, or quite rudimentary ones.

The Mouse Lemurs are so called from their small size; one of them was called by Buffon the Madagascar Rat. The food is chiefly fruit, insects, and probably small birds. Most of them build nests, and some of them æstivate, or indulge in a long sleep during the hot season. Just before they retire for their slumber, a large quantity of fat accumulates at and round the base of the tail, and the tail itself is enlarged. This fat nourishes them during their summer rest, and when they wake its loss is shown by the small size of their tails. Like the Galagos, the Mouse Lemurs have the bones of the ankle very long, thus giving great leverage to the muscles of the leg and increasing the jumping power.

The Galagos are confined to the continent of Africa. Some of the species are no larger than Mouse Lemurs, from which, however, they may be readily distinguished by their curious folding ears. The tail is long and bushy.

Garnett's Galago, from Eastern Africa, though it is often called the Black Galago, has dark-brown fur, fading into yellowish on the under parts. These animals have been exhibited in the Zoological Gardens,

F

and of one to which he gave its liberty in his room, Mr. Bartlett, the superintendent, writes :—

"Judge my utter astonishment to see him on the floor, jumping about upright like a kangaroo, only with much greater speed and intelligence. The little one sprang from the ground on to the legs of tables, arms of chairs, and, indeed, on to any piece of furniture in the room ; in fact, he was more like a sprite than the best pantomimist I ever saw. What surprised me most was his entire want of fear of dogs and cats. In bounding about on the level ground his jumps, on the hind legs only, are very astonishing, at least several feet at a spring, and with a rapidity that requires the utmost attention to follow. . . . He eats fruits, sweetmeats, bread, and any kind of animal substance, killing everything he can pounce upon and overpower. This strong and active little brute thus eats his prey at once, as I had proof in an unfortunate sparrow, which he unmercifully devoured head first."

COMMON LORIS.
(From Sketch by Col. Tickell.)

Other species are the Great or Thicktailed Galago, from West Africa, with a bushy tail longer than the head and body together ; the Senegal Galago, which has been known longer than any other ; and the tiny Demidoff's Galago, from the West Coast, with small ears and slender tail.

The Slow Lemurs, or Loris, have the tail short or absent. The thumb and great toe stand out from the other digits, and the latter is directed backwards. In those which live in Asia the first finger is small, while in those from West Africa it is reduced to a mere pimple. All are nocturnal and live among trees, amongst which they climb, and do not jump or run.

The Common Loris is found in India, to the east of the Bay of Bengal ; it lives also on the north-east frontier, and ranges southward through the Malay Peninsula and Siam to Borneo and the neighbouring islands. Colonel Tickell says : " It inhabits the densest forests, and never by choice leaves the trees. Its movements are slow, but it climbs readily and grasps with great tenacity. If placed on the ground, it can proceed, if frightened, in a wavering kind of trot, the limbs placed at right angles. It sleeps rolled up in a ball, its head and hands buried between its thighs, and wakes up in the dusk of evening to commence its nocturnal rambles." The total length is a little over a foot, the fur is ashy-grey, with a chestnut stripe on the back and dark rims round the eyes.

The Slender Loris, from Southern India and Ceylon, is much smaller, and has dark-grey fur with a reddish tinge. The Singhalese use the eyes of this creature in love-charms and philtres. To obtain them Sir Emerson Tennent says they hold the little animal to the fire till the eye-balls burst. Before we say hard things about the Singhalese it may be well to remember that living shrews were formerly plugged into ash-trees in this country, and that some North Country fisher-folk "after having caught nothing for many nights, keep the first fish that comes into the boat, and burn it on their return home as a sacrifice to the Fates."

The African species of Slow Lemurs are called by their native name —Pottos. Bosman's Potto was discovered on the Guinea coast early in the eighteenth century, and then lost sight of for a hundred and twenty years. The tail is short, and two or three of the vertebræ of the neck have long processes, which form little prominences, and are only covered by a thin skin. In habits it resembles the Common Loris, but is said to be even slower in its movements. Another species— the Awantibo—is found at Old Calabar.

The Tarsier, which lives in the forests of many of the islands of the East Indian Archipelago, is about the size of a squirrel, which it resembles in sitting up and holding its food in its hands while eating. The face is round, with sharp muzzle, large ears, and staring eye. The hind limbs are longer than the front pair, the tail is tufted, and the general colour of the fur is fawn-brown. It owes its name to the great elongation of the bones of the ankle, technically called the tarsal bones. These animals are nocturnal and arboreal, leaping from bough to bough in pursuit of insects and lizards. The natives regard them with superstitious dread, and if the people in some parts of Java see one near their rice-grounds they will leave them uncultivated.

The Aye-Aye is confined to the bamboo forests of Madagascar, where it lives solitary or in pairs. This animal is about the size of a large cat, has long, loose, dark-brown hair, with a woolly undercoat, and the long tail is bushy. With the exception of the great toe, which is opposable and bears a nail, all the digits are armed with long claws, and the middle finger of each hand is so thin that one writer has compared it to a piece of bent wire. The Aye-Aye builds a kind of nest of dried leaves in a fork of a tree, with an opening at the side, at which to go in and out. This creature is remarkable from the fact that its true position was long misunderstood. From the number and the peculiar character of its teeth, it was formerly placed with the Gnawing Animals. The resemblance was not confined to the number of the teeth. The incisors grow from persistent pulps—that is, they are pushed up from behind as fast as they are worn away in front, as is the case in the rabbit and the mouse. Specimens have lived in the Zoological Gardens, but have been seen by few visitors, for during the day they sleep in the little box at the top of the cage, only coming out when the house is cleared at dusk. Dr. Sandwith, who kept one of these creatures for some time, gives the following account of its habits in captivity :—

"The thick sticks I put into his cage were bored in all directions by a

THE AYE-AYE.

large and destructive grub. Just at sunset the Aye-Aye crept from under his blanket, yawned, stretched, and betook himself to his tree, where his movements are lively and graceful, though by no means so quick as a squirrel. Presently he came to one of the worm-eaten branches, which he began to examine most attentively ; and bending forward his ears and applying his nose close to the bark, he rapidly tapped the surface with his curious second digit, as a woodpecker taps a tree, though with much less noise, from time to time inserting the end of the slender finger into the worm-holes as a surgeon would a probe. At length he came to a part of the branch which evidently gave out an interesting sound, for he began to tear it with his strong teeth. He rapidly stripped off the bark, cut into the wood, and exposed the nest of a grub, which he daintily picked out of its bed with the slender tapping-finger, and conveyed the luscious morsel to his mouth. But I was yet to learn another peculiarity. I gave him water to drink in a saucer, on which he stretched out a hand, dipped a finger into it, and drew it obliquely through his open mouth. After a while he lapped like a cat, but his first mode of drinking appeared to me to be his way of reaching water in the deep clefts of trees."

CHAPTER VII.

BATS AND INSECTIVORES.

THE Bats are nocturnal animals having the fore-limbs specially modified for flight. We can easily trace out on our own bodies the general plan of this flying apparatus. If we stand upright, with the feet a little apart, the extended arms bent into a **V**-shape from the shoulder, with the thumb pointing upwards and the fingers downwards, we shall have some idea of the framework, so to speak, of the apparatus by which Bats fly. From the point of the shoulder to the thumb there stretches a thin sensitive membrane, the fingers are immensely elongated, and between these, and extending from the little finger to the heel, and running thence along the side of the body to the arm-pit, is the wing-membrane. Most Bats have, between the legs, reaching nearly or quite to the heel, a membrane, supported by the hind-limbs, and often by a bony spur which runs backward and downward from each heel. The wings are spread for flight by stretching out the arms, and opening the fingers, which, as

HEAD OF A BAT (MEGADERM).
Showing Earlets and Nose-Leafs.

the bones of the palm are free, seem to start directly from the wrist, something like the sticks of a fan. Bats have teeth of three kinds like our own; the number varies, but never exceeds thirty-eight. The ears are large, especially in the insect-eating Bats, which also have an inner ear, or earlet; and they generally have a "nose-leaf"—an outgrowth of skin on and round the nose. The ears, nose-leaf, and wing-membranes are extremely sensitive, and serve as delicate organs of touch.

On the ground Bats walk badly, owing to the fact that the hind-limbs are weak and the knee bent backwards. By means of the claws on their toes and their thumbs, they can climb up sloping and upright surfaces if there be any small projection for them to take hold of. When at rest, they hang by their feet in caves, or old buildings, or to the branches of trees, and sleep head downwards. In temperate regions Bats take a long winter sleep. Even in India the insect-eating Bats are rarely seen abroad in the cold season, though the fruit-Bats are as active then as at other times.

There are two sub-orders—the Large, or Fruit-eating Bats, from the tropical and sub-tropical regions of the Old World, and Australia; and the Small, or Insect-eating Bats, from the tropical and temperate regions of both hemispheres.

FLYING-FOXES AT REST.

The Flying-foxes of the East, which owe their popular name to their long, sharp muzzle, are good representatives of the Fruit-eating Bats. The average length is a little over a foot, and the wing-spread is about four times as much. There is no tail. The general colour of the fur is reddish-brown. Jerdon says: " During the day they roost on trees, generally in large colonies,hanging head downwards, wrapped in their wings, and resembling large dead leaves. Towards sunset they begin to get restless, move about along the branches, and by ones and twos fly off for their nightly rounds. If water is at hand—a tank, a river, or the sea—they fly cautiously down and touch the water; but I could not ascertain if they took a sip or merely dipped part of their bodies in. They fly vast distances occasionally to such trees as happen to be in fruit, returning from their feeding-grounds at early morning." Colonel Tickell says that, on their arrival at their roosting-places, " a scene of incessant wrangling and contention is enacted among them, as each endeavours to secure a higher or better place, or to eject a neighbour who presses too close. In these struggles the Bats hook themselves along the branches, striking out with the long claw of the thumb, and shrieking and cackling without intermission. Each new-comer is compelled to fly several times round the tree, being threatened from all points, and when he eventually hooks on has to go through a series of fights, and is probably ejected two or three times before he makes good his place."

The Kalong, or Malay Flying-Fox (Plate II., No. 7), is the largest Bat known, having a wing-spread of quite 5 ft. Wallace says: "These ugly creatures are considered a great delicacy, and are much sought after. At about the beginning of the year they come in large flocks to eat fruit, and congregate during the day on some small islands in the bay, hanging by thousands on the trees, especially on dead ones. They can then be easily caught or knocked down with sticks, and are brought home by basketfuls. They require to be carefully pre-pared, as the skin and fur have a rank and powerful foxy odour ; but they are generally cooked with abundance of spices and condi-ments, and are really very good eating, something like hare."

HEAD OF HAMMER-HEADED BAT.

The Grey-headed Fruit Bat. from Australia ; the Egyptian Fruit Bat, which lives among the ruins of the ancient buildings and in the dark chambers of the Pyramids ; and the Fulvous Fruit Bat, from India, Ceylon, and Burmah, are closely allied. The last is sometimes found in caves, near the sea-shore, and is said to feed on molluscs.

White's Fruit Bat ranges over Africa, from the Northern tropic to Senegal. It represents a group, in which the males of most of the forms have, near the shoulder, pouches. from the mouth of which long yellowish hairs project, whence they are called Epaulet Bats. They live principally on figs. To this group belongs the Hammer-headed Bat, discovered by Du Chaillu in Western Africa. There are many other species, but they differ little in habit from those described.

The Small, or Insect-eating Bats, fall into two groups, which may be distinguished by the character of the tail and the hair. In the first the tail is generally long, and enclosed within the thigh-membrane. In the second, the tail, when present, usually comes through the membrane. The character of the hair is pretty constant in the two groups. The figure *a* on the next page shows the hair of one of the Covered-tail Bats, while that marked *b* shows a hair of one of the Free-tailed group.

The Horseshoe Bats are so called from the fact that the nose bears leaf-like appendages, of which the front part surrounding the nostrils is not unlike a horseshoe in shape. The Greater Horseshoe Bat is found in the southern counties of England, and ranges over Central and Southern Europe, part of Asia, and the whole African continent. The total length is nearly 4 in., and the wing-spread 13 in. The fur on the upper surface is reddish-grey; below, the red tint is lost. Its favourite haunts are deserted quarries, old buildings, and natural caverns, preferring the darkest and most inaccessible parts. In some such situations it passes the winter in a torpid state, coming forth at the approach of spring. Cockchafers, which do so much damage to farm crops and plantations, are said to be its chief food.

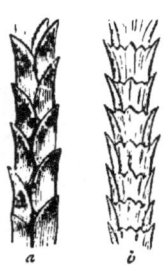

a   b

HAIR OF BATS.

The Lesser Horseshoe Bat is also British, but has a wider range, and spreads to Ireland. Dr. Leach described it as "a very cautious animal; very easily tamed, but fond of concealing itself." It probably flies higher than its larger relation, and there is great difference in its manner of alighting from that of other Bats. Mr. Bell, who turned one loose in a room, says that "instead of adhering by its claws against an object, it invariably sought for something from which it could hang freely suspended. The leaf of a table which was let down was often tried, but the polished surface not furnishing a suitable hold for its claws, was as often relinquished for some fringe over a window, from which it would hang suspended by one foot for some time, swinging about, and twisting itself round, to watch those who were observing it."

The Mourning Horse-shoe Bat is a native of India and the Asiatic Islands. The fur is long and thick, and black in colour, whence the popular name. The total length is a little under 6 in. There are several other Eastern species; and one, the Australian Horseshoe Bat, with mouse-coloured fur and very large nose-leaf, from Australia.

The Diadem Bat, from India, Ceylon, and Burmah, is from $3\frac{1}{2}$ in. to 4 in. long, with a wing-spread of about 2 ft. The fur is light brown. Captain Hutton says that "this species may be heard during its flight cracking and crunching the hard wings of beetles, which in the evening are usually abundant among the trees; the teeth are strong, and the general aspect is not unlike that of a bull-dog." This genus also has many Eastern species.

In the Megaderms and the Nycterine Bats the ears are enormously developed; the earlet also is very large. These Bats live in the warmer parts of the Old World. The best known of the Megaderms is the Lyre Bat from India and Ceylon, from 3 in. to 4 in. in length, with a wing-spread from 14 in. to 19 in. The fur is ashy-blue above, and yellowish below. Sterndale says they are very abundant in old buildings, and, undoubtedly, bloodsuckers. Blyth noticed one fly into his room one evening with a smaller bat, which it dropped. The latter was weak from loss of blood, and the next morning, the Lyre Bat having been caught and both Bats put into the cage, the little one was again attacked and devoured. Sterndale also records the killing of two canaries by this Bat, and Sir William Flower thinks that the Megaderms feed, when they can, upon the smaller Bats and other small mammals.

The Nycterine Bats, of which there are seven, inhabit Africa and

BARBASTELLE, WALKING.

Arabia, with the exception of one species found in Java. In all these the nose-leaf is absent, but in its place there is a deep groove extending upward from the nostrils. The Desert Bat, about 4 in. long, with grey fur. lives in the desert regions of Egypt and Abyssinia, and owes its popular and scientific names to the fact that it is found in the Thebaïd, the home of so many hermits in early Christian times.

In the True Bats the nostrils are at the end of the muzzle, and are not surrounded by leaf-like appendages. They are widely distributed throughout the temperate and warm regions of both hemispheres, and here belong most of the European Bats and all the British Bats, with the exception of the two already mentioned.

The Long-eared Bat, pretty generally distributed all over the country, and found throughout the greater part of Europe and Asia, is nearly 4 in. long, including the tail. The fur is rather long, thick, soft, and silky, lightish-brown above and brownish-grey beneath. According to

Bell, it may be easily tamed. Its favourite retreat is in the roofs of houses in towns and villages, and in the towers of churches. In moving along the ground the foreparts are raised, and the body thrown forward by successive jerks, first on one side, then on the other.

In the Barbastelle the ears are of moderate size. The total length is less than 4 in., of which the tail forms nearly a half. The long, soft fur is brownish-black. The Barbastelle ranges over great part of Europe and the southern and midland counties of England. When kept as a pet, not only does it shun its owner, but declines to make the acquaintance of other Bats that may share its cage.

The Pipistrelle is very common in Britain, and ranges to the Himalayas. The total length is a little less than 3 in.; the fur on the upper surface is reddish-brown, and dusky beneath. From the middle of spring to October this Bat may be seen after sunset, skimming the water like a swallow in search of insects, and instances are on record of its darting at a fisherman's fly and becoming hooked. Its close ally—the Serotine, some 4½ in. long, with rich chestnut-brown fur above, and yellowish grey beneath—is the only Bat found in both hemispheres. In England it is confined to the south-eastern counties.

The Great Bat, or Noctule, is widely distributed in the Eastern hemisphere. With us it is found as far north as Yorkshire. The head and body are 3 in. long, and the tail about half as much: the fur is reddish-brown. A naturalist, who kept some of these Bats in captivity, speaks of their voracious appetite. A female passed the day suspended by the hind feet at the top of the cage, coming down at evening to feed. Her weight was just two drachms, yet she managed to consume a whole half-ounce of cockchafers—just four times her own weight. She was careful in cleaning herself, using the feet as combs, with which she parted the hair on each side down the middle. A young one was born, but though the mother died the next day, the little thing lived for eight days on milk. The Hairy-armed Bat occurs in the Midlands and in Ireland. One specimen of the Parti-coloured Bat (from North-eastern Europe) was found at Plymouth. whither it had doubtless been brought by some ship.

The Tube-nosed Bats, of which there are seven or eight species, are found in Java, Japan, Tibet, and the Himalayas. The nostrils project on each side the muzzle, like tubes; and the lower part of the wings and the thigh-membrane are hairy.

The Painted Bat is found in the forests of Tropical Asia. It is of small size, but is very remarkable for its coloration—orange and black —which is probably protective: that is, it harmonises so well with the creature's surroundings as to render it difficult of detection. Of another species of the same genus, Swinhoe says: "The body of this bat was of an orange-yellow, but the wings were painted with orange-yellow and black. It was caught suspended head downwards on a cluster of the round fruit of the longan tree. Now, this tree is an evergreen, and all the year through some portion of its foliage is undergoing decay, the particular leaves being in such a stage partially orange and black. This bat can therefore, at all seasons, suspend itself from the branches and elude its enemies by its resemblance to the leaf."

The Brown Pig-Bat, from South and Central America, is nearly 3 in. long, including the tail. The fur is cinnamon-brown above, paler beneath, and the wings are dusky. This animal has suckers on the feet and hands, something like those on the arms of the cuttle-fish. These Bats are thus enabled to climb over smooth upright surfaces; and it is supposed that they capture the insects on which they feed, while crawling over the branches of trees.

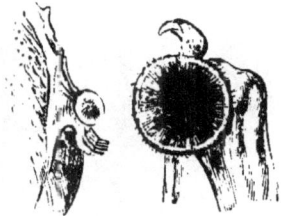

FOOT AND THUMB DISCS OF BROWN PIG-BAT (*Enlarged*).

The Thick-legged Bats are chiefly confined to the tropical and sub-tropical regions of both hemispheres. The Sack-winged Bats derive their name from a pouch or sac on the lower surface of the arm-membrane, near the elbow. It secretes a reddish substance with a strong smell. There are six species from Central and South America.

The Tomb Bats owe their name to the fact that the first species known was found in the ancient tombs of Egypt. There are some ten species spread over the Eastern hemisphere. The Egyptian Rhinopome is also a tomb-haunting Bat. The long slender tail is produced beyond the thigh-membrane. Owing to the length of the hinder limbs, and the fact that the wing-membrane does not extend the whole length of the leg, these Bats can walk much more freely than do others.

In the Mastiff Bats the muzzle is short and thick and the tail stout. Of the Smoky Mastiff Bat, which spreads from South America to Jamaica, Mr. Osburn says that they swarmed in the roof of his house, and passed out under the eaves. Frequently small parties of them would come in through the windows and take a short flight round the room. In hibernating, the males and females form separate groups, and this habit is common in most species of Bats. The strangest of the group is the Short-tailed Bat from New Zealand, which resembles the Brown Pig-Bat in the possession of special organs for climbing. It seems to go pretty well on all-fours. The thumb-claw bears a sharp tooth, which probably increases its clinging power. The lower surface of the hind limbs and the soles are covered with a soft, loose, wrinkled skin, almost certainly adhesive. Dr. Dobson believes "that this species hunts for its insect food not only in the air, but also on the branches and leaves of trees, amongst which its peculiarities of structure most probably enable it to walk about with security and ease."

The Javelin Bat, from Tropical America and the West Indies, is about 5 in. long, with a wing-spread of nearly 2 ft., and fur of a uniform brown hue. Wallace charges this Bat with blood-sucking, and other writers support him. On the other hand, the stomachs of

HEAD OF BLAINVILLE'S BAT.

many of these Bats have been examined, and found to contain insects, but no traces of blood. One of the strangest-looking of this family is Blainville's Bat, the head of which surpasses any demon-mask seen in a pantomime. There are allied species which have the leaf-like appendages on the chin.

The Vampirine Bats, long accused of blood-sucking, have been proved to be fruit eaters. Of the Great Vampire Bat (Plate II., No. 8), Bates says : "Nothing in animal physiognomy can be more hideous than the countenance of this creature when viewed from the front —the large, leathery ears standing out from the top of the head, the erect spear-shaped appendage on the tip of the nose, the grin, and the glistening black eye—all combine to make up a figure that reminds one of some mocking imp of fable." He opened the stomachs of several of these Bats, and found the contents to be fruit and insects.

The Soricine Bat, from the warmer parts of South America, may be taken as the type of a small group in which the tongue is long, thickly set with hairs, and capable of being protruded to some distance from the mouth. These Bats were formerly considered to be bloodsuckers, and the tongue was believed to be used in some way to increase the flow of blood. These Bats have been kept in confinement, and have been seen to use the long tongue to lick out the soft pulp of fruit.

In the Stenoderm Bats, which have a lance-shaped nose-leaf, springing from a regular horseshoe, the molar teeth have sharp points and a cutting edge. They feed chiefly, if not entirely, on fruit. The Mont-serrat Stenoderm, first described in 1894, is said to do great damage to the cacao-plantations in that island.

The blood-sucking Vampires, or Desmodonts, have the teeth and stomach fitted for a blood diet—a state of things found in no other mammals. The Common Desmodont is some 4 in. long, and nearly four times as much in wing-spread. The fur is brown, but the tint

DESMODONT, OR BLOOD-SUCKING VAMPIRE.

varies considerably in different individuals. This and an allied species seem to be the only Bats habitually guilty of blood-sucking. Horses, cattle, and man himself are the victims. The wound, which is difficult to heal, is probably inflicted with the sharp cutting teeth, the skin being shaved away till the small vessels are exposed and a constant supply of blood kept up.

Some persons are particularly annoyed by these Bats, while others are free from their attacks. Wallace tells of an Indian girl who was bitten again and again, till she became quite weakened from loss of blood, so that it was found necessary to send her to a distance where these bloodthirsty animals did not abound.

INSECTIVORES.

These animals are in many ways related to the Bats, but the limbs are organised for walking or burrowing, and in some few cases for swimming. There are generally five digits, armed with claws, on each limb; and in walking, the soles and palms are placed flat on the ground. Some have an external resemblance to some of the Rodents, and are often wrongly called by names that properly belong to that order. Thus, the Common Shrew is often called the Shrew-*mouse*, and some of the Indian Shrews are called Musk-*rats*. The teeth are of the ordinary three kinds—incisors, canines, and molars, the latter furnished with sharp points (see p. 85). Their chief food consists of insects and their larvæ, but some forms devour worms and molluscs, and shell-fish; while others attack frogs, snakes, fishes, small birds, and even small mammals. Some of them are found all over the temperate and tropical regions of both hemispheres, with the exception of South America and Australia.

For a long time naturalists were uncertain where to put the Colugos. They have been classed with the Lemuroids and called Flying Lemurs —a name which contains as many errors as words, for these creatures cannot fly, and are not Lemurs. They possess a parachute-membrane, or patagium, extending from the wrists along the sides of the body to the heels, as well as a thigh-membrane, like that of the Bats. But on comparing the body and limbs of the Colugo with our own in the same fashion that we compared those of the Bat (p. 69), we shall see in a moment that true flight is impossible for these animals. Motion of the fore-limbs would never raise them from the ground.

The Common Colugo is a native of Malacca, Sumatra, and Borneo. The general length is from 18 in. to 20 in. It has been said to live principally on leaves, but it also relishes insects, and it frequently captures and devours small birds. Wallace, in his "Malay Archipelago," says that the Colugo "rests during the day clinging to the trunks of trees, where its olive or brown fur, mottled with irregular whitish spots and blotches, resembles closely the colour of mottled bark, and no doubt helps to protect it. Once, in a bright twilight, I saw one of these animals run up a trunk in a rather open place, and then glide obliquely through the air to another tree, on which it alighted near its base, and immediately began to ascend. I paced the distance from one tree to the other, and found it to be 70 yards; and the amount of descent I estimated at not more than 35 ft. or 40 ft., or less than one in five. This, I think, proves that the animal must have some power of guiding itself through the air, otherwise in so long a distance it would have little chance of alighting exactly upon the trunk." Another species, of similar habits, lives in the Philippine Islands.

The Tree-Shrews are natives of South-eastern Asia and the islands of the Eastern Archipelago. They live among the branches of trees, and are active in the daytime. In form and size they closely resemble squirrels, for which they have been often mistaken. Their diet consists of fruit and insects. The Burman Tree-Shrew is about 14 in. in length, the tail being as long as the head and body together. The general hue of the fur is a dusky greenish-brown. It is a harmless little animal, in the dry season living in trees, and in the rainy season entering the houses. One that lived in a mango-tree, near the house of an Indian missionary, made itself nearly as familiar as the cat. It would take up its quarters on the bed, and was very fond of putting its

nose into the tea-cups immediately after breakfast, and acquired a taste
for tea and coffee. But at last it lost its life by walking into a
rat-trap.

In the same family is the Pen-tailed Tree-Shrew a native of Borneo
and Sumatra. Its general colour is blackish-brown above, and yellowish
on the under surface. Its great peculiarity is in its tail, which is hairy
at the base, then black and scaly for some distance, and for about a
third of the length at the end furnished with white hairs arranged on

COLUGOS.

each side like the wings of an arrow or the plume of a feather. Its
habits resemble those of the Tree-Shrews.

In the next family are the Elephant-Shrews, from Africa. The
snout is prolonged into a kind of proboscis, which accounts for the
popular name. The hind-legs are more developed than the fore-limbs,
and they advance by a succession of leaps, thus resembling the Jerboas,
and causing some writers to call them Jumping Shrews. The Common
Elephant Shrew, from South Africa, is about 8 in. long, of which
the tail takes up 3 in. The colour is tawny-brown, becoming whitish
on the limbs. It is active by day, and lives in burrows, to which it
retreats on being disturbed. There are several other species.

The Hedgehogs are small, stoutly-built animals, with pointed snouts and very short tails, and in most of them the hair on the upper surface is so thickened as to form spines. In this family we meet for the first time with a mode of defence, or means of protection—that of rolling into a ball, common among many of the lower Mammals.

The Common Hedgehog (Plate II., No. 10), when full grown, is about 10 in. in length; the spines of the upper surface are dirty-white ringed with black, and about an inch long. The face is black, and the hair on the spineless parts yellowish-white. It is spread over Europe, except in the extreme North, and ranges into the South-west of Asia. It generally sleeps by day, coming out to hunt at night, and hibernates in the winter, a habit which is not shared by the Indian, nor probably by the African species. Its chief diet consists of insects and beetles; and these creatures are sometimes kept in houses to kill cockroaches (which despite their popular name are not beetles at all), and so fond are they of this diet that some have died of over-feeding. Earthworms, slugs, and snails are also eaten, as are frogs, toads, snakes, vipers, and mice. Strange to say, the Hedgehog does not fear the bite of the viper, and will crunch up the Blistering Beetles or Spanish Flies as if they were sweetmeats; but the secretion from the skin of the toad is disagreeable, and the Hedgehog rubs its muzzle on the ground after each bite. From its habit of devouring slugs and snails, the Hedgehog must be reckoned as the gardener's friend; but he has a bad habit of feeding on birds' eggs, and sometimes attacks poultry. A writer in the *Field* says: "I was going to bed, I heard a tremendous outcry from a hen in an adjacent orchard. Hurrying out with a friend to the rescue, and picking up some handy stones, we made for a coop under which was the hen with her chicks. Overturning the coop, we found a large Hedgehog hanging to the hen's throat. The stone, of which I expeditiously delivered myself, caught the Hedgehog between the eyes, and he rolled over dead." When disturbed, the Hedgehog rolls himself into a ball, with the head and legs tucked in, and only the spiny surface of the back exposed. A jet of water poured on the part where the head is concealed will cause it to unroll; and it is said that foxes and some dogs will push a Hedgehog into a pond or ditch, so that he must either unroll and be eaten, or be drowned. The young are born blind and naked, but possessing the rudiments of spines, which are then soft and flexible. There are several allied species.

The Collared Hedgehog, a native of India, "has the habit, when touched, of suddenly jerking up the back with some force, so as to prick the fingers or mouth of the assailant, at the same time emitting a blowing sound like that from a pair of bellows."

The Bulau, discovered by Sir Stamford Raffles in Sumatra, is a Shrew-like Hedgehog, "with the body, and especially the head, more elongated than in the Common Hedgehog, with flexible hairs, and furnished with a tail that is nearly as long as the body." The general colour is greyish-black; the head and body are about 14 in. long, and the tail 12 in. It is more active by night than by day, and lives in holes among the roots of trees. There is another closely-allied species

The Shrews constitute a numerous family of mouse-like or rat-like creatures, spread over the Old World and North America. The

snout is long and pointed, the body mouse-like, and the tail thick and tapering, and more or less densely set with hairs. Many of them are furnished with glands which secrete a strong-smelling fluid.

The Common Shrew is about 2¾ in. long, with a tail of rather more than 1½ in. It feeds on insects, worms, small snails, and slugs; and it is preyed upon

HEDGEHOG AND YOUNG.

by barn owls and weasels. It is said that cats will kill but not eat them, owing to their strong-smelling glands. In the autumn great numbers of these little creatures are found dead, without apparent injury, on roads and footpaths in the country—probably starved.

Some old superstitions still linger round the Shrew, which is, or was till very recently, credited with causing cattle to fall lame if it ran over their backs, while its bite made them "swell at the heart and die." The only cure was to stroke the part affected or bitten with a twig from a shrew-ash—that is, an ash-tree, into which a hole had been bored with an auger, and a Shrew plugged up alive in the hole.

The Pygmy Shrew, which is also British, is rather smaller, though the tail is longer in proportion to the body. There is also more white on the under-parts.

The Garden Shrew is common over nearly all Europe. The total length is about 4 in., of which the tail occupies somewhat less than 1½ in. The fur is mouse-grey, shading into light ash-grey below. In

G

BRITISH SHREWS.

habit it resembles the Common Shrew, and the same story of its injuring cattle is told of it.   Our illustration shows the three British Shrews.   To the left is the Common Shrew; the Water Shrew is on the right; and the little creature at the top is the Pygmy Shrew.

The Tuscan Shrew, with ashy-red fur above and ash-grey beneath, is probably the smallest living mammal. From the snout to the tip of its tail (about 1 in.), it measures from $2\frac{1}{2}$ in. to $2\frac{3}{4}$ in.

The Indian Musk Shrew has bluish-grey or mouse-coloured fur; the head and body together measure from 6 to 7 in., and the tail nearly 4 in., so that, compared to the British Shrews, this is quite a giant. It has a strong musky odour, which it was (falsely) said to impart to wine and beer by running over the bottles in which these liquors were contained. Mr. Sterndale champions these little creatures on account of their insect-eating habits, and proved by experiment that the mere passing of a Musk Shrew over a substance does not necessarily impart a musky odour. While dressing for dinner one day he saw a Musk Shrew in his room. Placing a clean white handkerchief on the floor, he chased the Shrew till it had crossed the handkerchief five times.   At mess he asked his brother-officers if they could perceive any peculiar smell about the handkerchief, but none of them could.   "Well, all I know is," said he, "that I have driven a musk-rat five times over that handkerchief just now." From which it seems that the Musk Shrew emits no odour except at certain seasons, or when irritated.

The Water Shrew is a little more than 3 in. long, with a

tail of rather more than 2 in. The fur is black, or nearly black, above, and white below, the two being sharply marked off from, and not shading into, each other; but there is great diversity in the coloration. On the under-side of the tail is a long fringe of hair, and there are comb-like fringes of stiff hairs on the feet and toes, thus making the limbs and tail good swimming-organs. The Water Shrew forms a burrow in the banks of ponds or streams; its prey consists of freshwater shrimps, insects, larvæ, and the fry of small fishes; and it has been seen feeding on a rat that had been killed in a trap. It is fairly common in Britain, and ranges through Europe to the Altai Mountains. The Himalayan Water Shrew is somewhat larger; and another species is found in Japan.

The Tibetan Water Shrew is about 8 in. long, of which the tail counts for half. The feet are webbed, and furnished with sucker-like discs, which probably enable the animal to cling to the stones in the river-bed. It is said to feed on small fishes.

The Tailless Shrew, also from Tibet, is a burrowing animal. Like the Mole, it has the fore-feet broader and stronger than those of the hind-limbs. It is about 4 in. long, clad in grey fur with a greenish-brown tinge.

The Desmans and True Moles are confined to the temperate parts of Europe, Asia, and North America. The eyes are very small, and in some forms covered with skin; the ears are short, and hidden in the fur; and in most cases the fore-limbs are modified into shovel-like organs for burrowing.

The Desmans, which are aquatic in habit, are not unlike big rats, but the nostrils are very long, and form a tube-like snout; the toes are webbed, and the scaly tail is flattened from side to side to aid the animals in swimming. They frequent standing water and slow streams, in the banks of which they form their burrows, which are only used as resting-places, the greater part of their time being spent in the

PYRENEAN DESMAN.

water. They feed on worms, pond-snails, and insect larvæ ; and probably no small aquatic animals come amiss to them. The Common
Desman is a native of Southern Russia and South-western Asia. Its
length is about 18 in., of which the tail forms a little more than
a third. The fur is reddish-brown on the back, ashy-grey below,
with a silvery lustre in certain lights. The Pyrenean Desman is
about two-thirds the size of its relative, and has chestnut-brown
fur on the back, and silvery-grey on the under-surface. It is said

COMMON MOLE.

to feed principally on trout. Both species have a strong musky
smell.

The Mole Shrews are small animals that connect the Shrews with
the True Moles. The Hairy-tailed Mole Shrew, common in mountain
regions of Southern and Eastern Japan, but rarer in the north, is about
the size of the Water Shrew, with velvety-brown fur. It burrows like
a Mole, but does not throw up heaps of earth. Gibbs' Mole Shrew,
from North America, is closely allied. The Tibetan Mole Shrew differs
from the other species in not making a burrow. The fur is slate-
coloured, and the general appearance Shrew-like, but the skull is like
that of a Mole.

The True Moles have the collar-bone (*c*) and the bone of the upper
arm (*b*) from the shoulder to the elbow, very short and broad, and,
consequently, of great strength ; and from the inside of each wrist there

projects a stout sickle-shaped bone (*f*), which adds to the width of the hand and renders it more serviceable for digging and burrowing.

The Common Mole (Plate II., No. 9) is found in Britain, but not

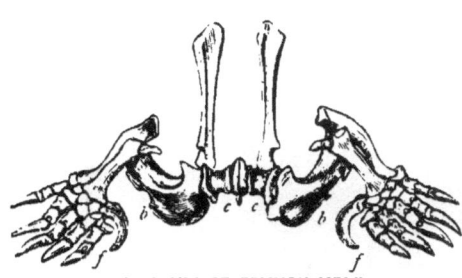

FORE-LIMBS OF COMMON MOLE.

in Ireland, and ranges eastward through Central and Southern Europe to India. It is about 6 in. long, of which the tail counts for a little more than 1 in. The body is long and cylindrical, with sharp-pointed muzzle. There appears to be no neck, owing to the fact that the fore-limbs are set so far forward. The fur is soft and velvety, black or blackish-brown in colour, with more or less of a whitish tinge in certain lights ; and the feet and hands are naked and flesh-coloured. There are five clawed digits on each limb, and the shovel-shaped hands are turned outwards.

The figure below shows the teeth of the Mole, and the sharp points of the back teeth are capital tools for cutting up the animal food on which the creature lives. The burrowing habits of the animal, and the heaps of earth it throws up in its travels beneath the surface, are well known. The beautifully regular plans of a Mole's dwelling as generally given are probably imaginative, and have been copied into most books on Natural History from the work of a French writer, who " figured the habitation of some mole of genius, or improved on Nature from his love of symmetry." The underground home of the Mole consists of a central chamber, just below the surface, and has several passages or runs branching out from it. These are generally connected by cross-runs, but Mr. Alston, who has paid much

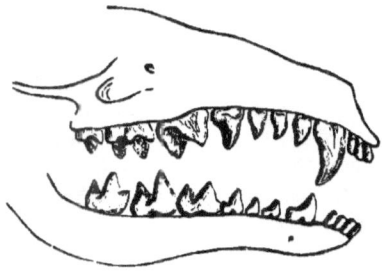

TEETH OF COMMON MOLE.

attention to the subject, says he has never seen anything like the regular system of circles generally shown as the approaches to the dwelling of the Mole. Moles are extremely voracious. The food consists of worms, insects and their larvæ, snails and slugs, frogs, and even snakes. One kept in confinement ate within twenty-four hours a large slow-worm,

a large snail, two chrysalids, and a snake 32 in. long. Of the latter
it only left the bones and skin.   It seems also to be a provident
creature.   A writer in the *Gardener's Chronicle* says that " Previous
to the setting in of winter the Mole prepares a sort of basin, forming it in a bed of clay, which will hold about a quart.   In this basin
a great quantity of worms is deposited, and, in order to prevent their
escape, they are partly mutilated, but not so much as to kill them.

THE CONVENTIONAL PLAN OF MOLE'S DWELLING.

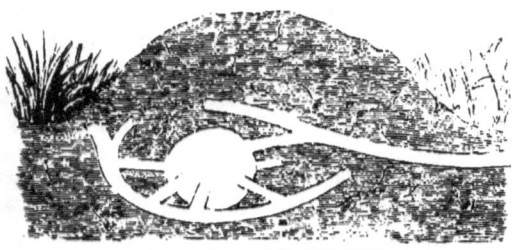

SECTION OF MOLE'S DWELLING.
(From Nature.)

On these worms the
Moles feed in the
winter months." This
account was con-
firmed by several
mole-catchers near
Southampton, where
Moles are very
numerous.   Many
farmers consider
Moles do injury to
pasture-land.   It is
probable that the in-
jury they do by eat-
ing the roots of
grasses is compen-
sated by the insects
and insect larvæ they
devour.   On this sub-
ject Frank Buckland
questioned a man in the employment of a large dairy farmer in
Hampshire.   The man's reply is worth quoting : " The Moles are of
great service ; they eat up the worms that eat the grass ; and wherever
the Moles have been, afterwards the grass grows there very luxuriantly.
When the Moles have eaten all the grubs and the worms in a certain
space, they migrate to another, and repeat their gratuitous work.   The
grass where Moles have been is always the best for cows."

The Blind Mole lives in Southern Europe, and differs little from
the common species except that the eyes are covered by a membrane
pierced by a tiny hole, so that the animal can have but a faint
perception of light and no distinct vision.   In India are found
several species nearly allied to our Common Mole, but they throw up
no mole-hills.   The Woogura Mole, from Japan, has tawny fur, and
the snout is greatly elongated.

The Star-nosed Mole, a native of the United States and Canada,

is about 8 in. in length, of which the tail counts for 3 in. The fur
is brownish-black, and the hairless parts flesh-coloured. The snout
is elongated, and at its tip is a fringe of long, fleshy filaments surrounding the nostrils, and probably
serving as organs of touch. It generally forms its runs near water or
swampy places.

The Shrew Moles of North America
have the hind feet webbed, and were
supposed to be of aquatic habit. This
is now known to be a mistake, for
their runs are scarcely ever found

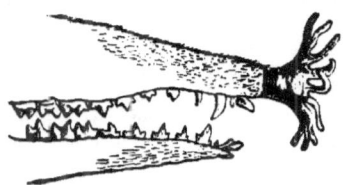

SNOUT OF STAR-NOSED MOLE.

near water, and in habit they agree with our Common Mole. Two
other species, also from North America, resemble the Shrew Moles in
habits, but are probably more closely akin to the Star-nosed Mole,
though the snout has no fleshy appendages.

The West African River Shrew, discovered by Du Chaillu, is about
2 ft. long, of which the tail forms a half. The fur is dark brown
above and white beneath, the tips of the hairs on the back showing
violet metallic reflections. It burrows in the banks of streams, and
though the toes are not webbed, is well fitted for an aquatic life,
the strong tail forming a powerful swimming-organ. The nostrils can
be closed by valves when the animal is in or under the water. Like
the Otter, it feeds on fish, and lands its prey before eating it. In
the family of River Shrews is included the Madagascar Shrew, a small
mouse-like form, with skull and teeth like those of the West African
River Shrew. Nothing is known of its habits.

The Agouta is a native of St. Domingo, and the Almique of Cuba.
Both are about a foot long, of which the tail forms more than half,
and differ chiefly in the colour and quality of the fur: in both the
snout forms a kind of proboscis, the tail long and naked, and the
feet are adapted for running rather than burrowing. In habit they
are nocturnal, and feed on insects, and probably birds and small
mammals. The Agouta is brown on the upper parts, lighter on the
head and neck, and darker behind and on the thighs, the belly and
feet tawny brown, the tail greyish with a white tip. In the Almique
the head, neck, and chest are a tawny yellow, and the rest of the body
dusky brown.

The Tanrecs are confined to Madagascar. In some the hair of the
back is mixed with spines. The Common Tanrec is of a tawny colour,
about 15 in. long, of which the head counts for a third: there is no
tail. These animals feed on worms, in search of which they turn up

the ground with the snout like a pig. The flesh is eaten, and by some it is esteemed a delicacy, though the musky odour is distasteful to others. It is hunted with dogs trained for the purpose, and great numbers are taken for the table. It is the largest Insectivore, and probably the most prolific mammal, twenty being produced at one birth. The Streaked Taurec is much smaller, and is marked with black and yellow.

The Tendrac has spines like those of the Hedgehog on the upper surface and short tail, but it is probable that it cannot roll itself into

TENDRAC.

a ball. The spines are black, with white or reddish tips, so that the general hue is dusky. The Rice Tendrac is greyish-brown in colour and mole-like in form. Its popular and scientific names refer to the damage it does to the rice-crops by burrowing in the fields and rooting up the young plants.

The Golden Moles are small burrowing animals from Eastern and Southern Africa. The fore-limbs bear four digits, and the hind-limbs five. The head is conical, the tail rudimentary; the limbs are short, and the eyes covered with skin : and, as in the Common Mole, there are no external ears. The fur is fine and close, and in nearly all the species (seven or eight) on the upper surface there is a brilliant metallic lustre of golden bronze, green, or violet.

The Cape Golden Mole, from Cape Colony and Caffraria, has brown fur with metallic reflections. Its habits resemble those of the British Mole, and the farmers wage war on it, alleging that it does damage in fields and gardens by its burrowing. The Blunt-nosed Golden Mole is a native of Mozambique and Caffraria.

CHAPTER VIII.

CARNIVORES. THE CAT FAMILY.

CARNIVOROUS animals are, for the most part, large and fierce, and generally feed on the flesh of warm-blooded creatures which they have killed. They are armed with strong claws and sharp teeth, admirably adapted for seizing and tearing their prey. Their senses are keen, as one would expect in beasts that live by rapine, and are, in their turn, hunted by man. There are never less than four digits on each limb, the canine teeth are well developed, and some of the back teeth are fitted for cutting flesh. There is no part of the world in which some of the Flesh-eaters are not found, with the exception of Australia, where the Dingo, or wild dog, was probably introduced by man.

The Cat family consists of a single genus, with about fifty species, the most highly specialised of the Beasts of Prey. They do not hunt, like the dogs, but lie in wait for and spring upon their victims, which they seize and hold with their claws. There are five digits on the fore limbs and four on the hinder ones—that corresponding to our great toe being absent; and the claws can be drawn back, so as to preserve them from being blunted in walking.

We may get a good idea of the structure of these animals by examining the family cat, which, for several reasons, should be done as gently as possible. If we take pussy into our lap, and hold her in the hollow of the left arm, the right hand will be at liberty for the examination. The spine, or backbone, and its connection with the skull or brain-box, may be *felt*, and we may assure ourselves that the limbs are also connected with the spine. We shall not be able to trace anything corresponding to our collar-bones, owing to the small size of these bones in the cat. We shall notice that the head is short and round, very different from the long head of a dog. The whiskers are exquisitely delicate organs of touch, and serve to warn the animal of any obstruction, so that the popular notion that where a cat can get its head it can get its whole body is founded on a scientific fact. If the lips be turned up, the small size of the cutting or *incisor* teeth (*i*) will be noticed, and the fact

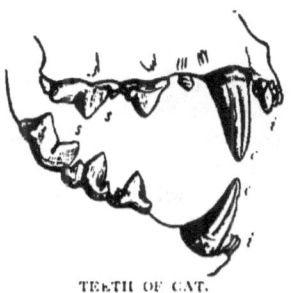

TEETH OF CAT.

that between them and the canines (*c*) in the upper jaw is a space into
which the canines (*c*) of the lower jaw bite.   In the upper jaw on each
side are four teeth, the first and last of which are very small ; while
below, there are three teeth on each side.   The tusk-like canines seize
and hold the prey—a mouse or a bird, while it is torn to pieces with the
claws, or cut with the last large tooth on each side, which for this reason
are called *sectorial* or *carnassial* teeth (*s*), and close on each other like the
blades of a pair of scissors.

If the finger and thumb of the left hand be inserted *behind* the back
teeth, the mouth may be held open so as to afford a good look at the
tongue.   The surface will be seen to be covered with rows of horny spines,
like tiny claws, bent backwards and in the larger cats these serve to rasp

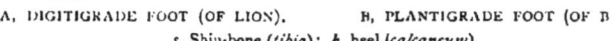

A, DIGITIGRADE FOOT (OF LION).   B, PLANTIGRADE FOOT (OF BEAR).
s, Shin-bone (*tibia*) ; *h*, heel (*calcaneum*).

the flesh from the bones of their prey.   You may easily test its roughness
by allowing the cat to lap a little milk from the palm of your hand.

The eye of the cat is that of a nocturnal animal.   It cannot, of course,
see in absolute darkness, but the pupil of the eye can be dilated so as to
catch the faintest beams, which are reflected and intensified by the
brilliant, golden-yellow lining of the eyeball, called the *tapetum*.   In a
strong light the pupil contracts to a mere slit—a fact you may verify
for yourselves.

It is an easy matter to examine the limbs.   The shoulder-blades may
be readily made out, and the bones of the arms and legs.   The elbow
and knee are situated close to the trunk, and what we call a cat's "foot"
corresponds to our fingers and toes, not to the palms of our hands or
the soles of our feet.

This will be better understood from Figs. A and B.   The terms (1)
Digitigrade and (2) Plantigrade were formerly used to denote (1) the
Cats and (2) the Bears and Dogs, because the first group walked on
their digits, and the second on their (palms and) soles.   But as it was

found that there were very many intermediate forms linking the two together, the names have been abandoned.

The "foot" of a cat deserves careful attention, for it shows how well fitted the members of this family are for their work of destruction. Underneath each fore-paw are seven soft fleshy pads or cushions, one below the end of each digit, a large one in part corresponding to the palm of our hand, and a small one on the outer side of the wrist. The hind "foot" carries only five of these pads, there being no great toe to need one, and the outer one being absent. These soft cushions enable the animal to approach its prey noiselessly, and serve to break a fall. It is owing to these that the domestic cat can jump from a great height to the ground without sustaining injury.

The great characteristic of the family is the possession of claws that can be drawn back or put forth at will. If the fore-paw be held lightly between the finger and thumb, and the latter moved forwards and backwards upon the skin of the animal, the claws will move forwards and backwards also. Then we shall see that each issues from a horny sheath

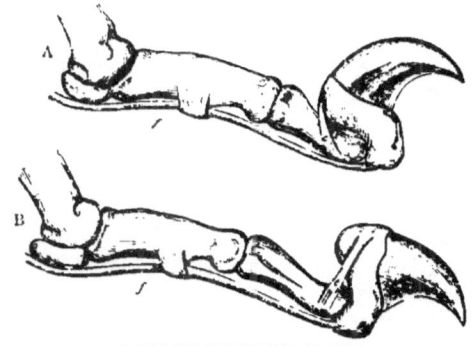

DIGIT OF FORE-PAW OF CAT.
A, With claw drawn back ; B, with claw put forth.

or hood on the last joint of each digit. The bone marked m is the last in one of the four rows of bones in the human palm ; 1, 2, 3 mark the joints of the finger, the last with its horny sheath enclosing the root of the claw. When the animal wishes to draw back its claws, the tendon (f) is relaxed, and the strain being thus taken off the elastic ligament (e) that ligament contracts and draws the third joint, carrying the claw, backwards till it fits into the hollow on the second joint. The contraction of the tendon (f) has, of course, a directly contrary effect, stretching the ligament (e), and bringing the first joint to a nearly upright position, so that the claw is put forth ready for use.

The colouring of the Domestic Cat may be black, white, tabby (that is, grey striped with black), tortoise-shell (fawn-colour spotted with black), sandy, and grey. There are numerous varieties, as the tailless Manx cat, a capital mouser, the long-haired Angora and Persian cats, generally kept as pets, the bluish-grey Carthusian cat, and many others. A good deal has been written about a Chinese breed, with long ears hanging down like those of a lop-eared rabbit, but Père David, the French missionary, who has travelled so long in China, was never able to find one, and thinks the story a fable.

The habits of the Domestic Cat are too well known to need description. Wherever civilised man is found, the cat is an inmate of the house, though rarely made such a pet of as the dog. It is quite distinct from the Wild Cat (Plate III., No. 4), which is extremely rare in Great Britain, occurring only in the Scottish mountains, and unknown in Ireland; it is still found in Central Europe. Our cats probably descended from the Egyptian cat, a native of North Africa, smaller than the Wild cat of Europe, with yellowish fur, darker on the back, and fading into white below. There are some obscure stripes on the limbs; and stripes or spots occur on those of most cats, large and small, at least in early life, which seems to show that they are all descended from a striped ancestor.

The Wild Cat is larger and more stoutly built than the Domestic Cat, and the tail, instead of tapering to a point, is of the same thickness throughout. The colour is grey, striped with black, so that it looks like a large, fierce tabby. Its disposition is extremely savage, and when hard pressed it is said to attack man, and inflict severe wounds with its teeth and claws. The female makes a kind of nest in hollow trees or clefts of the rocks, where she brings up her young—for cats are excellent mothers.

Having got some idea of the "make" of the cats—for our common species is a very good example of the family—we may go on to the larger species. In the Old World are found the lion, the tiger, the leopard, the ounce or snow leopard, the cheetah or hunting leopard, a number of smaller kinds known as tiger cats, and lynxes. Some tiger cats and lynxes are found also in the New World, which has only two large species —the puma and the jaguar, and some smaller cats. The colouring may be sandy, as in the lion and puma, striped as in the tiger, or spotted as in the leopard and jaguar. Black varieties are said to occur in many species, but there is no black race of any wild form, though black individuals often occur in the same litter with cubs of the ordinary kind. This unusual colour of the skin is called *melanism*, and occurs chiefly in leopards and jaguars, but even in these the spots appear of a deeper black than the rest of the skin.

The Lion (Plate III., No. 1) is a native of Africa and South-western Asia, but in both continents is being driven back by the advance of civilisation. In classic times it roamed over South-eastern Europe, and remains have been found in bone-caves in this country which show that at one time a lion closely allied to, if not identical with, the living species, was native in Britain. The lion is distinguished

LION AND GIRAFFE.

from all other cats by the presence of a large, thick mane in the adult male. A full-grown animal will measure rather more than eight feet from the nose to the end of the tail, which counts for nearly half, and is furnished at the end with a tuft of hair, in the centre of which is a small horny prickle, the use of which is unknown. The lion certainly does not employ it, as was once thought, to excite himself to fury by pricking his sides with it when he lashes his tail. The lioness is smaller than her mate and without a mane. She bears from two to four cubs at a litter, and native hunters often steal them to sell them to the dealers in wild beasts who supply the menageries, for the capture of a full-grown lion is rarely effected. The sire and dam both watch over their young, and train them to hunt prey. Thus young lions are more destructive than old ones ; the former kill for the sake of killing, the latter only to satisfy hunger and provide for their mate and her cubs.

Lions generally lie in wait for their prey, concealed in the reeds near some place where other animals come to drink, and then, springing from their lair, leap upon the victim, striking it down with the paws. The neck is generally broken with a violent wrench of the powerful jaws, and the carcase is carried off to be devoured at leisure. The lion does not disdain the flesh of animals killed by the hunter. Gordon Cumming frequently saw lions feeding on antelopes that had fallen to his rifle ; and Stevens, who was sent by the *New York Herald* to find Stanley, saw three " bunched up inside the capacious carcase of a rhinoceros, and feeding off the foulest carrion imaginable." When pressed by hunger the lion will approach a native village by night and carry off goats and calves, but fires and torches will scare him away.

The lion has been called the King of Beasts, and a good deal has been written about his courage and magnanimity. The former has been exaggerated ; the latter he does not possess. He will generally fight savagely if brought to bay, and the lioness, when with cubs, is still more dangerous ; but as a general rule, the "king of beasts," if not molested, will bolt on sighting a man.

There are several varieties, differing somewhat in size and greatly in colour—from red chestnut to a grey, so pale that at one time it was said that a race of white lions existed at the Cape. The Asiatic lion was formerly supposed to have no mane, but one that lived in the Zoological Gardens, Regent's Park, from 1854 to 1857, could show as fine a mane as any lion from South Africa.

The roar of the lion is extremely grand and striking, and at times a troop may be heard in concert, one taking the lead and three or four others chiming in like persons singing a catch.

The lion is easily tamed if taken young, and is capable of strong attachment. Most of the so-called " lion-tamers " who exhibit performing animals maintain their authority by the terrorism of the whip. This was not the method of Van Amburgh, who subdued his troop of wild beasts by personal influence, nor of the celebrated Frenchman, Henri Martin, who never took a whip into the cage, and whose influence over

TIGERS ON THE PROWL.

savage creatures was so extraordinary that many people attributed to him the power of fascination.

The Tiger (Plate III., No. 2) is confined to Asia, ranging from the shores of the Caspian Sea to the island of Saghalien, but is not found in Borneo or Ceylon. It equals, if it does not exceed, the lion in size, and is an exceedingly beautiful animal. The fur is a reddish-tawny, with black stripes on the body, and bars of the same colours on the limbs and tail. The under surface is white, and there is some white about the face and ears. Tigers from the hot plains of India are larger and have smoother fur than those from northern localities. The unworn canine teeth in the upper jaw show faint serrations, like those of the teeth of the extinct sabre-toothed tigers, one of which formerly lived in Britain. It is these jagged teeth that make the tiger's bite so dangerous as to give some colour to the native notion that it is poisonous.

The tiger is a very cowardly animal, but extremely savage and dangerous when roused or wounded : in 1893 the Commander-in-Chief of Madras was killed by one that he had wounded. It delights in marshy places and thickets near rivers, and its striped marking is a protection to it among the reeds of the jungle. It lives chiefly on antelopes and domestic cattle, usually seizing the prey with the teeth and using the paws to hold it and give a purchase for a terrific wrench, by which the neck is broken. Like the lion, it will feed on animals it has not killed ; and Jerdon tells a story of a wounded tiger being dragged off and devoured by another tiger. In old age, when unable to take stronger and more active prey, some tigers become "man-eaters," and will then prowl round the villages, carrying off victim after victim, till some English sportsman puts an end to their career.

Tiger-hunting is a favourite sport with native princes and Europeans in India. The sportsmen are usually seated in howdahs—constructions something like the body of a carriage—on the backs of elephants. Beaters are sent forward to rouse the game, and the elephants follow in line. But shooting a tiger from an elephant's back is not an easy task, and if the game be missed the tiger often carries the war into the enemy's camp and charges the nearest elephant, sometimes inflicting severe wounds upon the head and trunk before it is trampled to death by the mighty beast, or despatched with a shot. Cases are known in which the infuriated tiger has sprung upon the elephant and carried off the sportsman. The tiger is also shot from platforms in trees, a cow or goat being tied up near as a bait to attract it within range ; and the natives take it in traps and pitfalls, or destroy it with poisoned meat.

If not interfered with the tiger will generally run from, rather than attack man. A recent writer in the *Field* tells a story of how, having sent his guns and luncheon on before him, his attention was arrested by a rustle in the jungle, and, looking that way, he saw a tiger, crouched low, coming rapidly towards him. He says : " My first feeling was one of horror, for it seemed all up with me the tiger being very close and in his rush. Of course, it was not me but the pony that he wanted ; but had he knocked over the latter, his own fears at finding a man under him would have made him maul me, too. There was but one thing to be done—viz. to put a bold front on it and try to frighten him : and I therefore instantly wheeled the pony's head

directly towards him, snouting at the same moment. The tiger stopped short and stared at me, but he did not offer to retreat. I then moved the pony towards him, shouting loudly as I did so, and the tiger then turned his tail to me and, having retired about thirty yards, he sat bolt upright on his haunches and stared at me. I was naturally desirous of withdrawing from an interview so unpleasant to me in my unarmed condition. I therefore rode straight in at the tiger, waving my arm and sternly ordering him off, and before I reached him he

LEOPARDS IN THE FOREST.

decided to remove himself, this time somewhat hastily, and in marked contrast to his previous orderly, not to say dignified, retreat; and, having at last routed him, I lost no time in cantering over the remaining portion of the jungle cart-track until it emerged upon the high road."

The Leopard (Plate III., No. 3) is found in Africa and the warmer parts of Asia. It is about 6 ft. long, of which the tail forms a little less than half. The fur is reddish-fawn, marked on the body with dark rosettes; the tail is tinged with black, and the under-surface is whitish. It is arboreal in habit—that is, it lives much more on trees than on the ground; in this respect differing from the lion and the tiger, which rarely climb trees—so rarely, indeed, that some writers have doubted if these larger cats have the power to do so. It is a very destructive animal, and preys upon sheep, goats, antelopes, and calves. Donkeys it leaves severely alone, because, to quote a recent

H

writer on Eastern Equatorial Africa, " it knows well that a donkey, like an English football-player, is generally a good kick, and so prefers to give it a wide berth."

It has a strange liking for dog-meat, and is always ready to dine off a dog provided he be not too large. Dr. Pruen, in " The Arab and the African," tells an amusing story of the experiences of a leopard with two English mastiffs. His servant chained up the dogs in the verandah at dusk, and little time elapsed before a leopard, who had smelt dog from below, jumped in between them. He was evidently surprised at their size, and still more so at the treatment he received, for " one dog got him by the head, the other by the tail, and the two quickly bowled him over. He lay perfectly still, astonished at the unexpected turn which events had taken, whilst the dogs, evidently puzzled at his quiet behaviour, simply held him there and growled, but offered him no further violence. Before the men who had been standing near could return with their guns, the leopard had taken advantage of the dogs' indecision to suddenly wriggle away and disappear in the darkness, leaving them without even a scratch."

It sometimes carries off old women and children, but rarely attacks man, though when wounded it fights with great fierceness, and sometimes succeeds in killing its foe. In 1892 a high official in India wounded a leopard, as he thought mortally, when the beast sprang upon him, threw him down, and badly mauled his left arm. Fortunately, a native hunter came up and pinned the brute to the ground with a spear, when the Englishman scrambled to his feet, and killed the leopard with a shot through the head.

The Ounce, or Snow Leopard, bears the same relation to the true leopard that the extinct mammoth did to the elephants of our own day—that is, it is fitted for existence in cold climates. The wide spreading feet, that seem disproportionately large, are admirably suited for travelling over an expanse of yielding snow. It lives in the highlands of Central Asia and the Himalayas, rarely descending much below the snow-line, and is clothed in thick fur as a protection against the cold. The colour is yellowish-grey with irregular dark spots, and the bars on the tail do not form rings. It is said to live on sheep, goats, and dogs, and has never been known to attack man. Very little is known of its habits in a wild state ; and only two specimens have been brought alive to England. The first was exhibited at the Zoological Gardens, Regent's Park, in 1891 ; but, unfortunately, it only lived in confinement for about a month. A second specimen was received in 1894. For some little time, when quite a cub, it was in the possession of Mrs. Tyacke, and fed with milk from a spoon. In " How I Shot my

Bears," she says :—"He was never happy unless in my lap or arms, and would cry bitterly if I left the room without taking him with me." But the animal passed into other hands, and when Mrs. Tyacke saw him three months later "he was an exceedingly graceful little beast, sleeping on the Mem Sahib's bed, and answering to the name of 'Moti' (pearl). He followed her everywhere, even running along the road on a fifteen-mile march. But he was already developing the natural ferocity and treacherousness of his race, and learning to make pretty free use of his claws. We gave him a wide berth, except when tired out with a long march, and disinclined to mischief, for he had left the marks of his poisonous natural weapons on the hands of his master, and even of his mistress . . . He had already bitten a boy, and had taken to knocking down goats, so that in a few months we saw that it would be necessary to consign him for life to an iron cage, or sell him to some rajah with a fancy for a menagerie." His owner sold him to the Zoological Society, and it is only fair to say that his behaviour towards his keepers in London is much better than it seems to have been towards his friends in India.

The last of the Old World large cats is the Clouded Tiger, found in some parts of South-eastern Asia. It is a very beautiful and graceful creature, with brownish-grey fur, irregularly marked and striped with black. The total length is about 6 ft. or a little more, and of this the tail makes up some 30 in. This animal is arboreal in habit, and preys upon sheep, goats, dogs, and pigs.

The Puma, sometimes called the American lion, is found as far north as Canada and as far south as the Straits of Magellan. In shape it is not unlike a small lioness, with short legs, and the colour varies from reddish-brown to silvery-grey, so light as to be almost white. The young are spotted, but when the cubs are about a year old the spots disappear. The prey consists of monkeys, deer, calves, pigs, and wild colts ; indeed, it destroys so many of the last-named animals, that wherever the puma abounds wild horses are sure to be scarce. Hudson, in his "Naturalist in La Plata," says that a native told him that while driving a troop of horses through the thicket, a puma sprang out of the bushes and seized a colt that was following behind. The puma alighted directly on the colt's back, with one foot grasping its chest, while with the other it seized the head, and, giving it a violent wrench dislocated the neck. The colt fell to the earth as if shot, and the natives affirmed that it was dead before it touched the ground.

The same author believes that there is no authenticated instance of a puma making an unprovoked attack upon any human being, though he relates a case when it defended itself when attacked, and severely injured its assailant before it was killed. The Gauchos call it by a Spanish name

H 2

which means " the friend of man," and to the native Indian the beast was sacred. The Jesuit missionaries found little difficulty in making converts of the Indians, but when they tried to destroy the pumas, that preyed on their domestic cattle, they met with great opposition from their spiritual children, who thought that instant death would overtake anyone rash enough to interfere with the sacred animal. But early in the eighteenth century, when matters had become unbearable, owing to the depredations committed by the pumas on the farm-stock of the mission, a Jesuit Father, who had just come from Spain, determined to take strong measures to remedy the state of affairs. After exhorting the Indians to hunt and destroy the puma to no purpose, he waited till he encountered one, and, having killed it, he put it on his mule, and brought it to the station. The Indians expected to see him fall dead ; but as no harm befell him, their faith in the belief that some terrible harm would happen to him who killed a puma was shaken, and they were at last persuaded to hunt the beasts that so terribly thinned the flocks and herds of the mission-station.

The puma not only refrains from attacking man, but some writers tell us that it is unable to defend itself when attacked. " For then," says Gay, in his " Natural History of Chili," " its energy and daring at once forsake it, and it becomes a weak, inoffensive creature, and trembling and uttering piteous moans, and shedding abundant tears, it seems to implore compassion from a generous enemy."

Between the puma and the jaguar a deadly feud seems to rage. Whenever they meet there is sure to be a battle-royal ; and though the latter is the larger and heavier beast, he generally gets the worst of the encounter. Hudson, whom we have quoted before, says that the puma harasses the jaguar as the tyrant-bird harasses an eagle or a hawk, moving about it with such rapidity as to confuse it, and when an opportunity occurs, springing upon its back and inflicting terrible wounds with teeth and claws.

Kingsley, in his "Standard Natural History" (of America) is responsible for the statement that the puma of the Northern States wages as deadly war with the grisly bear as the Southern variety does with the jaguar, and generally comes off victor. The puma is easily tamed, and there are numerous instances of its having been kept as a pet.

The Jaguar is somewhat larger than the leopard, and, like it, is a spotted cat. In the Jaguar, however, the spots form dark rings with smaller spots within them. It ranges from Texas and Louisiana to Patagonia, chiefly haunting the wooded banks of rivers, and feeding on the capybara, the largest living rodent, though it by no means confines itself to this diet. Wallace ("Travels on the Amazon ") says that " it appears to approach very nearly in fierceness and strength to the tiger, and that many persons are killed every year by these animals, though they rarely attack man when they can procure other food. The Indians say that the jaguar is the most cunning animal in the forest: he can imitate the voice of almost every bird and beast so exactly as to draw them towards him; he fishes in the rivers, lashing the water with his tail to imitate falling fruit, and when the fish approach, hooks them up with his claws. He

JAGUAR.

catches and eats turtles, and I have myself found the unbroken shells, which he has cleaned completely out with his paws. He even attacks the manatee in its own element, and an eye-witness assured me he had watched one dragging out of the water this bulky animal, weighing as much as a large ox.

"A young Portuguese told me he had seen (what many persons assured me often happened) a jaguar feeding on a full-grown live alligator, tearing and eating its tail. On leaving off and retiring a yard or two, the alligator would begin to move towards the water, when the jaguar would spring upon it and again commence eating at the tail, during which time the alligator lay perfectly still. We had been observing a cat playing with a lizard, both behaving in exactly the same manner, the lizard only attempting to move when the cat for a moment left it; the cat would then immediately spring upon it again, and my informant assured me he had seen the jaguar treating the alligator in exactly the same way."

Bates and his party once disturbed a jaguar that had come to a water-hole to drink, and as they went on they found the mangled remains of an alligator. "The head, fore-quarters, and bony shell were the only parts which remained; but the meat was quite fresh, and there were many footmarks of the jaguar round the carcass, so that there was no doubt this had formed the solid part of the animal's breakfast."

OCELOT.

The Indians believe that this animal has the power of fascination. Wallace was told by a person who said he had witnessed the scene, how a jaguar stood at the foot of a high tree, looking up into the branches, where was a howling monkey making a piteous noise, and gazing steadily at the great cat below. The monkey descended slowly, branch by branch, still uttering mournful cries, till at last it fell down at the very feet of the jaguar, which seized and devoured it. The same author adds:—"Many incidents of this kind are related by persons who have witnessed them, but whether they are exaggerated or altogether imaginary, it is difficult to decide."

In both hemispheres there are found smaller striped, spotted, and in a few cases, self-coloured animals of the Cat genus, which are generally called tiger-cats, though some of them have distinctive names. Among the most important are the Ocelot, a savage arboreal form, with a wide range in America; the Margay, from Central America: and the Fishing Cat, from Southern and Eastern Asia. This species, which is very fierce, and which has been known to carry off children, lives chiefly on fish, which it dexterously hooks out of the water with its paw. The Serval is a large long-legged African cat, of tawny colour spotted with black. Of uniform coloration are the Jaguarondi and the Eyra, of weasel-like shape, with a very long tail, from the New World; and the Flat-headed Cat, with a long body and short legs and tail, the small Bornean Bay Cat, and the Egyptian Cat, from the Old World. Besides these, several wild cats are found in both hemispheres.

The Lynxes are cats with short tails, long limbs, especially behind, and a tuft of long hair at the tip of the ears. They are found in all four continents. The Northern Lynx (Plate LII., No. 5) is about 40 in. long, exclusive of the tail. Its colour is reddish-grey, more or less spotted, and there is a fringe of hair on the cheeks. It is a native of Scandinavia, Russia, and Northern Asia, and a few linger in the mountainous districts of Northern Europe. In winter it loses its bright colour, and the fur becomes longer and thicker, as a protection against the cold. Two or three varieties of this species are found in America. The Pardine Lynx is a native of the South of Europe, from Spain to Turkey. Lynxes may be considered as large wild cats; they prey upon small mammals and birds, doing great damage to poultry-yards. The Caracal from Asia and Africa is more like the tame cats in form. The tail reaches to the heels; the colour is reddish or yellowish-brown, generally uniform on the back and sides, but sometimes with darker spots. The under-surface is paler and occasionally white. This animal is used by the Arabs and Persians for hunting small antelopes.

Last of the family comes the Cheetah or Hunting Leopard, which differs so much from all the rest that some authorities put it in a genus by itself—the Dog-like Cat. It is found in Africa and Asia, and in the latter continent it is kept in a half-domesticated condition, and trained to hunt antelopes and deer. The legs are long, and the general appearance, with the exception of the head, dog-like, and the claws are only partially retractile. The length is about 4½ ft. exclusive of the tail, which is 2½ ft. to 2¾ ft. The ground colour is reddish-fawn, marked with black spots.

CHEETAH.

CHAPTER IX.

THE CIVET, HYÆNA, AND DOG FAMILIES.

THE Civet family occupy a position between the True Cats and the Hyænas. They have long, thin bodies, short limbs, a long tail, and a sharp-pointed snout; and are clothed with stiff, harsh fur. There are usually five digits on each limb, but those corresponding to our thumb and great toe may be wanting; and in walking the wrist and ankle are brought much nearer the ground than is the case with the Cats. The claws in most species can be but partially drawn back. The skull is longer than that of the Cats, and there are more teeth; the canines are smaller, and the back-teeth less scissor-like—bearing, especially in the Palm Civets, little blunt projections, so as to crush or grind. There is a pouch under the tail, in which an odorous substance is secreted. These animals are confined to Asia and African regions, with the exception of one species that is European.

The Foussa, from Madagascar, is about 5 ft. long, and has soft, close fur of a reddish-brown hue. Its reputation is a very bad one. In habit it is nocturnal, and strange stories have been told of its fierceness when wounded or molested. It feeds on flesh, and is said to carry off kids and to attack and prey upon goats and sheep. The illustration was sketched from life from the first specimen brought alive to England (in 1890), and exhibited at the Zoological

FOUSSA.

Gardens. Since then some young ones have been exhibited. The claws can be drawn back like those of the cat.

Next come the True Civet Cats, from which the musky perfume called civet is obtained. The fur is coarse, yellowish-grey in colour, more or less spotted or striped with black, and forming an erectile mane on the back. They feed chiefly upon flesh, but also on fruits and roots.

The African Civet is a native of those parts of the Dark Continent lying between the tropics. It is somewhat larger than a Common Fox, and, like some other species, is kept in confinement for the sake of its strong-smelling secretion, which is used in the manufacture of perfumery. The odour is far too strong to be pleasant, unless the civet is diluted with oil or spirit. The Asiatic Civet, about the same size as the African, has a wide range in the East, where the natives keep it in cages in which it can hardly turn round, so that they may conveniently extract the contents of the pouch. The Tangalung, from Java, Sumatra, Borneo, and the Philippines, and the Burmese Civet, are of smaller size, but similar in habits. The spots of the latter are large and distinct.

The Lesser Civet Cat, or Rasse, found over the greater part of India and in the Malay Peninsula and China, is about 40 in. long, of which the tail counts for 16 in. or 17 in. They are generally solitary, a pair being rarely seen together, and feed on small mammals, eggs, snakes, frogs, insects, fruit, and roots. The Chinese eat the flesh, but the musky odour is generally disagreeable to Europeans. They often rob poultry-yards; and one kept by Dr. Jerdon caught rats, squirrels, and small birds. The Fossa—a very different animal from the Foussa—is an allied species from Madagascar.

AFRICAN CIVET.

The Genettes are smaller than the Civets, less stoutly built, and with shorter limbs. They emit a musky odour, but there is no pouch in which the product of the scent-glands is stored up. The soft grey fur is spotted with brown or black. All the species are African, but one, the Common Genette, is also found in the South of France, Spain, and South-western Asia. It is often domesticated as a mouser.

The Linsangs are beautiful and graceful cat-like animals, with three species from Asia and one from Africa. The body is long and slender, the limbs short, the tail long and round, and ringed with black. In the Asiatic Linsangs the ground colour is rich buff or greyish-white, marked with oblong black patches. The African Linsang, by far the largest, is marked with spots and small blotches. They are as much at home in trees as on the ground, and prey on small mammals and birds. They become gentle in confinement and are easily tamed.

The Palm Civets, of which there are nine or ten species, are confined to Southern Asia and the Malay Archipelago. They are about the size of a domestic cat, and can draw back their claws. They are nocturnal and arboreal in habit, are capital climbers, and feed more on vegetable food than the True Civets do. The long tail, which is ringed in the Indian species, is not prehensile, as was formerly supposed, but it can be partially coiled, and in specimens kept in confinement, this coiled condition sometimes becomes permanent. The pouch is represented by a fold of skin, and the secretion has no musky smell. The general coloration is dull. Mr. Sterndale kept one as a pet, and thus describes it : " It used to sleep nearly all day on a bookshelf in my study, and would, if called, lazily look up, yawn, and then come down to be petted, after which it would spring up again into its retreat. At night it was very active, especially in bounding from branch to branch of a tree which I had cut down and placed in the room in which it was locked up every evening. Its wonderful agility on ropes was noticed on board ship. Its favourite food was plantains, and it was also very fond of milk. At night I used to give it a little meat, but not much ; but most kinds of fruit it used to like. Its temper was a little uncertain, and it seemed to dislike natives, who at times got bitten ; but it never bit any of my family, although one of my little girls use to catch hold of it by the forepaws, and dance it about like a kitten." Those in the Gardens at Regent's Park will come down to the front of their cages if called, and a small piece of apple or banana will serve to begin an acquaintance.

The Binturong somewhat resembles a raccoon in appearance, though but distantly related to that animal. It is about 5 ft. long, of which the tail, which is prehensile, counts for half ; with stiff harsh black or dark-greyish fur, and the ears bordered with white. In habit it is nocturnal and arboreal, and eats anything that comes in its way. One that has lived for some years in the Zoological Gardens, Regent's Park, knows its friends well, and howls reproachfully if they pass its cage without noticing it. It ranges from Nepaul eastward to Sumatra and Java.

The Cynogale, from Borneo and the Malay peninsula, is an "otter-like animal, with very broad muzzle, clothed with long bristles." It is as much at home in the water, feeding on fish and crustaceans, as it is climbing trees and feeding on birds and fruit. It is about 30 in. long, of which the tail counts for 6 in., and has yellowish-brown fur.

The Ichneumons are chiefly African, only the True Ichneumons ranging into Asia and Europe. They vary in size from that of a large cat to that of a weasel, which animal many of them resemble in form.

They live mostly on the ground, and feed on small mammals, birds, reptiles and their eggs, and insects. Some are domesticated as mousers and snake-killers. The Common Ichneumon (Plate II., No. 20) of North Africa, found also in Spain, was a sacred animal among the ancient Egyptians. It is commonly domesticated at the present day, and makes an affectionate pet, and a capital servant in killing rats, mice, serpents, and lizards. The Indian Ichneumon, or Mungoose, is much smaller, with pale-grey fur. It is noted as a snake-killer; and while

INDIAN ICHNEUMON.

some maintain that it is proof against snake-poison, others declare that when bitten the mungoose rushes away to feed on some herb that acts as an antidote. Sterndale kept one as a pet for some time, and says:—"It travelled with me on horseback in an empty holster, or in a pocket, or up my sleeve; and afterwards, when my duties as a settlement-officer took me out into camp, 'Pips' was my constant companion. He was excessively clean, and after eating would pick his teeth with his claws in a most absurd manner. I do not know whether a mungoose in a wild state will eat carrion, but he would not touch anything tainted; and, though very fond of freshly-cooked game, would turn up his nose at high partridge or grouse. He was very fond of eggs, and, holding them in his fore-paws, would crack a little hole at the small end, out of which he would suck the contents. He was a very good ratter, and also killed many snakes against which I pitted him. His way seemed to be to tease the snake into darting at him, when, with inconceivable rapidity, he would pounce on the reptile's head. He seemed to know instinctively which were the poisonous ones, and acted with corresponding caution. I do not believe in the mungoose being proof against snake-poison, or in the antidote theory. Their extreme agility prevents their being bitten ; and the stiff, rigid hair which is excited at such times, and a thick loose skin, are an additional protection. I think it has been proved that if the poison of a snake is injected into the veins of a mungoose it proves fatal."

The Crab Mungoose, extending from the South-east Himalayas into Assam and Arakan, is iron-grey in colour, and about 18 in. long, of which the tail counts for a third. In habit it is partially aquatic, and feeds on crabs and frogs. The glands under the tail secrete an ill-smelling fluid, which the animal can squirt to a considerable distance. There are a few other forms, of which the best known are the Kusimanse—a small burrowing animal with dark brown fur—from West Africa, and the Suricate, or Meerkat, from South Africa. We know little about their habits in a wild state, but both soon become tame in captivity. The Suricate is about the length of a Crab Mungoose, but more stoutly built and stands higher on the legs. The fur is greyish-brown, barred with dark stripes along the back. They are commonly kept as pets at the Cape ; but the Malays use them for charms and love philtres, treating them in much the same way as the Cingalese do the Loris. Of their habits when kept as pets, Mr. Morgan Evans wrote recently in the *Field :*—" They quickly avail themselves of any opportunity of being nursed—climbing up one's legs and nestling in the lap. If they have the chance and are allowed, they creep on to the breast and seek warmth and comfortable quarters underneath the coat or inside the waistcoat. whence they will probably push on until they curl themselves up at your back, if your garments are loose-fitting. I have several times carried one about with me of winter nights in this fashion or in my pocket, to the houses of friends who wished to interview one of my pets. They have a ' curious, low, throaty' cry—a kind of croak which is by no means disagreeable."

The Aard-Wolf (Earth-Wolf) is not much unlike a small Striped Hyæna in form and colour, with a sharp muzzle and large ears. Down the back runs a line of stiff hair, which can be erected like a mane when the animal is angry. South Africa is the home of this animal :

it lives in a burrow, is more active by night than by day, and feeds on carrion, insect larvæ, and termites.

In the Hyænas the teeth are large and strong, and the muscles of the jaws are so powerful that these animals can crush the thigh-bone of a horse or an antelope. The hind limbs are shorter than the front, and all have four digits, with stout claws that cannot be drawn back ; the tail is short, and a pouch receives the unpleasant secretion from the scent-glands.

HYÆNAS IN A BURYING-GROUND.

The Striped Hyæna (Plate II., No. 19) is about the size of a large dog, with dirty grey fur, marked on the body with stripes and on the limbs with bars of a blackish hue, and stiff and mane-like on the back and shoulders. It is found in North Africa and Southern Asia, and is thus described by Mr. Sterndale :—" This repulsive and cowardly creature is yet a useful beast in its way. Living almost exclusively on carrion, it is an excellent scavenger. Most wild animals are too active for it, but it feeds on the remains left by the larger cats, and such creatures as die of disease ; and it can, on a pinch, starve for a considerable time. The African hyæna is said to commit great havoc in the sheep-fold. The Indian one is very destructive to dogs, and constantly carries off pariahs from the outskirts of villages. The natives declare that the hyæna tempts the dogs out by its unearthly cries, and then falls upon them. Dr. Jerdon relates a story of a small dog, belonging to an officer of the 33rd Madras Native Infantry, being carried off by a hyæna whose den was known. Some of the Sepoys went after it, entered the cave, killed the hyæna, and recovered the dog alive, and with but little damage done to it. The hyæna is of a timorous nature, seldom, if ever, showing fight. Two of them nearly ran over me once as I was squatting on a deer-run waiting for sambur, which were being beaten out of a hill. I flung my hat in the face of the leading one, on which both turned tail and fled."

Closely allied is the Spotted or Laughing Hyæna, from Africa, south of the Great Saharan Desert. The fur is yellowish, marked with dark-brown spots. The tail is not so bushy as that of the Striped Hyæna, and there is no mane. In habit it resembles its relative from Asia and North Africa ; but while that animal is solitary, silent, or nearly so, this hunts in packs, and utters an unearthly cry, which, according to the late Professor Kitchen Parker, is "enough to wake the dead and madden the living."

The Brown Hyæna, or Strand-Wolf, is also South African, but less common than the Spotted Hyæna, and of smaller size. The ground-colour is reddish-grey, with blackish stripes and spots.

The Dog family contains, besides the creatures we commonly call by that name, the Wolves, Wild Dogs, Jackals, and Foxes. Some of these are found all over the world, with the exception of a few islands ; and wherever Man has settled, there the domesticated Dog is sure to be. The animals of this group stand midway between the Cats and the Bears. The skull is long, and the muzzle sharp and pointed ; the limbs are rather long, and the Dogs, like the Cats, walk on the tips of their fingers and toes : the former are five, the latter four in number, armed with curved blunt claws, which cannot be drawn back. The great toe is absent, except in some Domestic Dogs, where it is represented by the bones and claw, which hang loosely in the skin, unconnected with the bones of the foot. This is what fanciers call the "dew-claw." The tail is rather long, and generally bushy. The Dogs are less carnivorous than the Cats, and the teeth are not so well adapted for cutting and tearing raw flesh. The larger Dogs feed by choice on the flesh of animals they have themselves killed ; many of them hunt in packs, and most are, to some extent, sociable. Many of the smaller forms feed on carrion, insects, fruit, and other vegetable substances.

The Wolves are the largest wild members of this family, though some breeds of Domestic Dogs exceed them in size. They are found in Europe, Asia, and America, from the extreme north as far south as Mexico ; but from South America and Africa they are absent, their place being taken by Jackals and Foxes. With such a wide range, in such different climates, they vary much in size, colour, and character of the coat, and many have been described as distinct species which have no real claim to that character. The Common Wolf (Plate II., No. 17)

is still found in Europe, being most abundant in the northern and
eastern parts. It is also widely distributed over Asia, where it runs
into varieties; and the Common Wolf of North America is probably
identical with it, though some naturalists give it a different name.
The fur is generally yellowish-grey; but specimens almost white and
others nearly black have been met with. Wolves from the northern
regions have longer and closer fur than those from the south.

WOLVES ON THE TRACK.

The Wolf was formerly common in the British Isles; and, as we
might expect, lingered longer in Scotland and Ireland than in England.
It is well known that King Edgar commuted the money-tribute of the
Welsh for three hundred wolf-skins, or heads, delivered year by year.
But as late as the reign of Edward II. these beasts infested the royal
forest of the Peak in Derbyshire; and they do not appear to have
been exterminated in this country till the reign of Henry VII. (1485–
1509). Wolves probably lived on in the North of Scotland till about
the middle of the eighteenth century, disappearing a little later from
Ireland—according to tradition, somewhere about 1766.

The Common Wolf is about 5 ft. long, of which the tail forms nearly a third, and stands from 30 in. to 32 in. high at the shoulder. Its habits are those of the group. Wolves generally hunt in packs, and pursue their prey in the open, thus differing from the Cats, which lie in wait, or creep up, and attack with a sudden spring. Deer, antelopes, and cattle are relentlessly run down; and in former days, the mighty bison was often separated from the herd, and pressed by his pursuers till he dropped from sheer exhaustion, and was torn to pieces by the yelling pack. Where their numbers have been thinned by man, they are obliged to hunt singly or in couples, and to content themselves with smaller game. When they break into a sheepfold they kill, as it would seem, for the sake of killing, and not merely to satisfy their hunger—a bad characteristic that also distinguishes the larger Domestic Dogs that develop a taste for mutton. Wolves rarely attack man, unless first molested, or hard-pressed by hunger; but in India, according to Sterndale, "hundreds of children are carried off annually, especially in Central India and the North-west Provinces."

Wolves are tamed without much difficulty. Two cubs of the Indian Wolf, kept by an officer, recognised him after an absence of nearly three months, fawning on him, and licking his face and hands, though always ready to growl and snap at a stranger. They soon became so tame that they were turned out in the camp, where they lived on excellent terms with the dogs.

The Prairie Wolf of America ranges from about 55° N. to Mexico. It is smaller than the Common Wolf, the head and body together measuring about 3 ft., and the tail about a foot. The hair is of shades of grey, darker along the spine; the tail is bushy, and the short, erect ears are white in front and brown behind.

Next to the Wolves come the Jackals. The head and body are about 30 in. long, and the tail rather less than a foot; the height at the shoulder is from 16 in. to 18 in. The fur is dark yellowish or reddish-grey, lighter on the under-parts; and the tail, which is moderately hairy, is reddish-brown, and ends in a darkish tuft.

Like the Wolves, they hunt in packs. Sterndale says that the Jackal is well known "both as a prowler and a scavenger, in which capacity he is useful, and as a disturber of our midnight rest by his diabolical yells, in which peculiarity he is to be looked upon as an unmitigated nuisance. He is mischievous, too, occasionally, and will commit havoc among poultry and young kids and lambs; but, as a general rule, he is a harmless, timid creature, and when animal food fails, he will take readily to vegetables." The Jackal sometimes feeds on dead bodies, which it digs out of the shallow graves made by the natives; and Sterndale says: "I once came across the dead body of a child in the vicinity of a jungle-village, that had been unearthed by one." The Indian species is also found in Asia Minor and North Africa. South of the Sahara Desert it is replaced by the Black-backed Jackal, of somewhat larger size. The Senegal Jackal has bright tawny fur and dark bands on the back, sides, and chest.

Very closely related to the Jackals are the Wild Dogs of Asia. The Common Indian Wild Dog ranges through Burma to the Malay Archipelago. It is somewhat larger and more dog-like in appearance than the Indian Jackal, of a fierce disposition, and is said to be un-tamable. These Wild Dogs hunt in small packs, and endeavour to seize the quarry by the flanks and pull it down. They have often been seen chasing deer and antelopes, and the natives assert that a pack will run down and kill a tiger.

The Dingo, or Australian Wild Dog, has been partially domesticated by the natives, and breeds freely with the dogs kept by the colonists. The general colour is yellowish-brown, though this may be tinged with red, or deepened till almost black. Dingoes hunt in small packs, and commit great damage in sheepfolds and poultry-yards. On this account the farmers do their best to exterminate them.

The Eskimo Dog is wolf-like in appearance, and, like the wolf, does not bark. It is but partially domesticated ; for though it is used to draw sledges and carry burdens, and performs these duties very well, it often lapses into savagery, and its master can keep none of the smaller domestic animals lest his dogs should prey upon them.

The Hare Indian Dog takes its name from the tribe of American Indians who domesticated it. It is found round the Great Bear Lake, and seems to have been derived from the Prairie Wolf.

The Pariah Dogs of the East are domestic dogs that have escaped from their masters and run wild ; they have gone back to a partially savage condition. They are, however, protected to some extent from the fact that they act as scavengers in clearing the streets of offal.

The Domestic Dogs run into more breeds than can be easily counted. A recent writer estimated them at nearly two hundred. Mr. Harting classes them all in the following six divisions :

| Wolf-like Dogs, | Spaniels, | Mastiffs, |
| Greyhounds, | Hounds, | Terriers, |

and thinks that cross-breeding will account for the forms that do not fall readily under any one of the classes.

Among the Wolf-like Dogs he places the half-domesticated Eskimo and Hare Indian Dogs, the Pariahs, the Sheep Dog, the Drover's Dog, the Collie, and the Spitz.

I

In the Greyhound Class are our Coursing Dog, the Italian and Persian Greyhound, the Scotch Staghound, the Irish Wolfhound, and the Lurcher.

The Spaniels form a very numerous class. Many of them are sporting dogs, used to flush game or bring it out of the water. Here also belong the Retriever, the Newfoundland, and the St. Bernard— the last two well known for their life-saving services to man—and the King Charles and Blenheim lap-dogs.

Chief among the Hounds is the Bloodhound, with its magnificent power of scent. In this group are also the Staghound, with which the red deer of Exmoor is hunted, the Fox-hound, the Harrier and Beagle, the Otter-hound, the Dachshund and its relation the Turnspit. These hunt by foot-scent; the Pointer depends on body-scent, and the Dalmatian, or Plum-pudding Dog, is purely ornamental.

Besides the English Mastiff there is a Tibetan breed. To the Mastiffs belong the German Boarhound, the Great Dane, the Bull Dog, and the Pugs.

The Terriers form the last group, and perhaps the most numerous. We can only mention the Fox Terrier, which makes such a capital companion, the rough-coated Irish Terrier, the long-haired Skye, with its varieties, the short-haired English Terriers, and the Poodles.

These various breeds of Dogs need no description. The Dog was probably brought into subjection by Man before the dawn of history, and is one of his most useful conquests. When Man was a hunter, the Dog helped him to track and secure his game; when he moved upward to the pastoral stage, these animals guarded his flock and herds; and as civilisation advanced, they became his companions and friends. In some parts of Europe dogs are still used, as they formerly were in England, to draw small carts; and they look none the worse for their labour. In Continental armies they are trained to carry despatches and to act as sentinels. During the German army manœuvres of 1893 five trained dogs—two Scotch and three German sheep-dogs—carried despatches from outposts to headquarters and back again with speed and certainty, distinguishing themselves especially by night. As searchers for wounded men their keen scent and instinct were particularly valuable. The Scotch dogs displayed the greater fidelity, while the German dogs excelled in speed, one of them having covered nearly two miles in seven minutes.

The Common Fox (Plate II., No. 18) would soon be extinct in Britain were it not strictly preserved for the sport of fox-hunting. It is a good type of the second group, distinguished by their smaller size, erect, pointed ears, and brush-like tail. The length is about 3 ft., 2 ft. of which is taken up by the head and body. The fur is reddish above and white beneath, and the end of the brush—for so the tail is called—is white. The back of the ears and the front part of the limbs are black. The muzzle is long and pointed, and the animal has a sharp, wide-awake look, which is quite warranted by its character; indeed, the Fox is proverbial for its cunning. And it has need of all its skill and sagacity, for it lives by plundering game-preserves, poultry-yards, and rabbit-warrens, and when taken in the fact is only spared that it may become the prey of the hounds. Fox-hunting is a national sport, and

FENNECS.

has been introduced abroad by Englishmen wherever possible. In America the Fox is shot. Its habits are nocturnal, and the day is passed in the burrow, which is called its " earth." There are several allied species spread over the world.

The Arctic Fox has a very large, bushy tail, and the palms and soles covered with fur. In summer the fur is bluish-grey, changing to white in winter. Some of these animals may be seen in the Zoological Gardens, Regent's Park, and their mottled appearance when assuming the darker summer coat is very strange. At that time no two of these creatures are alike in colour.

The Fennecs are small African Foxes, with very large ears. Their colour is fawn or sandy, as is that of most desert animals. Their food is chiefly vegetable, and they are very fond of fruit. Some that were sent to the Zoological Gardens in 1894 were very timid at first, but with a little coaxing they would come down to the wires of their cage and take a piece of fruit.

The Cape Hunting Dog, which is found in South and East Africa, is about the size of a mastiff, and is marked somewhat like a Spotted Hyæna, with white, yellow, and black. It hunts in large packs.

The Bush Dog, from Brazil and Guiana, is about the size of a fox; the Long-eared Fox, from South Africa, is somewhat smaller, and its bushy tail is very short.

CHAPTER X.

BEARS AND BEAR-LIKE CARNIVORES.

THE Bears form a well-known group. They are more stoutly built than the Cats or the Dogs, have short limbs all with five digits, armed with strong curved claws that cannot be drawn back, and well adapted for digging. In walking, the soles and palms, which are generally hairless, are planted flat on the ground; the tail is very short. The young are born blind and naked, but there is no truth in the story that their mothers lick them into shape. Flesh forms the principal part of the diet of the larger forms, but many are, to a great extent if not entirely, vegetable feeders, and their teeth are modified accordingly.

The Polar Bear (Plate II., No. 12), a native of the Arctic regions of Europe and America, is one of the largest of the group, attaining a length of nearly 9 ft. Unlike the Arctic Fox, it is white all the year round. The head is small, but long, and the soles and palms are hairy, giving the animal a firm grip as it walks over the ice. This Bear takes a winter sleep, and the young are born in a cave or hole beneath the snow. It has a bad character for ferocity, but is rarely the first to attack. Dr. Robert Brown says, "It does not hug, but bites; and it will not eat its prey till it is dead, playing with it like a cat with a mouse. I have known several men who, while sitting watching or skinning seals, have had its rough hands laid on their shoulders: their only chance then has been to feign being dead and to shoot it while the bear was sitting at a distance watching its intended victim. Though Eskimo are often seen who have been scared by it, yet unless attacked or rendered fierce by hunger, it rarely attacks man."

Like most other wild animals, Polar Bears are diminishing in numbers before the advance of man. They are hunted for the sake of their skins, which make excellent rugs, their flesh, and their fat. They feed on seals, walrus, and fish, and in summer their diet consists largely of seaweed, grass, and lichen.

The Brown Bear (Plate II., No. 11) is found in Northern and Central Europe, and is much more abundant in Asia north of the Himalayas; it formerly lived in Britain, but it is uncertain when it became extinct, probably about the time of the Conquest. Bear-baiting was formerly a popular sport in England, and lingered on till the beginning of the eighteenth century. Bear-gardens were places of public entertainment where these animals were baited with dogs. At Berne, in Switzerland, the name is still applied to the public bear-pit.

A large specimen will measure as much as 7 ft. in length, and 3 ft. in height at the shoulder; but 5 ft. may be taken as the length of an ordinary European bear. In Europe the Brown Bear feeds

BEAR-HUNTING IN NORWAY.

more largely on flesh than the Asiatic form does. In the Himalayas the animal subsists principally on insects and vegetable food, sometimes plundering the gardens and orchards of the natives, while in the north of the continent it lives for a great part of the year on fish. The Syrian Bear and the Isabelline Bear are local varieties.

It was, and perhaps still is, supposed that these bears attack by "hugging." In his "Mammals of India," Mr. Blanford says that this idea has probably arisen from the fact that a bear strikes round with its paws, as if grasping, and the blow of its powerful arm drives its claws into the body of its victim, causing terrible wounds; but the idea of its 'hugging' appears not confirmed by recent observers." But in a fight between brown bears some years ago at the Zoological Gardens, the conqueror undoubtedly killed his victim by hugging, as was shown by the nature of the fatal injuries.

The Grizzly Bear ranges from Alaska to Mexico. Mr. Shields puts the length at about 7 ft. from the snout to the tip of the tail. The fur is brownish-yellow, with a dark stripe along the back and one on each side; the muzzle is pale, and the legs are nearly black. These animals prey chiefly on moose deer, and often raid the farmer's stock-yards; but they will eat poultry, fish, roots, herbs, vegetables, fruit, honey, and insects. When the snow is deep, the grizzly retires to his winter quarters for his long sleep, during which the fat it has put on during the summer furnishes heat and nourishment sufficient to maintain life. Mr. Shields says that he has been "convinced that some grizzlies, at least, will attempt to make a meal off a man, even though he may not have harmed them previously." He also tells a story of a man being attacked by an old female grizzly, who "struck a powerful blow at his head . . . Her claws caught his scalp, and laid it open clear across the top of his head in several ugly gashes." He says nothing about hugging, and the blow seems to confirm what Mr. Blanford said about the Brown Bear's method of attack.

The so-called Cinnamon and Brown Bears of America are only varieties of the Grizzly Bear.

The Black Bear of America is now confined to some of the mountain ranges south of the St. Lawrence, the Great Lakes, and to the unsettled country east of the Mississippi. The length rarely exceeds 5 ft., and the fur is smooth and glossy-black in colour.

The Black Bear of the Himalayas has on its breast a white mark shaped somewhat like a half-moon. These animals, about 5 ft. in average length, frequent forests and wooded districts, and though they are principally vegetable-feeders, will sometimes attack sheep, cattle, and ponies. Sterndale says that, if cornered, they attack savagely, as all bears will, and the face generally suffers.

Of the habits of the Spectacled Bear, in its wild state, nothing is known. It is a small black bear, with yellowish rings round the eyes, and white on the throat and breast, and lives in the Peruvian Andes.

The Malayan Bear is also small, and has brownish-black fur, with a yellowish mark on the chest. Its tongue is long and extensile, and with it the animal scoops the honey from the nests of wild bees.

The Sloth Bear is found all over India, and a large specimen will measure from 5 ft. to 6 ft. in length. It is covered with long black, shaggy hair, except on the chest, where there is a white U-shaped mark. The Sloth Bear carries her young on her back. Mr. Sanderson shot one that was carrying a cub as large as a sheep-dog. Fruit, insects, and honey are the chief food of these bears, and they are especially fond of white ants. Colonel Tickell says that "the bear scrapes away with his fore-feet till he reaches the larger combs at the bottom of the galleries. He then with violent puffs dissipates the dust and crumbled particles of the nest, and sucks out the inhabitants of the comb so forcibly as to be heard at two hundred yards' distance or more. Large larvæ are in this way sucked out from great depths under the soil."

The Black-and-White Bear is a native of Eastern Tibet; the length is a little less than 5 ft., and the height at the shoulders rather more than 2 ft. The ground colour is white, there are black rings round the eyes; the limbs are black, and from each fore-limb a band of the same colour goes up across the back. This animal is said to feed entirely on vegetable food, but nothing is known of its habits.

RACCOON.

The Panda, or Red Cat-Bear, a link between the True Raccoons and the Bears, is the only one of the family found in the Old World. It is a native of the South-eastern Himalayas. In size it resembles a large cat; the fur is rusty-red; there is some white on the face, and the tail is light red with dark rings. The claws can be drawn back like those of a cat, but are only used for climbing, not for taking prey or defending itself. Its food is chiefly vegetable, and it is said to steal milk and butter from the villagers.

The True Raccoons are natives of North America, and are generally found in wooded places near running water. They are flesh-eaters, but do not object to birds' eggs, shell-fish, fruit, and grain. The Common Raccoon is from 30 in. to 3 ft. long, of which the tail measures nearly a foot. The fur is greyish-brown, and the tail is ringed with black. The palms and soles are naked; and though these animals cannot grasp, they will hold food between the palms closely pressed together. They are excellent climbers, and those at the Zoological Gardens will generally climb up the wires of their cages for biscuit, but fruit will always tempt them. When at liberty, they are said to dip their food in water before eating it. They will often do this in confinement, but quite as often they will eat it without moistening or washing it, especially if two of them are struggling for some coveted dainty. But even in this case I have occasionally seen the lucky possessor shuffle off to the basin in the middle of the den and dip his prize in the water, and rub it between his palms before eating it. The fur of these animals is commercially valuable.

The Crab-eating Raccoon—a native of South America—is of a larger size and has shorter fur.

The Coatis, from Central and South America, may be known by their long, tapering tail and upturned snout. The White-nosed Coati, a native of Mexico and Central America, has long reddish-brown fur. Its total length is nearly 3 ft., the tail, which is sometimes ringed, counting for about half. The Red Coati lives in South America, has darker fur, and the tail is always broadly ringed with black; but the colour varies greatly in individuals of both species. These animals are often seen in captivity, where they appear to be quite comfortable. One now living in the Zoological Gardens knows those who feed him, and will climb up the wires of his cage for any dainty they may offer him.

The Kinkajou, from Southern and Central America, is not unlike a lemur in general appearance, though the limbs are shorter, and is clothed in soft yellowish-brown fur. It is arboreal in habit, and its tail is prehensile. Bates kept one as a pet for several weeks. It grew tame in a short time, and allowed strangers to caress it; but, as was natural, was much more friendly with its master than with them.

The Cacomixle (the *x* is pronounced as *st*), from Mexico and the United States, is about the size of a cat. The fur is greyish above and white below, and the tail is ringed with black. These animals live in woods and rocky places. They are said to do great damage in poultry-yards, but are often tamed and kept to kill rats and mice.

The Wolverene, or Glutton, is a small bear-like animal, with long brownish-black fur, a short, bushy tail, and a snout like that of a dog. It lives in the northern forests of Europe, Asia, and America, and preys upon birds and small mammals, and is said to attack reindeer, cows, and horses. It climbs trees with ease; but there is no truth in the stories formerly told of it, that it lay in wait on the branches and dropped down on large animals that passed below. It seems to be as fond of stealing and secreting articles for which it can have no possible use as is the magpie. Dr. Coues says that "A hunter and his family having left their lodge unguarded during their absence, on their return found it completely gutted—the walls were there, but nothing else. Blankets, guns, kettles, axes, cans, knives, and all the other paraphernalia of a trapper's tent had vanished, and the tracks left by the beast showed who had been the thief. The family set to work, and by carefully following up all his paths, recovered, with some trifling exceptions, the whole of the lost property."

The Martens live among woods and rocks, and prey on squirrels, rabbits, mice, snakes, lizards, and frogs, sometimes varying this diet with berries. They are bloodthirsty creatures but less so than the True Weasels, and, with one exception, are pretty much alike in size and

WOLVERENE.

colour of the fur. Thirty inches, of which the tail occupies about one-third, may be taken as the average length, and the colour is brown, varying from an orange-tint to nearly black. There is generally some white about the head. The Pine Marten (Plate II., No. 14) roams over the woods of Northern Europe and Asia. Probably some still live in the North of Scotland, though they are extremely rare, if not extinct, in England. There are several closely-allied species. The Sable, from Eastern Siberia, the fur of which is so highly valued, and the North American Sable, of which over one hundred thousand skins are imported every year, are probably only varieties of the Pine Marten.

The Beech, or Stone Marten, widely distributed over Europe, was probably the "cat" of the Greeks. Other animals of this group are still domesticated as mousers and ratters. The Pekan, or Fisher Marten, is the largest of the group, and has a fox-like appearance. It is a native of North America, but is becoming scarce east of the Mississippi. The fur is blackish, for which reason the trappers sometimes call it the Black Fox. It is said to steal the fish used to bait traps, whence its name of Fisher.

We have some of the True Weasels in Britain. They are smaller than the Martens, with shorter limbs and longer body. The Common Polecat (Plate II., No. 15) is abundant in Europe, but is becoming scarcer in Britain. The total length is about 2 ft., of which the tail counts for a quarter. The long, coarse, brownish-black fur changes little in the winter. These animals possess scent-glands of a most offensive odour. During the day they sleep in their holes and hiding-places, coming abroad at night, and preying on young rabbits, mice, birds, snakes, lizards, etc. Farmers and gamekeepers destroy the Polecat whenever possible, for it kills far more than it can eat. It generally carries off its prey to be devoured, but has been known to kill all the birds in a fowl-house, to gratify its love of killing.

FERRET.

The Ferret is an albino variety of the Polecat, that breeds true. It has been domesticated chiefly for its services in rabbiting, but is also employed to kill rats. Ferrets are muzzled and put into a rabbit-hole, whence they drive the rabbits, which, as they bolt, are caught in nets at the mouth of the burrows. There are four other Polecats, with the habits of the common species.

The Weasel is about 10 in. long, of which the tail measures 2 in. Its fur is reddish-brown above, becoming lighter in winter, and whitish below. It is quite as bloodthirsty as the Polecat, though its smaller size prevents its doing as much damage. In some respects it is a "farmer's friend," for it frequents rickyards and destroys large numbers of rats and mice. The female is smaller than her mate,

MINK.

and has been sometimes wrongly taken for a distinct species.

The Stoat is not so common in Britain as the Weasel, from which it may be distinguished by its larger size and the fact that the tail has a black tip. This never changes colour, though the rest of the fur becomes white in winter in Northern latitudes, and the creature is then known as the Ermine. It is an exceedingly active animal, and sometimes hunts in family parties of six or seven.

The Minks are Weasels of aquatic habits. One is found in Eastern Europe, and the American Mink is bred extensively. It is a capital ratter, and when dead the sale of its skin leaves its owner a good profit on its keep. The Grison and the Tayra are weasel-like animals, both from South America.

Next come the Badgers, which for the most part live on the ground and in burrows. The Common Badger (Plate II., No. 13) is found in many parts of Britain, on the continent of Europe, and in Asia. The body is stoutly built, the limbs are short and strong and armed with large claws. The length is from 30 in. to 3 ft., and the height at the shoulder some 12 in. The general

BADGER.

colour is grey, but the head is white, with a black band on each side. These animals feed on mice, snakes and frogs, insects, fruit, acorns, and roots. They are very fond of wasps' nests. The cruel sport of badger-baiting—now, fortunately, nearly extinct—consisted of putting a badger in a barrel and setting on dogs to pull it out. There are a few other species.

The Ratels, or Honey Badgers, are grey above and black below. One is a native of India, the other of Africa ; in the latter the two colours are sharply divided by a white line. These animals usually live in pairs, and eat rats, birds, frogs, white ants, and other insects. They are quiet in confinement, and have an amusing habit of running round their cage and turning somersaults at pretty regular intervals.

The Teledu, or "Stinking Badger," from Java and Sumatra, possesses very offensive scent-glands, as does the Cape Zorilla, from South Africa. These two creatures seem to link the Badgers with the Skunks.

The Skunks are small nocturnal burrowing animals, confined to America : their diet is similar to that of the Badger. The usual colour is black, varied with white spots or stripes, and the tail is long and plume-like. These animals have the disagreeable quality of being able to discharge the secretion of their scent-glands to a considerable distance : this secretion is so offensive as to cause nausea in many persons, and clothes tainted with it will retain the vile smell for a long period. The Common Skunk, about the size of a small cat, ranges from Hudson's Bay to Guatemala. A writer in the *Ibis* says that he once saw a bird of prey attack a skunk. It seized the tail with its claws and in an instant began "staggering about with dishevelled plumage, tearful eyes, and a profoundly woebegone expression. The skunk turned and regarded his victim, as who should say, 'I told you so,' and trotted off unconcernedly." The other species differ little from the Common Skunk.

The Otters are a small group of aquatic animals, with five webbed digits on each limb. Species occur in all the continents. Their home

OTTER.

is generally near water, and fish forms their principal food. The Common Otter (Plate II., No. 16) is a native of Europe and Asia, and is pretty plentiful in Britain. The total length is rather more than 3 ft., of which the tail counts for nearly half; the fur is a soft brown colour. These animals feed chiefly on fish. which they bring to bank to eat; they only consume the choicest parts, unless pressed by hunger. Kingsley, in "The Water Babies," makes the old otter say. "We catch them, but we disdain to eat them all; we just bite out their soft throats, and suck their sweet juice—oh, so good!- (and she licked her wicked lips)- and then throw them away, and go and catch another." This habit very naturally makes enemies of those interested in fishing, and the otter is trapped and killed without mercy. Otter-hunting with dogs was formerly common. The beast was brought to bay by the dogs, then speared by the huntsman, and after being lifted up on the spear was broken up by the pack. Otters are readily tamed, and make affectionate and playful pets. In India they are trained to drive fish, and in China they fish for their masters.

There are some other species, the largest of which is the Canadian Otter, regularly hunted for its fur. These Otters are remarkable for indulging in a game of sliding. English Otters are equally fond of play, and merit the description of being "the merriest, lithest, gracefullest creatures you ever saw." The Sea Otter, which is larger and more stoutly built than the English species, is a native of the shores of the North Pacific Ocean, and lives principally on shell-fish and sea-urchins. The fur is extremely valuable, and on that account these animals are so hunted that in a short time the species will probably become extinct.

CHAPTER XI

MARINE CARNIVORA

THE Fin-footed Carnivores are chiefly dwellers in the sea or on the coast. Some ascend large rivers, and a few others inhabit inland seas, but all bring forth their young on shore. There are five digits on each limb, united by a web.

The Eared Seals have, as their name denotes, a small external ear; and when moving on land the hind limbs are turned forward, and aid the animals in their progress. These Seals are called Sea-Bears or Sea-Lions, or Fur-Seals and Hair-Seals, according as their skins do or do not yield the sealskin used for jackets, etc. This consists of a fine close under-fur, which is left on when the longer hairs are removed, by shaving the under-side of the skin. The longer hairs are more deeply set in the skin, and consequently this shaving process cuts off their roots without touching those of the fur.

The largest of the Eared Seals is the Northern Sea-Lion, from the North Pacific, which attains a length of about 10 ft. But the best known is the Patagonian Sea-Lion (Plate IV., No. 15), the first of these creatures brought alive to England. Lecomte, a French sailor, captured one near Cape Horn in 1862, tamed it, and brought it to England, where it was bought for the Zoological Gardens. The animal was taught to perform many tricks, and when it died Lecomte was sent out by the Zoological Society to procure other specimens. This creature has a decided mane, which is not noticeable when the animal is wet, and some writers have denied its existence. The feeding of the Sea-Lion at the Zoological Gardens is always an attraction, the docility with which the creature obeys the commands of its keeper, the certainty with which it catches the fish thrown it, and its evolutions in the water, are generally witnessed by large crowds. The Californian Sea-Lion is often brought to Europe.

From the Common Sea-Bear or Fur Seal of the North Pacific most of the sealskin is procured, and to attain it

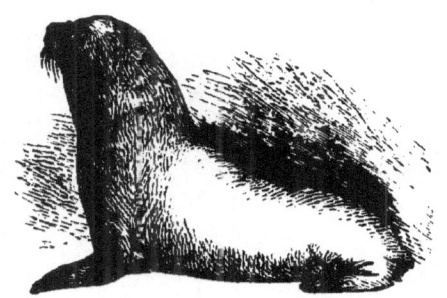

SEA-LION.

WALRUS BULL.

these animals are slaughtered in such numbers that, although a close
time has been adopted and a limit put to the quantity allowed to be
killed, the number of fur seals visiting the Pribyloff islands is growing
less year by year.

These islands are their breeding-places, to which they come in
summer, the winter being spent in following the fish in their southward journey. Those killed by the sealers are young, unmated males,
or "bachelors." These herd by themselves, and are driven away to a
convenient place, where they are slaughtered by a blow on the head
with a club, and immediately skinned. So quickly is this work done,
that forty-five men drove away, killed, and skinned more than 72,000
seals in four weeks, or about 400 seals a week for each man.
The Southern Fur Seals are far less important from a commercial
point of view. There are three or four species. The Cape Fur Seal
is at present living in the Zoological Gardens, where he affords visitors
great amusement by his ceaseless gambols in his tiny pond.

The Walrus (Plate IV., No. 14) is confined to the Arctic region. It
has no external ear, but its hind limbs are turned forward, though less so
than in Eared Seals, and they form less effectual organs of locomotion.
There are long, stiff whiskers on each side of the short, blunt muzzle.

The upper canines are developed into large tusks, used principally for digging up the molluscs from the sea-bottom, on which the Walrus feeds; but they form terrible weapons of defence, which the animal can use with great effect. Some writers say that they are also used to aid the animal in climbing up rocks and on to ice-floes, but this has been denied. The adult male is of great bulk, especially about the fore-quarters, and its total length may be from 10 ft. to 11 ft. The hair is yellowish-brown above and darker on the under-surface.

The Walrus is inoffensive unless first attacked, when a herd will join in the defence. The mothers are affectionate and very courageous in defence of their young. To the Eskimo and the neighbouring tribes the Walrus is very important. Captain Scammon says, "The flesh supplies them with food, the ivory tusks are made into implements used in the chase and for domestic purposes, as well as affording a valuable article of barter, and the skin furnishes the material for covering their summer habitations, harness for their dog-teams, and lines for their fishing-gear." The Americans hunt the Walrus for its ivory, and probably it will soon be numbered among the animals that were.

The True Seals have no external ear, and the hind limbs are directed backwards, affording no assistance in progression on land, which is effected by short jerking movements, aided in some species by the fore-limbs. The palms and soles are hairy.

The Common Seal, from 4 ft. to 5 ft. long, yellowish-grey with dark spots above and whitish beneath, is widely distributed. The larger Grey Seal is confined to the North Atlantic. It is plentiful in suitable localities round our coast, where it has not been driven away by man. That it is becoming rare in some places is not to be wondered at, if the following account (taken from the *Field*) of a Pembrokeshire seal-hunt is correct :—

"The hunters are armed with bludgeons or mattocks. Where there is a likely cave left high and dry, they land on the sand or shingle, a gun is fired off in the mouth of the cave, and if there are any seals there, they come tumbling out after each other in close line, when they are killed by a blow between the eyes. It is said they are so close together that if you miss the first it is difficult to stop any of them. There is no element of danger in the sport, except the possibility of being struck with the pebbles scattered by the flippers of the seals as they hurry down towards the sea for safety." It is wrong to call such slaughter "sport."

The Greenland Seal is regularly hunted, as are some of the other species, for its hides and oil. It occurs in vast numbers in the ice-fields of the North, and the annual catch round Jan Mayen island is estimated to be not far short of a quarter of a million. The colour is very variable, differing in the males and females, and the young undergo many changes of hue as they grow up. The whalers recognise several stages in the life-history of this seal, each of which has its own particular name. In old males the ground colour is tawny grey, with dark head, and two dark bands on the back, whence comes the name of Harp or Saddleback Seal.

Other species are the Bearded Seal, from the North Atlantic ; the

SEALS.

Ringed Seal, extending to the North Pacific : the Caspian Seal, from the Caspian Sea and the Sea of Aral ; and the Siberian Seal, from Lake Baikal.

There are two species of Monk Seal : one from the Mediterranean, the Talking-fish of showmen, the other from the West Indies. The Sea-Leopard, or the Leopard Seal, from the Southern Seas, is about 12 ft. long, and owes its name to its spotted skin. The Crested, or Bladder Seal, is a native of the Arctic and North Atlantic Oceans, coming as far south as Norway in the Old, and Newfoundland in the New World. It is the boldest and fiercest of the Seals, and will often turn upon the native hunters. The head of the adult male is furnished with a kind of sac, which can be inflated, so as to form a kind of crest.

The Elephant Seal, from the Antarctic Ocean and the coast of California, is the largest marine Carnivore—an adult male measuring from 15 ft. to 16 ft. in length. The general hue of the fur is greyish with a darker tinge above. The snout of the male is produced into a kind of trunk or proboscis, which can be expanded at will.

CHAPTER XII.

HOOFED MAMMALS.

THIS Order contains a large number of animals possessing the common character that the toes are enclosed in hoofs. Most of them live on the ground; none burrow; but the Hippopotamus is aquatic, and some of the Coneys are quite at home among the branches of trees. Nearly all feed on vegetable substances, though some few—like the Pigs—will eat anything that comes in their way. They are arranged in four groups: the Elephants, the Coneys, Odd-toed Mammals, and Even-toed Mammals (the reference being to the number of digits on the hind limbs).

ELEPHANTS.

These are the largest living quadrupeds—of massive build, walking softly and silently on the tips of their digits, of which there are five on each limb, united by a cushion-like pad that forms a flat sole. The head is large and joined to the body by a short neck, and the bones of the skull are filled with air-spaces divided by thin partitions, thus securing lightness. The brain is comparatively small; and the nose is produced into a flexible proboscis, or trunk, divided down the middle so as to form two tubes, at the ends of which are the nostrils. There are no canine teeth, and no incisors in the lower jaw; those of the upper jaw are very long, and are popularly called "tusks." They grow from behind as fast as they are worn away in front, like the incisors of a mouse or a rabbit. Only one molar tooth is in use on each side in each jaw at the same time.

The skin is very thick, and scantily covered with coarse, bristly hair. The eyes are small; but the senses especially those of hearing and smell, are acute. The limbs are set on to the immense trunk almost perpendicularly, and the great length of the thigh brings the knee almost in the position of that of the horse's hock. From this formation of the limbs, it follows that elephants cannot jump either over or across an obstacle. A trench 7 feet wide is impassable to an elephant, though the stride of a large one is about $6\frac{1}{2}$ feet.

The trunk serves many of the purposes of a hand, and its extremities act as lips. With it these animals gather food and convey it to the mouth; into it water can be drawn up and may be then blown into the mouth or scattered over the body shower-bath fashion. Sand and dust

J

are strewn over the body by the same means; and by forcibly driving air through the trunk, elephants make the noise known as trumpeting.

The Indian Elephant (Plate IV., No. 1) is found also in Ceylon (where most of the males are tuskless), Burma, Siam, and the islands of Sumatra and Borneo. The height of a large male is about 10 feet, while the female is about a foot less, and has smaller tusks. The head is oblong, the forehead concave, the ears small, the trunk ends on its upper surface in a finger-shaped lobe, and there are four nails on the hind foot. The enamel of the molar teeth forms parallel folds (Fig. A).

EAR, TRUNK, AND TOOTH OF THE ELEPHANT.
A, Indian.   B, African.

This Elephant has long been partially domesticated in the East; but the supply is kept up by driving wild ones into enclosures, called kheddahs, and taming them, rather than by breeding. Elephants are used as beasts of draught and burden, and are kept by princes and nobles for purposes of display. They are also employed in tiger-hunting, the sportsmen being mounted in a kind of carriage on the back; and the animals sometimes take part in the sport, by trampling the tiger, kicking it to and fro, flinging it to a distance, or pinning it to the ground with the tusks.

The newly-captured elephant is first led between two trees, and rubbed down by a number of men with long bamboos. It lashes out furiously at first; but in a few days it ceases to act on the offensive, and, as the natives say, "becomes ashamed of itself." Ropes are tied round the body, and it is mounted for several successive days as it stands captive. It is next taken out for exercise between two tame elephants, a man going before with a spear to teach it to halt when ordered to do so. When the tame elephants wheel to the right or the left, the driver presses its neck with his knees, and taps it on the head with a stick to train it to turn. It is taught to kneel by taking it when the sun is hot, into water about 5 feet deep, and pricking it sharply on the back. Partly to avoid the pain, and partly from a fondness for bathing, the beast kneels down, and the lesson is repeated in water that is shallower day by day, till the elephant learns to kneel down on land. It is taught to pick up anything from the ground by a piece of wood being dangled over the forehead by a rope. This strikes against the trunk and fore-feet, and to avoid the annoyance, the elephant takes the wood in its trunk and carries it.

The so-called white elephants, which are only albinos, are sacred

CAPTURING WILD ELEPHANTS.

animals in Siam. One belonging to Barnum was exhibited in the Zoological Gardens, Regent's Park, in 1883, before being shipped to America.

The African Elephant stands a little higher at the shoulder than the Indian species. The forehead is arched; there is a lobe above and below at the tip of the trunk, and these close like finger and thumb; the ears are of immense size, and there are only three nails on the hind foot, and the enamel ridges on the molar teeth are lozenge-shape (Fig. B). The Indian elephant uses his tusks only as weapons of

J 2

offence, while the African elephant uses his in ploughing up the ground in search of roots and bulbs. These animals were formerly tamed and used in war, and were brought to Rome to take part in the games of the circus. They are now being rapidly exterminated for the sake of the ivory of their tusks.

European hunters of course shoot these animals, and so do some of the natives. But pitfall traps are still common, and the unhappy beasts that tumble into them are soon killed with lances. Parker Gillmore says that if an elephant has escaped from one of these traps, it never forgets it, and is always on the look-out for a similar danger. He believes in the wonderful stories told about the memory of these animals, and says that after the lapse of years they will at once recognise a person who has treated them with cruelty or even unkindness. A young elephant belonging to Mr. Gillmore was burnt on the trunk by a Kaffir blacksmith. Many months afterwards he returned to the neighbourhood where this wanton act was performed, to have some repairs done to his waggons. His pet followed him to the forge. He found the same man employed in his trade ; and while its master was explaining his wants, the elephant gave the blacksmith a blow with its trunk that sent the man reeling for several yards.

CONEYS.

These are the "feeble folk" of the Bible, "that make their houses in the rocks." In general appearance they are not much unlike rabbits, but are larger and more stoutly built. The molar teeth resemble those of the rhinoceros in pattern, and the toes bear short broad nails. There are several species, chiefly from Africa, though the one longest known is from Syria. Some live on the ground in rocky places, and others live in trees. Both of them have the power of clinging to upright surfaces by the soles of their feet, which are furnished with suckers. The Dassie, or Cape Coney, is kept as a pet in South Africa, though it is difficult to capture on account of the almost inaccessible places in which it lives. Mrs. Martin says that "It is as pretty, soft-coated, and gentle as you could wish, and in its mild, placid way gets very tame." Some of the many species are generally to be seen in the Zoological Gardens. They make friends very readily with anyone who will notice and feed them, and will come to the front of the cage to be petted and to take biscuit or fruit, showing their joy by a little whistling cry.

ODD-TOED MAMMALS.

The principal character distinguishing this group, and giving it its name, is that the third digit in each limb forms the chief support of the body. The second and fourth digits may help in this task, as in the Tapir and the Rhinoceros, or the limbs may end in a single useful digit, as in the Horse, the Donkey, and the Zebra.

All the animals of the Horse family belong to the Old World. Those that are wild, or have escaped from domestication, live in herds or troops, and feed principally on grass. There is but a single perfect digit on each limb, encased in a solid broad hoof.

The Horse has a flowing mane and tail, the latter being hairy from the root, and there is a wart on the inner side of each limb. Horses have been known from the earliest times, and in the Polished Stone Age were hunted for their flesh by the men who dwelt in Europe in that period. From carvings discovered in the same cave with that figured on p. 13, we know that these horses were of small size and heavy build, and had shaggy manes and tails. Horses must have been domesticated before historic times; for the earliest literature speaks of them as used for beasts of burden, and drawing war-chariots and carrying armed men.

PREJEVALSKY'S HORSE.

It is very doubtful whether there are any really wild horses, though it is said that the Russian traveller Prejevalsky found one specimen in the sandy desert of Central Asia, near Zaisan. It seems to stand between the horses and the asses, having some of the characters of both; and may possibly be a hybrid. Two other Russian travellers also saw the animal in the desert of Dzungaria, and they brought back with them four skins and a skeleton. Other so-called wild horses are the descendants of animals that have escaped from captivity. Large herds exist in Tartary, where they are called "tarpans"; in America they are known as Cimarones or Mustangs; and similar herds exist in Australia.

The horse was introduced into Europe from the East; the Arab blood which has so improved our native breed was introduced at the time of the Crusades; and from Arab sires imported in the seventeenth century our racehorses, hunters, and steeplechasers are descended. The cart-horse was originally a Dutch breed. The name "pony" is given to horses less than thirteen hands (4 feet 4 inches) high at the shoulder. Shetland ponies are generally much less than this.

Asses are smaller than horses; the tail is covered with long hair for only about half its length, and there are no warts on the hind legs. Our common Donkey, or Domestic Ass, is said to have been first domesticated in Egypt. It was introduced into England as early as the reign of Ethelred, though it did not become common till after the time of Elizabeth. The horse and the ass breed together. The offspring, which are called mules, partake of the qualities of both parents, and are very serviceable to man. The Asiatic Wild Asses, the Kiang, the Onager (Plate IV., No. 4), and the Syrian Wild Ass, though

QUAGGA.

differing somewhat in size and colour, closely resemble each other in habit, and probably belong to the same species. The African Wild Ass is thought to be the stock from which our Donkey descended.

Sir Samuel Baker says: "Those who have seen donkeys only in their civilised state have no conception of the beauty of the wilder original animal. It is the perfection of activity and courage, and has . . . a high-actioned step when it trots freely over the rocks and sands, with the speed of a horse when it gallops over the boundless desert."

The Striped Asses are all African. The Quagga, now nearly, if not indeed quite extinct, as there is much reason to fear—having been exterminated for the sake of its hide—had the body reddish-brown, marked on the head and neck with dark stripes which grew fainter and fainter till they disappeared on the flanks. The limbs and under-surface were white.

The Zebra (Plate IV., No. 5) is white marked with black stripes, and stands about 4 feet high at the shoulder. It is a native of Cape Colony, but is becoming rare owing to the advance of civilisation. It is said to be untamable, and in 1894 one in the Jardin des Plantes, at Paris, attacked a workman who entered the enclosure where it was confined, and would probably have killed him had not help arrived. There is a warning notice, "THESE ANIMALS BITE," on their stalls in the Zoological Gardens, Regent's Park. For all that, they seem on good terms with visitors, and will follow one all round the paddock for a biscuit. In Cape Colony they have been broken to saddle and ridden by a lady, and to harness, and a team of them driven in a coach. The Dauw, or Burchell's Zebra, is somewhat larger and stronger. The ground colour is yellow, and the limbs and tail are not striped, but individuals vary greatly in this respect.

The Tapirs are large animals, of pig-like form, with the snout produced into a short trunk, which is of great use in pulling down and breaking off shoots of trees, etc., and in collecting roots or plants from the ground. There are four digits on the fore-limbs, and three on the hind limbs. Tapirs are solitary, nocturnal animals, frequenting the neighbourhood of water, for they are very fond of bathing. The American Tapir (Plate IV., No. 2) is found in Brazil and Paraguay, and the north of the Argentine Republic. It is about the size of a small donkey, dark brown in colour when full grown, but when young it is marked with yellowish stripes and spots, as are all the other species. It has a small, stiff mane. It is sometimes domesticated, and becomes tame and familiar. Its flesh is eaten, and its skin makes excellent leather. Roulin's Tapir is found at great elevations in the Cordilleras, and there are two other American species. The Indian or Malayan Tapir, from Sumatra, Borneo, and Malacca, is rather larger than the American Tapir, and has no mane. Like the common species of the New World, it is often domesticated. Its head, neck, and limbs are glossy black, and the back, rump, and sides white, the two colours meeting without shading into each other.

The Rhinoceroses are only exceeded in size by the Elephant, and have three digits on each limb. They are of timid disposition, but when irritated or attacked become very formidable, using the horns on the nose with great effect. Their diet is entirely vegetable; they are more active towards evening and at night than during the day, and are fond of bathing, and wallowing in mud. These are now confined to Asia and Africa, and it is from the former continent that the rhinoceroses in menageries and zoological gardens are chiefly procured. They are distinguished from the African forms by the skin being raised into folds, called "shields," which make these creatures look somewhat as if they were clad in armour. The "horns" are composed of fibres, bound together in a solid mass.

The Indian Rhinoceros (Plate IV., No. 3), with a single horn, is often seen in zoological collections. "Old Jim" has lived at the Regent's Park Gardens since 1864. He is said to be 12 feet long, as much round, and about 5 feet high at the shoulder. The Javan Rhinoceros, found from Calcutta to the Malay Peninsula, and in Java, Sumatra, and

Borneo, is smaller, and the folds of the skin are not so strongly marked. The Sumatran Rhinoceros, with two horns, is the smallest of the family, and there is a variety from Chittagong called the Hairy-eared Rhinoceros.

There are two African Rhinoceroses, both with two horns. They are generally called the Black and the White Rhinoceros, though the so-called White animal is the darker-coloured of the two. It is better, therefore, to call it the Square-mouthed Rhinoceros. Mr. Nicholson,

AFRICAN RHINOCEROS.

an old African hunter, says that the name "White" was given because albinos are very common, and that he himself shot three of a light yellow or cream colour. A specimen of the Black Rhinoceros lived in the Regent's Park Gardens for twenty-three years. The latter is almost extinct, and it is scarcely probable that another specimen will be brought alive to this country.

Mr. R. T. Coryndon, who shot two Square-mouthed Rhinoceroses in 1893, and brought home their skins and skeletons, says that the Black Rhinoceros has a prehensile upper-lip and a small head, and feeds entirely on leaves and twigs. The calf always follows the mother. The Square-mouthed Rhinoceros has a disproportionately large head, with a jaw that looks as if it were cut off square in front, and feeds entirely on grass. The calf always runs just before the cow, which guides it by the pressure of her horn upon its flank.

Other forms have been described as species, but these have been only varieties, or individuals showing some peculiarities in the horns. Sir John Willoughby shot a Black Rhinoceros with three horns. These were brought to England and exhibited at a meeting of the Zoological Society.

The natives take these animals in "game pits." Sir John Willoughby fell into one, and thus describes his mishap: "I was swinging along a pass between two small hills where the grass was dry and smooth, and the path apparently well trodden by game. Suddenly the ground gave way under me, and I found myself supported by my arms, with my legs dangling in space, and vainly struggling to reach something more solid. The gun-bearers rushed to my assistance, and soon extricated me from my undignified and uncomfortable position. This pit, unlike many others, was luckily free from spikes and stakes, but the way in which the mouth was concealed by the smooth and well-trodden grass was most creditable to the artist who had planned and arranged it, and I should imagine quite capable of deceiving an animal with four legs as well as one with only two."

EVEN-TOED MAMMALS.

In this group the number of digits on each limb is even. In most of these animals where four toes are present, two of them (the second and fifth) are useless for walking on, as in the Ox and the Pig. A line drawn down the middle of the limb would pass between the third and fourth digits. In the Odd-toed Mammals, a similar line would pass down the third digit. Most of the Even-toed Mammals "divide the hoof and chew the cud," and in all the stomach is complex. Here belong the Pigs, Hippopotamuses, Cattle, Sheep and Goats, Antelopes, Deer, Giraffes, and Camels, with their allies from the New World.

The Pigs possess a short, generally cylindrical snout, at the end of which the nostrils are situated, used to turn up the ground in searching for food. There are four digits on each limb, two only of which touch the ground in walking. The teeth are of three kinds, and the incisors are developed in wild males into formidable tusks. The Domestic Pig, which has run into many breeds, is too well known to need description. The Wild Boar, from which it is descended, is still found in many parts of Europe, the North of Africa, and Asia Minor, but has been extinct in England for more than two hundred years. It is driven and shot by European sportsmen. The Indian species is larger and fiercer ; it is hunted by mounted men armed with spears, and the sport is called "pig-sticking." Instances are on record of tigers being killed by wild boars. Sterndale says that "an old boar which has been driven from the herd is generally a match for a tiger ; in fact, few tigers, unless young and inexperienced, would attack one." There are some other species, and in all the young are dark brown, striped from head to tail with a paler colour, as are those of the Pigmy and River Hogs.

The Pigmy Hog, from Sikkim and Nepaul, is about the size of a hare. These are said to go in herds, and the fierceness with which they attack those who disturb them is out of all proportion to their size.

The River Hogs and Wart Hogs are natives of Africa. The Wart Hogs are so called from the large fleshy lumps on the face. The snout is broad and flat, and the upper canines are bent upwards, in both sexes. It is said that these teeth are used to dig up roots and plants for food. Ælian's Wart Hog, from North Africa, is ashy-brown, and the Ethiopian Wart Hog from South Africa is dark grey in colour.

The Babirusa, from Celebes and Buru, is nearly hairless. The upper incisors in the male do not grow down into the mouth, but upwards through the skin of the face, and curve backwards like horns. Old writers suggested that these teeth served as hooks by which the creature could rest its head on a branch. Others have supposed that they were developed to protect the eyes while the animals hunted for their food in the spiny thicket. But this explanation is not satisfactory, for the females, which must seek their food in the same way as their lords, do not possess these teeth.

The Peccaries, found only in the New World, are small pig-like animals, with but three toes on the hind limbs, the fifth being absent ; and the stomach is something like that of the Ruminants. In the middle of the back is a gland that secretes a fatty substance with a musky smell. If this is not removed immediately the animal is killed, the flesh is rendered unfit for food. The White-lipped Peccary is met with in large herds between British Honduras and Paraguay. In disposition it is fierce and pugnacious, and it is no uncommon occurrence for hunters who meet with a herd to be obliged to take to a tree as their only chance of safety. Peccaries do great damage to cultivated lands.

The Hippopotamuses, or River Horses, are confined to Africa. There are four digits on each limb, all used in walking ; the head is long, and the muzzle broad and rounded : the body is massive and hairless, and the limbs and tail are short. The Common Hippopotamus is widely distributed in African rivers and lakes, on the

banks of which it lives in small herds, spending the day in the water,
and coming on shore at night to feed, doing more damage to crops by
what it treads down and destroys than by what it consumes, though
this is no trifling matter, for the stomach will hold from five to six
bushels. The general colour is reddish-brown, and large specimens are
said to measure 14 feet in length, and 5 feet high at the shoulder.
" Obaysch," the first specimen brought to Europe in modern times, which
lived for about thirty years in the Zoological Gardens, Regent's Park,
measured 12 feet from the
snout to the tip of the
tail. " Guy Fawkes," born
in the Gardens (Nov. 5th,
1872), is nearly, if not
quite as large.

HIPPOPOTAMUS.

Dr. Hugh Rayner, of
the Grenadier Guards,
writing in the *Field* (Aug.
5th, 1893), says that " A
hippopotamus lives in
the water by day, and
only comes out, after the
sun is down, for purposes
of grazing. Then in the
early morning, before the
sun reappears, he once
more makes his way to
the water, and remains
there until feeding-time
comes round again. Therefore it is whilst the animal is in the water
that the sportsman's opportunity occurs. A hippo suddenly puts his
head above the surface, gives a ' blow ' of air, and as suddenly
disappears. When frightened by the presence of man, his head is
rarely above the surface more than a couple of seconds, and it is
during this short period of time that one must shoot. Unless, too,
the animal be hit in some part of the head by which the bullet can
penetrate the brain, apparently no injury is done, and frequently
bullets will glance off the flat part of his head and whizz into space.
. . . Once I thought an animal was coming out of the water to charge
me. He certainly looked like it, but he was either too badly hurt or
else he changed his mind." These animals can stay under water for
some time. Dr. Rayner says that he timed one, and " it remained
down fourteen minutes."

An infuriated hippopotamus has been known to bite a man literally in two; and the late Sir Samuel Baker wrote of one that charged a steamboat, and pierced two of her iron-plates as cleanly as if the holes had been made with a sharp pick.

The Liberian or Pigmy Hippopotamus, from the West Coast of Africa, is a much smaller species.

The Ruminants, or animals that chew the cud, consist of the True Ruminants (Sheep and Goats, Oxen, Antelopes, Deer, and Giraffes), the Camels, and the Chevrotains, or Water Deer. Except in the Camels, there are no incisor teeth in the upper jaw, their place being supplied by a horny pad, against which the lower incisors bite. The lower canines look like incisors; the upper ones are often absent, but in the Musk Deer and some few others they are large and tusk-like. The Hollow-horned Ruminants (Sheep and Goats, Oxen and Antelopes) never shed their horns, which are hollow, and borne on bony projections from the skull, called horn-cores. Both sexes generally bear horns, but in some forms the females are hornless.

The stomach of the True Ruminants has four divisions. While grazing, the animal fills the first stomach with partially masticated food, and when it lies down to ruminate, some of this food is forced into the second stomach, where it is formed into a solid mass, and returned to the mouth. After being thoroughly masticated, it passes through the second and third stomachs to the fourth, to undergo the process of digestion. The Camels and Water Deer have no third stomach.

The Domestic Sheep is extremely valuable to man; its flesh yields excellent food, and its fleece is made into various articles of clothing. Its domestication must have taken place in prehistoric times, for it is mentioned in the oldest literature that has come down to us. The Domestic Sheep of the New World have been introduced since the Spanish Conquest. Moseley says that he saw sheep used as beasts of burden in South America—carrying barrels of water from a well to houses in the town. The Tibetans also employ them to transport goods over the mountains, and their load is from 20 to 30 lbs.

Asia is the home of wild sheep, and thence probably came the stock from which our breeds have sprung. Many of them are of large size; Marco Polo's Sheep, from the Thian Shan Mountains, is said to stand nearly 4 feet high at the shoulder, and to weigh nearly 600 lbs. The Ammon, or Argali, from Tibet, is nearly as large, and its horns are as much as 4 feet along the curve, and nearly 2 feet round at the base. The Barbary Wild Sheep has the lower part of the neck and the knees fringed with long hair. In Europe there is one wild sheep, the

Mouflon of Sardinia and Corsica, and one in America, the Bighorn, allied to the Argali, by which name it is sometimes called.

Goats differ from Sheep in having the horns flattened, and curving backwards, or twisted in a spiral. There are several European Wild Goats, of which the best known is the Alpine Ibex, which lives about the snow-line, coming lower down at night to graze. The Grecian Ibex, ranging as far east as Persia, is probably the parent of our domestic breeds. The Sinaitic Ibex has knobs on its large horns, which must be formidable weapons of defence.

There are several Asiatic species, of which the Markhore, with

MUSK-OX.

spirally twisted horns, and the Thar, with short thick horns, ridged in front, are the most remarkable.

The Domestic Goat is very serviceable to man. Its milk has valuable properties, and is often made into cheese. In the East its flesh is eaten, and the fine hair of the Angora and Cashmere goats is made into cloths and shawls.

The Musk-ox is related to the Sheep and the Oxen: it is a native of Arctic America. It is about the size of a small Welsh Ox, stoutly built, with short legs, covered with long brown hair that reaches nearly to the ground, and beneath this is a thick under-fur shed in summer. It has a strong musky odour, the source of which is not ascertained. The artist of the Schwatka Expedition tracked a herd with some Eskimo and their dogs. He says:

"Standing huddled together, with their heads outwards, the

Musk-oxen defied the attacks of the dogs; and when one dog, more daring than the rest, tried to seize an old bull by the nose, he paid dearly for his rashness, and was savagely tossed in the air. But though the oxen could keep off the dogs, the guns of the hunters soon thinned their numbers, and in a few minutes four out of the small herd had fallen, and the rest galloped off over the broken ground at a pace that left the dogs behind."

The Antelopes are chiefly found in Africa, though their numbers are diminishing; Asia has many species; there are two in Europe, while America has only one true Antelope and the Pronghorn, which is Deer-like in some of its characters.

The Elk-like Antelopes, of large size, and generally of uniform colour, are natives of Africa and Syria. The head is long and narrow, broadening out at the extremity. Both sexes are horned. The Hartebeest, which stands about 5 feet high at the shoulders, is well known in zoological collections.

The Common and the White-tailed Gnu (Plate IV., No. 12) are strange-looking creatures, ungainly enough when at rest, but really graceful when in rapid motion, for their large heads and manes give the necks the appearance of being arched, like that of a well-bred horse, while the flowing tails stream well out behind. They have a bad reputation for ferocity, but the cows in the Zoological Gardens will come up when called, and follow one round the paddock for biscuit or apple. A young White-tailed Gnu was born in the Gardens in 1894.

The Duikerboks (Plate IV., No. 10) are small African Antelopes, one of them no bigger than a rabbit. The Four-horned Antelope of India belongs to this group; its horns are in pairs, those in front being much the smaller.

In the next group, entirely African, only the males are horned. Among these are the Royal Antelope from Guinea; the smallest known ruminant, and Salt's Antelope, from Abyssinia, which is very little larger; the Rehbok, from rocky districts of South Africa; and some others. The Water-bucks are large Antelopes with large horns and long dark hair. They have bred in the Zoological Gardens.

The next group may be called Desert and Steppe Antelopes, from the places in which they are found. Here belongs the Saiga, a sheep-like Antelope, with very large swollen muffle: it is a native of Eastern Europe and Western Asia. Here also belong the Gazelles, of small size, generally fawn or sandy-coloured, with some white on the face, and a white mark on the rump, that serves the same purpose as the upturned tail of the rabbit. The Springboks of South Africa are well-known Gazelles.

GEMSBOK AND LEOPARD.

The Equine Antelopes are African. They are large animals with long horns, straight or curving backwards. The Gemsbok, the Beisa, and the Oryx are rusty-grey with black markings, and have straight horns; and to these three the name Oryx is often applied. The Leucoryx is uniformly coloured, and has curved horns. These horns—straight or curved—are terrible weapons. In attacking, these antelopes lower the head and strike sideways and upwards with great force. Gordon Cumming says of a wounded Oryx: "Lowering her sharp horns, she made a desperate rush towards me, and would have inevitably run me through, had not her strength at that moment failed her, when she staggered forward and fell to the ground." Andersson tells a story of a fight between a Lion and an Oryx: the Antelope pierced the Lion with his horns, but was killed by the weight of his dying foe. Boer hunters have also met with the skeletons of these animals and of the great Carnivora, lying in such a position as to show that similar conflicts must have taken place.

The Bovine Antelopes are African, with the exception of the Nilghai, which is Indian. Here belong the Harnessed Antelopes, pale bay, marked with white stripes; the Koodoo, slaty-grey, also marked with white stripes; the marsh-loving, mouse-coloured Speke's Antelope, with long, spreading hoofs; and the gentle, fawn-coloured Eland, the largest living antelope. Livingstone's Eland, a variety with white stripes, was first brought to England in 1894.

In the last section are the Goat-like Antelopes, with horns in both sexes. The Serow and the Takin are Asiatic. The Chamois (Plate IV., No. 11) is found in the mountains of Europe, from the Pyrenees to the Caucasus. It stands about 2 feet high at the shoulders, has long chestnut-brown hair, and horns bending backwards at the tips. Chamois live in small herds, and post sentinels, which give an alarm on the approach of danger. Chamois-hunting is attended with great danger, on account of the almost inaccessible heights to which these animals betake themselves when pursued. The Rocky-Mountain Goat, with long, white hair, is closely allied. The Pronghorn, from the prairies of North America, bears horns with a single branch, which are shed annually. It is about the size of a Fallow Deer.

The Buffaloes are confined to the Old World. The Anoa of Celebes is a small animal, with black hair and short horns. It resembles the young of the Cape Buffalo.

The Indian Buffalo is a large beast, with dull black skin, almost hairless, and long horns that sweep upward and backward. The domesticated race has been introduced into the South of Europe as beasts of draught. Wild buffaloes exceed the domestic race in size,

AMERICAN BISON (1875).

standing 6 feet high at the shoulder, and measuring over 10 feet from the nose to the root of the tail. A pair of horns of the Wild Buffalo in the British Museum measure 6 feet 6 inches each; this in the living animal would give a curve of nearly 14 feet. Jerdon says that a wounded wild buffalo will charge the elephant, and he had it on the authority of many sportsmen that the charge was sometimes successful. The domesticated animals are often used to drive a wounded tiger out of cover, and Sterndale records an incident, from his own knowledge, of a herdsman being seized by a tiger, when his buffaloes, hearing his cries, rushed up, and so saved his life.

The Cape Buffalo is a larger and much fiercer animal than the domestic Indian race. Its horns are very broad at the base, where they almost meet, and they curve downwards and backwards. Varieties occur in the more northern parts of Africa.

There are two Bisons, one American, the other European, and the few that still survive are carefully protected, but in a few years both species will be extinct. The head is carried low, and there is a hump.

K

The horns are short and rounded. There is a thick shaggy mane, a long beard, and a fringe of long hair on the forequarters. The height at the shoulder is about 5 feet 8 inches, and the general colour is brown. Within the memory of living men bison existed in America in countless myriads. The numbers of the great southern herd killed from 1872 to 1874 are estimated in millions, and Hamlin Russell says that "Buffalo Bill and his kind, with English 'sportsmen' and American army officers, vied with each other in the wanton slaughter." How the great northern herd disappeared is not certainly known, but since 1883 herds of bison—or "buffalo," as the Americans call them—have disappeared. One consequence of their disappearance will be the rapid extinction of the Red Indian. "With his food supply cut off, the Indian became suddenly tame and easy to handle."

The European Bison (Plate IV., No. 13), though somewhat larger than the American, is not so heavily maned. It is preserved in the forests of Lithuania, Moldavia, Wallachia, and the Caucasus.

The Yak, from the higher regions of Tibet, is allied to the Bisons, but though the withers are high, there is no true hump. The size is a little less than that of the Ox, and the body is fringed with long hair, which, Mr. Bartlett thinks, serves the purpose of a mat, on which the animal can lie amid the snow. There is a domesticated race, used as beasts of burden.

There are three Asiatic Wild Oxen, all with high withers and a ridge down the back. The Gaur is the largest, standing 6 ft. at the withers. Like the Gayal, it is dark coloured, with white "stockings," but there is a difference in the shape of the horns. The Banteng from Java is much smaller ; the females and young are bright chestnut, but the bulls become black with age.

Of our Domestic Cattle, with their numerous breeds, and their importance to man, there is no need to speak. Everybody has drunk milk, and eaten butter and cheese and veal and beef; and most people have seen oxen ploughing and drawing carts. The hair, skins, horns, hoofs, bones, and blood all have commercial value. Some breeds of cattle said to be descendants of a wild race, are preserved in English parks.

Closely allied are the Humped Cattle of India, some of which are much larger than any English Ox, while others are smaller than a donkey. They are used for ploughing, drawing carts, and riding on. The white humped cattle are sacred to Siva, and Hindus think it wrong to kill any of whatever breed.

Deer are found pretty plentifully in three of the continents. Africa has but one species, which lives along its Mediterranean coast. The males of nearly all bear antlers, which differ from horns in being bony outgrowths from the skull, and are shed and renewed year by year. The females of the reindeer alone carry them, and theirs are smaller and slighter than those of their lords. The young of most Deer are spotted.

The Red Deer is found in many parts of Europe, is still wild on Exmoor, in the North of Scotland, and the West of Ireland, and ranges into the North of Africa, those living there being sometimes called Barbary Deer. A full-grown Stag, for so the male is called, will stand four feet high at the withers ; the body-colour is reddish-brown, the neck greyish, and there is a white patch on the rump, present also in most of the allied forms, and serving as a mark or signal. The females, which are smaller, are called hinds, and the young calves. The flesh of deer is called venison. In Scotland Deer are stalked and shot ; on Exmoor they are still hunted with hounds, and killed by the huntsman when brought to bay. One took refuge in a room where some ladies were just sitting down to dinner. One of the hunters described the scene

YOUNG SAMBUR.

as resembling a football scrimmage by men in boots and breeches, a frantic, struggling deer taking the place of the ball. The Persian and Cashmere Deer are closely allied.

The Wapiti is a North American Deer resembling the Red Deer, but of larger size and with finer antlers. Attempts are being made to acclimatise these noble creatures in England. There is a small herd at Haggerston Castle in Northumberland, and in 1893 six males and fourteen females were brought to Osmaston Castle, near Derby. Wapiti breed freely in the Zoological Gardens, and the small herd there are very tame, coming up to the bars to be fed by visitors. The stag, however, will drive the hinds away, and they in their turn will keep the calves at a distance, so as to secure the biscuit or bread for themselves.

The Sambur is a large Indian Deer, standing about 5 feet 8 inches at the shoulder, dark brown in colour, with an orange-yellow recognition mark on the rump. The ears are large, and the neck is heavily maned. Sterndale describes this deer, "with his proud carriage, and shaggy, massive neck, sauntering slowly up the rise, stopping now and

K 2

then to cull a berry or to scratch his sides with his wide-sweeping antlers," and applies to the animal the merited epithet "noble." Sambur are driven or shot as they come to drink, but the sportsman-like way is to stalk them.

The Axis Deer, also Indian, stands 3 feet or a little more at the shoulder. The colour is reddish-fawn, spotted with white. The Hog Deer is closely allied. The Swamp Deer, smaller and lighter than the Sámbur, is found near water and in the open country. A sportsman once had three herds in sight at the same time, the nearest not five hundred yards away. He says: "There must have been at least fifty of them—stags, hinds, and fawns feeding together in a lump, and outside the herd grazed three enormous stags." There are some other species.

FALLOW DEER.

The Muntjacs are small deer from the south and east of Asia, with antlers borne on bony stalks. They are solitary in habit, and are often called Barking Deer by sportsmen, from their alarm cry.

The Fallow Deer stands about 3 feet high at the shoulders, is marked with white spots, and has the antlers spread out at the top like the palm of one's hand. It lives in many English parks in a half-domesticated condition. There is also a Persian species.

The Reindeer (Plate IV., No. 8) is a native of the northern regions of Europe, Asia, and America (where it is called the Caribou). The general colour is deep brown, but individuals vary. The size is about that of the Red Deer, but the wild race is larger than that domesticated by the Laplanders. These animals feed in the summer on the branches of birch and willow, and in winter on lichens and reindeer moss, from which they sweep away the snow with their broad deeply-cleft hoofs. They make partial migrations in search of food and to avoid the attacks of a gad-fly. To the Laplander the Reindeer serves instead of the Horse, Cow, Sheep, and Goat. It will draw a sledge, weighing with its load 300 lbs., over the frozen snow at the rate of 100 miles a day. Its flesh and milk are used for food : its horns and bones are made into tools and weapons ; its skin, sewn together with the sinews, serves for clothing and tent-covers. The Siberian race is used as a beast of burden and for riding on.

The Elk or Moose is found in the regions inhabited by the Rein-deer. It is the largest living deer, adult males often standing 6 feet at the shoulder. The general colour is brown, and the hair on the neck

and throat is long. The muffle is large and prominent; there is a
swelling on the throat, and the antlers are very large, and spread out
into a kind of shovel-shape, the two together forming a sort of basin.
These animals are very shy and wary, and to hunt them successfully is
a difficult task. They generally lie with the tail towards the wind,
trusting to their senses of hearing and smelling to warn them of danger
from that quarter, using their eyes to warn them of danger to leeward,
where hearing, and especially smelling, would be of little use. These

ELK OR MOOSE.

animals are naturally timid, but when roused the antlers and hoofs are
formidable weapons, and a blow from the latter has been known to
kill a wolf.

The Roebuck stands about 2 feet at the shoulders, and is reddish-
brown in colour, with a white patch on the rump. It is widely distributed
over Europe and Western Asia, and is found wild in the north of Britain.

The Chinese Water-Deer is small, and has no antlers, but the upper
canine teeth of the males are developed into tusks. The female has
five or six young at a birth, unlike other deer, which have but one, or
rarely two.

America has several species of Deer of uniform colour and medium size. The odour of the Pampas Deer is said to be fatal to snakes. The Gauchos tie a strip of deerskin to their horses' necks when turned out to graze. All deer hate snakes, and have often been seen to leap upon them and cut them to pieces with their hoofs.

The Musk-Deer is found in the highlands of Central and Eastern Asia. It is a little smaller than the Roebuck, has no antlers, and the long upper canines of the males project downwards and backwards. The colour is brownish-grey, with patches of a lighter hue. The animal owes its name to the highly-scented secretion of the male, which is found in a sac, about the size of a small orange, on the abdomen. The sac is cut out, dried, and sold for making perfumes.

The Mouse-Deer, or Chevrotains, are the smallest living Hoofed Mammals. They are sometimes called Pigmy Musk-Deer, but wrongly. They resemble the true Musk-Deer in the absence of antlers, and in their long canines; and in little else. Sir William Flower tells us they are intermediate in structure between the Deer, the Camels, and the Pigs. There are four or five Asiatic species, with the habits and appearance of Agoutis, and an African form, found near water and having the habits of a Pig.

The Giraffe (Plate IV., No. 9) is a native of Africa. It is the tallest living mammal, adult males having reached a height of 18 feet and upwards. The colour is yellowish-buff, with large reddish-brown blotches. There are two bony "horns," covered with skin, and below these, in the middle of the face, a third "horn," also covered. The neck is immensely long, the nostrils can be closed at will, and the tongue is prehensile, and well adapted to pluck off the foliage on which these animals feed. Giraffes generally occur in small herds of about twenty, but larger ones have been met with.

Gordon Cumming thus describes his first adventure with these animals :—" Before me stood a troop of ten colossal giraffes, the majority of which were from 17 to 18 feet high. On beholding me they at once made off, twisting their long tails over their backs, making a loud switching noise with them, and cantered along at an easy pace, which, however, obliged my horse to put his best foot foremost to keep up with them. My senses were so absorbed by the wondrous and beautiful sight before me that I rode along like one entranced, and felt inclined to disbelieve that I was hunting living things of this world. The ground was firm and favourable for riding. At every stride I gained upon the giraffes, and after a short burst at a swinging gallop I was in the middle of them, and turned the finest cow out of the herd. . . . In a short time I brought her to a stand in the dry bed of a watercourse, where I fired at

GIRAFFE.

fifteen yards, aiming at where I thought the heart lay, upon which she again made off. Having loaded, I followed, and had very nearly lost her: she had turned abruptly to the left, and was far out of sight among the trees. Once more I brought her to a stand, and dismounted. There we stood together, alone in the wild wood. I gazed in wonder at her extreme beauty; while her soft dark eye, with its silky fringe, looked down imploringly at me, and I really felt a pang of sorrow in this moment of triumph for the blood I was shedding. Pointing my rifle towards the skies I sent a bullet through her neck. On receiving it she reared high on her hind legs and fell backwards with a heavy crash, making the earth shake around her."

The only mode of defence employed by these animals is kicking, and with their powerful legs they can deal terrific blows. The cows are said to keep off in this fashion the larger beasts of prey that attempt to carry off their young. But if, as is often the case, the lion springs upon a giraffe, all the victim's efforts to dislodge its assailant are unavailing, and it gallops off wildly with its cruel enemy holding fast with claws and teeth, till it sinks exhausted on the ground, and becomes an easy prey.

The first giraffes were brought to the Regent's Park Zoological Gardens in 1836. This small herd, consisting of three males and a female, bred freely, and seventeen giraffes have been born in the Gardens. The last of these animals, however, died in March, 1892, and the Society was for several years unable to procure other specimens. There is indeed great reason to fear that this noble animal is rapidly becoming extinct.

In 1893 a variety from Somaliland was described. A herd of seven was seen, and the bull was said to be "like a small Eiffel Tower." The female shot was marked with large patches of rich chestnut, separated by white lines; the legs were creamy white. In February, 1895, a female Giraffe from South Africa was brought to Southampton, and purchased by the Zoological Society. The animal, between two and three years old, is nearly 12 ft. in height. Thus for nearly three years there had been no Giraffe in the Society's menagerie.

The Camels and Llamas are hornless. The upper lip is deeply cleft; there are canine teeth above and below; the young have cutting teeth in both jaws, but in the adults only the outermost on each side remains. There are only two digits on each limb, the hoofs are small, and the animals walk on broad cushion-like pads that form the sole of the foot.

The Arabian or True Camel (Plate IV., No. 6), often called the "Ship of the Desert," is domesticated in the North of Africa, and South-west Asia, and eastwards to India. It is a bulky animal, with long neck and limbs, and a single hump on its back. The dromedary is a lighter, swifter variety of this species. "The Camel prefers such plants as a horse would not touch to the finest pastures. He is satisfied with very little, and if he should be stinted of this hard fare, the fat hump contains a store of nourishment to sustain him till he reaches an oasis of prickly bushes. If the best of liquids be there, he fills the water-tanks with which his interior is fitted up, and goes on his way rejoicing."

BACTRIAN CAMEL.

To the Arabs the Camel is not only valuable as a beast of burden ; the milk is made into butter, the hair is woven into cloth, the skin is made into leather, the flesh is eaten, and the droppings form almost the only fuel to be found in the desert.

Camels have been introduced into Italy, and in Spain there are even a few half-wild ones that have escaped from domestication. They have lately come into use in the sandy deserts of Australia. Camels are also used in the army as baggage animals, to carry light swivel guns, and infantry are mounted on them, each animal also carrying a driver.

The Bactrian Camel, with two humps, is a larger animal, with shorter legs and thicker coat of darker colour. It is domesticated pretty generally throughout Central Asia. The "wild" Camels discovered by a Russian traveller in 1879 are probably the descendants of animals that escaped from domestication. Mr. Littledale shot three near Lob Nor in 1893, and brought home the skins, one of which has been set up in the Natural History Museum.

The Llamas are confined to South America. There are four forms, two wild and two domesticated, all smaller than the Camel and more lightly built. The Guanaco (Plate IV., No. 7) ranges from Peru to Patagonia. Its general appearance combines some of the characters of the camel, the deer, and the goat. The body—deep at the breast, but small at the loins—is covered with long soft hair, brown above, fading to white beneath, and there is usually some black on the face. These animals are generally met with in herds, sometimes containing several hundreds. The Vicugna is smaller and slenderer, and is confined to the mountain regions as far south as Bolivia.

The Llama, a domesticated breed of the Guanaco, is found in the south of Peru. It is larger than its wild stock, and is white marked with brown or black, or altogether black. It is still used as a beast of burden, as it was in the days of the Spanish Conquest, when the ore from the silver mines was brought down from the mountains by these animals.

The Alpaca, probably another breed of the Guanaco, is smaller than the Llama. The colour is generally brown or black, and from its long hair the fabric known as alpaca is made.

LLAMA.

CHAPTER XIII.

RODENTS, OR GNAWING ANIMALS.

THIS order contains a larger number of animals than any other, and some of them are found in every country in the globe, for in the few places where native species do not exist others have been introduced by man. They are generally of small size, and live on or burrow under the ground, but some live among the branches of trees, and a few are aquatic; many of them take a long sleep in cold weather. Their food is principally vegetable, often bark and roots. There are no canine teeth; and the incisors grow from behind as fast as they are worn away in front. The limbs usually bear five digits, armed with claws or hoof-like nails. Their name refers to their method of eating. There are two groups, of which the Simple-toothed Rodents, with two incisors in the upper and two in the lower jaw, are by far the more numerous. Of these there are three divisions, of which the Squirrel, the Mouse, and the Porcupine are the types.

The true Squirrels are found in most temperate and tropical countries, but are absent from Madagascar and Australia. The largest and most brilliantly coloured forms are natives of the East, where some are as large as a cat, while others are no bigger than a mouse. The tail is long and bushy, and the pointed ears are generally tufted with hair.

Our English Squirrel (Plate III., No. 8) ranges over the greater part of Europe and Northern Asia, and has many near relatives in the latter continent. Its total length is about 9 inches, and the general colour is brownish-red above, and white on the under surface. Squirrels live for the most part among the branches of trees, and are exceedingly active, leaping from bough to bough so quickly that the eye can scarcely follow them. Their movements on the ground are also rapid. They do not often take to water, but can swim well, though there is, of course, no truth in the story that they use pieces of bark for canoes and their tails outspread for sails. They feed principally on nuts, acorns, buds, and bark, etc., and the food is held in the hands and gnawed with the incisors. They make a kind of nest called a "drey," and lay up stores of provision against the winter.

The Palm Squirrel, a common Indian species, has been known to "carry off bits of lace and strips of muslin and skeins of silk for its house-building purposes."

Though none of the true Squirrels has a flying membrane, yet by spreading their limbs and tail they in some measure supply the want. Mr. Burroughs, an American naturalist, while hunting, started a squirrel, which his dog drove to take refuge in a tree. He says: "To see what the squirrel would do when closely pressed I climbed the tree. As I drew near he took refuge in the topmost branch, and then, as I came on, boldly leaped into the air, spread himself out upon it, and with a quick tremulous motion of his tail and legs descended quite slowly, and landed upon the ground thirty feet below me, apparently none the worse for the leap, for he ran with great speed, and escaped the dog in another tree."

The Spiny Squirrels are natives of Africa, and are clad in coarse fur mixed with spines.

The Flat-tailed Flying Squirrels are chiefly Asiatic, but one is found in North America and another in eastern Europe. Like the Colugo (p. 78), they have a flying membrane. The Round-tailed Flying Squirrels, in which there is a membrane between the thighs, as in some Bats, are mostly Indian. All the Flying Squirrels are nocturnal.

The Ground Squirrels are natives of America, where they are called Chipmunks, but one species ranges through the North of Asia into Europe. They have cheek-pouches, in which food is stored and carried to their burrows.

The Pouched Marmots or Spermophiles are spread over nearly the same limits as the Ground Squirrels. Like the latter, they have cheek-pouches; but though their ordinary food is vegetable, they also prey on small animals, the bones of which have been found in their burrows. There is one European species, the Souslik, which is considered a delicacy by the Russian peasants.

The Marmots are stoutly built, have no cheek-pouches, and the tail is short. There are about a dozen species, which are confined to the northern hemisphere.

The Alpine Marmot (Plate III., No. 9), found in the Alps, Pyrenees, and Carpathians, is nearly 2 feet long. Sir William Flower, quoting from a German author, says that these animals " live high up in the snowy regions, generally preferring exposed cliffs, whence they may have a clear view of any approaching danger, for which a constant watch is kept. When one of them raises the cry of warning, the loud piercing whistle so well known to travellers in the Alps, they all instantly take to flight and hide themselves in holes and crannies among the rocks, often not reappearing at the entrance of their hiding-places till several hours have elapsed, and then frequently standing motionless on the look-out for a still longer period. Their food consists of the roots and leaves of various Alpine plants, which, like squirrels, they lift to their mouths with their fore-paws. For their winter-quarters they make a large round burrow, with but one entrance, and ending in a sleeping-place thickly lined with hay. Here often from ten to fifteen Marmots pass the winter, all lying closely packed together, fast asleep, until the spring." The Bobac, the only other European species, is confined to the eastern parts of the Continent.

WOODCHUCK.

The Woodchuck, the common American Marmot, shows great skill in forming his burrow on a hillside, guarding himself against being drowned out by making his sleeping-place, at the end of the hole, higher than the entrance. He digs in slantingly for 2 or 3 feet, then makes a sharp upward turn, and keeps nearly parallel with the surface of the ground for a distance of 8 or 10 feet further, and so sets rain and flood at defiance.

The Prairie Dogs are so called from their cry, which somewhat resembles the bark of a small dog. There are two species; the best known lives in the prairies east of the Rocky Mountains, the other in the western States as far south as New Mexico. The total length is about 1 foot, of which the tail counts for some 3 or 4 inches, and the fur is reddish-grey. Their chief food is the blades and roots of the buffalo grass. These animals live together in large groups, dwelling in burrows pretty close to each other, and an assemblage of these burrows is called a "town" or "village." Burrowing Owls and Rattlesnakes share these underground dwellings with their rightful owners—the former probably because they are spared the trouble of

PRAIRIE DOG.

excavating a burrow for themselves, as they do in some places; the latter because of the plentiful supply of their favourite food.

The Sewellels are also American burrowing rodents from the west of the Rocky Mountains. They are of similar habits to the Prairie Dog, but somewhat larger, and with darker fur.

The Anomalures, of which there are several kinds, live in Western and Central Africa. They have a flying membrane, which is supported by a gristly spur that projects from each elbow, and so gives the membrane a greater spread. At the root of the tail are scales that overlap each other, and act as support to these animals in climbing. Mr.

Adams thus describes their habits in the *Proceedings of the Zoological Society* (1894) :

"They come out of their holes in the trees some hours after sunset, returning long before daybreak. They are only to be seen on bright moonlight nights, and in fact the natives say they do not come out at all in stormy weather or on very dark nights. They live on berries and fruits, being specially fond of the palm-oil nut, which they take to their nests to peel and eat. The most I have seen in one hole is three, though, according to the natives, five or six are sometimes found. They pass from tree to tree with great rapidity, usually choosing to

BEAVER.

jump from a high branch to a lower one, and then climbing up the tree to make a fresh start."

The Beaver (Plate III., No. 11) is found in America between the Arctic Circle and the Tropic of Cancer, but its numbers are rapidly diminishing, owing to the fact that it is trapped for its fur, and it is being driven from some of its old haunts by the increasing population. These animals formerly spread over the northern hemisphere, including our own country, where their memory lingers in such place-names as Beverley, which means the Beaver-lake. In the Old World they are now confined to a few of the larger European rivers, where they live in pairs. A full-grown American Beaver is about 30 inches long, with a flat, oval, scaly tail that may measure 12 inches in length and 5 inches in breadth. The fur is reddish-brown above, paler below. The tail is used as a rudder, not, as some have thought, as a sledge to carry materials for building, or as a trowel to spread the mud with which beaver-dwellings are plastered. The fore feet are much smaller than the hinder ones, and the latter are webbed. The food consists of the bark of trees, the roots of the water-lily, and sometimes grass.

The chief interest of these animals lies in their social habits—for the American Beaver forms large communities—their lodge-like dwellings, and their dams for keeping back water. Some of them, however, inhabit burrows, differing from those of other aquatic animals in little except their size and in opening below the surface of the water.

Mr. H. P. Wells, writing in *Harper's Magazine* (January, 1889), says of beaver-dams that they "resemble a narrow pile of brushwood thrown together higgledy-piggledy. The largest poles are perhaps as thick as a man's wrist, the butt-ends sticking up in the air, with the brush ends inclined towards the bottom and up stream. On the up-stream side these branches and poles are weighted down with mud, mixed with grass and small stones, so as to form a solid and water-tight bank." He says that the

DORMOUSE.

general feeling, when one sees a beaver-dam for the first time, is one of surprise and disappointment at the rough character of the work; and he has but a poor opinion of the Beaver's engineering skill.

The tools of the Beaver are its sharp, strong, cutting teeth, with which it can fell trees from 12 to 18 inches in diameter. This is effected by gnawing a ring round the trunk, gradually working deeper and deeper till trunk and stem resemble two sugar-loaves point to point. In the Zoological Gardens, Regent's Park, you may see the short stems of trees which have been cut down in this fashion. Beavers work only at night, and will do much more when it is dark and rainy than in bright moonlight.

The Dormice are the first of the Mouse-like Rodents to be mentioned, because they are more closely allied than the others to the Squirrel-like forms. They are found chiefly in the temperate parts of

Europe and Asia, and a few live in Africa. They are small creatures, with large ears and eyes, and long bushy tail, living in trees or thickets, and many build nests. The common English Dormouse is reddish-tawny in colour, nearly 6 inches long, of which the tail forms a little less than half. It is often kept as a pet, and soon becomes very tame. Dallas describes the Dormouse as living in thickets and hedgerows, where it is as active in its way among the bushes and undergrowth as its cousin, the Squirrel, upon the larger trees. Among the small twigs and branches of the shrubs the Dormice climb with wonderful adroitness—often, indeed, hanging by their hind feet from a twig in order to reach fruit or nuts otherwise inaccessible, and running along the lower surface of a branch like monkeys.

The Lophiomys, from North-east Africa, is somewhat like a small Opossum in form and habit, and has a thumb-like great toe. The long hairs in the middle of the back can be raised into a kind of crest.

The Rats and Mice and their close allies are very numerous, and form more than a third of the whole order. In Britain we have two Rats—the Black, or Old English, Rat, about 16 inches long, of which the slender tail counts for rather more than half; and the Brown Rat, with stouter body and shorter tail. The latter was probably introduced from Asia, and is rapidly driving out the native species. The habits of these animals are well known. They form burrows, and in their burrowing often do considerable damage. They will eat almost anything that comes in their way, and they increase with marvellous rapidity. Many instances are on record of their undermining masonwork, and by making passages through dams letting out the water from reservoirs and canals to overflow the country. The White and Pied Rats that are kept as pets are varieties of the Old English Rat.

The common Mouse follows man in his wanderings all over the globe. Like the English Rat, it has white and pied varieties, well known to most boys as pets. But the ordinary Mouse can be tamed with a little trouble and patience, and is then just as amusing as his white brother.

The Wood or Long-tailed Field Mouse is about the size of the common Mouse, but the fur is reddish-grey above and white below. It lays up a store of seed and grains in its burrows, and rarely ventures into houses, though it will sometimes pass the winter in barns.

The Harvest Mouse, the smallest European Mouse, is yellowish-red, with short ears and tail. It builds an elegant nest, generally fastened to cornstalks above the ground, or hung on to the heads of thistles. Gilbert White describes one which he took as being " most artificially plaited, and composed of the blades of wheat, perfectly round, and about the size of a cricket-ball, with the aperture so ingeniously closed that there was no discovering to what part it belonged. It was so compact and well filled that it would roll across the table, without being discomposed, though it contained eight little mice that were naked and blind." These last two Mice feed also on insects and worms.

The Wood Rats, from Southern Asia, are closely allied to the true Rats and Mice. The largest is the Indian Bandicoot Rat, which often reaches 2 ft. in length, of which the tail forms a half.

The Voles differ from the true Rats and Mice in being of stouter build, and having shorter limbs and tails. They are found in cold and temperate regions of the northern hemisphere. There are three British Voles, the largest of which is the Water Vole, often miscalled the Water "Rat," and it is sometimes blamed for the misdeeds of the Rat. It lives in the banks of streams and ponds, feeding chiefly on vegetable substances, with insects, mice, and small birds as occasional delicacies.

The common Field Vole is much smaller, and ranges over

HARVEST MICE AND NEST.

Britain as far north as the Hebrides. Quite recently this animal has become so numerous in Scotland as to be a plague. A farmer holding 6,500 acres of land told the Agricultural Commissioners who were inquiring into the subject that he estimated there were three million Voles on his land during the great Vole-plague of 1891 and 1892 ; and putting the damage done by them at twopence a head, the amount would be £50,000 in two years. Greece was visited with a similar plague about the same time. The Bank-Vole is also British, but the fur, instead of being brown, as in the other two, is rusty red. Other species are found on the Continent, and in Asia and America.

There are several Lemmings, of which the Norwegian species is the

I.

best known. It is about 5 in. long, clothed in soft brownish-yellow fur marked with dark spots. The fore limbs are strong and well fitted for

LEMMING.

digging. This little creature is remarkable for its periodical migrations, probably in search of food. The Lemmings appear so suddenly and in such vast numbers in the cultivated districts of Norway that one cannot wonder at the old superstition, not yet extinct, that they drop from the clouds; and within recent times prayers were offered in the churches as a protection against their ravages. One strange part of this migration is that of the myriads that move onwards till they reach the sea, none ever returns. When they come to the seashore they plunge in as boldly as they crossed the streams that lay in their path, and continue swimming till they perish in the waves. Another strange circumstance is that though the advancing host is preyed upon by numberless beasts and birds, and shot and trapped by man, they actually increase while on the march, probably owing to the abundant food-supply.

The Musquash, or Musk-Rat, is an American species, ranging from the Barren Grounds to the Rio Grande. It is about 2 ft. long, the tail counting for a third; and the fur is dark brown above and grey below. It frequents swampy grounds and the banks of lakes and streams, feeding on roots, shoots and leaves, and fresh - water

MUSK RATS.

mussels. The fur is valuable, and large numbers of these animals are trapped or speared year by year. They generally live in a burrow in the bank, but sometimes build dome-like houses, like those of the beaver, rarely far from water.

The Hamsters are nearly allied to the true Mice and Rats, but have large cheek-pouches. There are many species, all confined to the Old World ; and the best known is the common Hamster (Plate III., No. 10) of Central Europe and Asia. It is about 1 ft. long, of which the tail forms a little less than a quarter. The fur is thick and shiny, yellowish-brown above, a reddish band on the neck, and a yellow spot on each cheek ; the lower surface, limbs, and a band on the forehead are black, and the feet white. It is a pretty little animal, and will defend itself vigorously if attacked. It feeds mostly on vegetable food, but does not object to small animals, and in confinement hardly anything comes amiss to it.

MOLE RAT.

The Tree Mice are small African forms, of which not much is known. They are said to resemble Dormice in habit.

The Gerbilles, or Kangaroo-Rats, of South-eastern Europe, Asia, and Africa, have very long hind limbs and long hairy tail. Most of them inhabit deserts and sandy plains, and in one the nostrils are furnished with a lobe, which, Blanford says, "is evidently to keep out sand and dust from the air passages." Of the common Indian Kangaroo-Rat Sir Walker Elliot says : "They do not hoard their food, but issue from their burrows every evening, and run and hop about, sitting on their hind legs to look round, making astonishing leaps, and on the slightest alarm flying into their holes." The species are numerous.

The Long-tailed Spiny Mouse, sometimes called the Spiny Dormouse, is a native of India, living in clefts of the rocks and hollow trees, and is said to hoard grain and roots. The hill-people call it the "pepper-rat," because it destroys large quantities of ripe pepper. The length is about 6 in., and the tail 3 in. The fur is light brown. The hairs on the tail stick out like a bottle brush.

The Mole-Rats are so called from their form and burrowing habits.

L 2

The fore-feet are well adapted for digging, and the eyes are small and sometimes hidden in the skin, as in the common Mole Rat of South-eastern Europe. The Indian species are often called Bamboo Rats, from their burrowing under old bamboo roots. The Strand Mole Rat, from the Cape of Good Hope, is about the size of a rabbit, with greyish-white fur. It lives in sandy places near the shore, and burrows to such an extent as to make the ground unsafe to ride over.

The Pouched Rats, widely distributed over America, have cheek-pouches that open outside the mouth. The Pocket Mice, with the hind limbs long and adapted for leaping, as in the Gerbilles and Jerboas, belong to this family.

The Jerboas constitute a family, with representatives in all the four continents, characterised by the wonderful way in which they progress by leaping, being better adapted for this than the Gerbilles, which in many ways they resemble.

The American Jumping Mouse ranges from Labrador to Mexico. The true Jerboas are confined to the Old World, and the

JERBOAS.

Cape Jumping Hare is from the Cape of Good Hope. It is like a tiny kangaroo, and is said to cover twenty feet at a bound.

The Porcupine-like Rodents consist of several families. In the first are a number of small rat-like animals, chiefly from South America. The Tuko-Tukos, of which there are four species, are named from their cry. They are small burrowing animals from South America, and on each of the hind feet is a kind of comb of bristles, used for dressing the fur. The Coypu is found on the shores of South American lakes and rivers. It is about 3 ft. long, of which the round hairy tail measures a third. Four of the five digits on the hind limbs are webbed. The fur is valuable, and is sold as South American Otter. The body is dull brown, the muzzle greyish, and there is a little warm brown on the side of the head. Mr. Aplin says that the Coypu "swims with the nose, the top of the head, and a narrow line of the back out of water, all on a dead level, or almost so ; the nostrils being very high up in the line of the skull, they are kept out of the water without the nose being poked up towards the sky."

The Porcupines form two groups, those from the Old World, which live on the ground, and the American forms, which live among the branches of trees. The hairs are converted into spines or quills.

The common Porcupine (Plate III., No. 12) is a native of Southern Europe and the North and West of Africa, and other species range through India to the Malay Archipelago. The length is somewhat under 3 ft., including the short tail. The skin is blackish-brown, hairy on the muzzle, and with a stiff bristly crest. The large quills are mostly ringed with black and white. These can be erected by a con- traction of the skin ; but the old notion that the animal could shoot them at its foes is erroneous, and probably arose from the fact that they drop out easily. Sterndale, who kept one of the Indian forms, says that the

COYPU.

porcupine attacks by backing up against an opponent or thrusting at him by a sidelong motion. When a dog or any other foe comes to close quarters, the porcupine wheels round and rapidly charges back. These animals are nocturnal and solitary, living in burrows during the day, and coming out at night to feed, often doing great damage to gardens. The flesh is excellent eating.

The Brush-tailed Porcupines, of which two are Malayan and one African, have the spines of the tail flattened at the base. The Tree Porcupines, with one exception, use the tail as a grasping organ, and in those of South America there is a pad on the hind feet, between which and the toes objects can be firmly held. These are more lightly built than the Ground Porcupines, and the short close spines are mixed with hair. They feed on fruit and the buds and leaves of trees.

The Urson or Canada Porcupine has a short tail, and the spines are almost hidden by long hair.

In the Chinchilla family are three species, all South American. The Chinchilla, valuable for its soft grey fur, is a squirrel-like animal from great elevations in the Andes. It is nocturnal in habit, and very shy. It is often seen in captivity, but rarely becomes tame and familiar. Cuvier's Chinchilla, a larger species, is found over the same range of country.

The Viscacha, which lives on the great plains from Buenos Ayres to Patagonia, is about 2 ft. long, of which the tail counts for a quarter. The general colour is mottled-grey, with some black and white on the cheeks. They are nocturnal in habit, and live in burrows in family parties, large numbers of these burrows being near each other. They feed at night on herbage, roots, and bark, and often do great damage in cultivated land. They post sentinels to give notice of danger, and on the alarm being raised they scamper away to their holes.

CHINCHILLA.

The Agoutis are natives of South America and some of the West Indian islands. They have been compared to small slender-limbed Pigs, and to Deerlets. The general colour above is a brownish-olive, due to the mixture of black and yellow hairs. Agoutis live in the forests or along the banks of rivers, and are solitary in habit. They are vegetable-feeders, and from their habits in confinement appear to store food. Those living in the Zoological Gardens generally have pieces of biscuit hidden under the straw, and I have repeatedly noticed that a piece of the same kind of biscuit given to these animals is added to the store, while sweet biscuit or fruit is eaten immediately.

The Paca, of similar habits, is more stoutly built, and its brown fur is marked with white spots forming lines on each side. It is found from Guatemala to Paraguay.

The Restless Cavy, the stock from which our Guinea Pig sprang, is a native of South America, and very common on the banks of the La Plata River. In Uruguay, according to Mr. Aplin, it makes runs among the grass, coming out chiefly about sundown to feed. He says that Cavies are almost as destructive as rabbits; and where the foxes, their chief natural enemies, have been killed down, they are apt to increase inconveniently. The grey mouse-coloured fur is long and pretty, but generally seems loosely attached to the skin.

The Patagonian Cavy is somewhat hare-like in form, with long slender legs. It is a burrowing animal, but where the Viscacha is found it utilises the dwellings of that animal in preference to making a burrow of its own. The long, thick fur is rusty-grey.

The Capybara is the largest living Rodent, being about 4 ft. in total length. The body is stoutly built and the hair long and coarse,

CAPYBARA.

reddish-brown above, lighter beneath. These animals are aquatic, and frequent the borders of streams and lakes in South America, feeding chiefly on water plants. Mr. Aplin says that "their skins tan into splendidly thick soft leather. Like other thick-skinned animals, they like to wallow in mud. They do not go to ground, but live in the banks of the rivers, in such cover as they can find. They can remain under water, proceeding for some distance below the surface; but when a herd has been disturbed, the members probably lie low, by putting just their noses above water, under the shelter of a bed of water-plants."

The Double-toothed Rodents owe their name to the fact that behind the incisors of the upper jaw is a pair of very small teeth (see figure on next page). To this group belong the Hares and Rabbits and Pikas or Calling Hares. The Common Hare is a native of Britain and the greater part of Europe, but is absent from Ireland. Its length is about 27 in., the tail counting for 3 in. ; the fur is reddish-grey above, and white below. It lies in its "form," a mere depression in fern or brushwood, or in the grass, coming out towards dusk to feed on green vegetables and root-crops, sometimes on the bark of trees. These animals

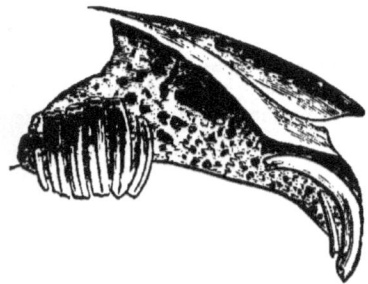

pair early in spring, and produce three or four litters every year. The young, called Leverets, can see, and are covered with hair when born. Hares are timid animals, but the males fight fiercely at certain seasons, standing upright on their long hind legs, and striking downwards at their opponent. This habit of theirs has been utilised by showmen and the keepers of "happy families," who often exhibit

SECTION OF UPPER JAW OF HARE.

hares that engage in mock combats and beat tattoo on the tambourine.

There are several closely-allied species. The Mountain Hare (Plate III., No. 13), from Cumberland, Scotland, Ireland, and the mountains of Central Europe and Asia, is of smaller size, and in cold climates the greyish-brown fur becomes white in winter, but the tips of the ears are always black.

The Rabbit is a burrowing animal ; its young are born blind and helpless, and it multiplies more rapidly than the Hare, from which it may be readily distinguished by its smaller size, greyer fur, and the absence or small size of the black tip of the ears. Spain is probably its native country, but it is now widely distributed. Some years ago it was introduced into Australia and New Zealand, where it has become such a pest that rewards have been offered for keeping down its numbers. There are various breeds of domestic rabbits, and in Belgium these are bred for the table.

The Pikas, or Calling Hares, are small animals not much unlike Guinea-pigs. The North of Asia is their chief home, but one lives in the Rocky Mountains and another in the South-east of Europe. They are mountain-dwellers, living in burrows, and laying up large stores of grass to serve for food in the winter.

SCHOOL OF PORPOISES.

CHAPTER XIV.

DOLPHINS AND WHALES.

THESE are aquatic mammals of fish-like shape, and generally of large size. The body is spindle-shaped, and tapers from the head backwards to the tail, which forks into two "flukes," one on each side, thus differing from the tail of a fish, which spreads out above and below. The fore limbs are like paddles, and there are no hind limbs, though traces of them sometimes exist in the shape of small bones; and many of them have a fin on the back. The body is hairless, except round the mouth, but beneath the skin is a thick layer of blubber, which keeps these animals warm. All feed on other animals; most of them on small crustaceans, some on cuttles, squids, and fish; while one attacks and devours its fellows. In habit they are social, and are found in schools. They come to the surface to breathe, and the vapour expelled with the air condenses into spray, and sometimes water is carried up also. The "spouting" of whales and dolphins is, however, nothing more than their breathing. There are two groups, one with teeth, and one in which teeth are replaced by whalebone. The Toothed Whales have but a single blow-hole—that is, the nostrils form but one aperture; in the others the nostrils are distinct.

The Dolphin (Plate IV., No. 16), a native of the North Sea and the Mediterranean, is also found in the Atlantic, and has near relations in many other seas. It is from 6 to 8 ft. long, black above and white below, with the head produced into a kind of snout, and a fin on the back. This is the animal which is said to have borne Arion on its back through the waves, and Dolphins were believed to have drawn the car of Amphitrite over the foaming waters. It must be distinguished from the so-called Dolphin which exhibits a beautiful play of colour when dying, for this last is really a fish.

The Porpoise is well known all round our coasts, and it or some closely-allied form is found in nearly every sea. From 4 ft. to 5 ft. is the average length; the colour is bluish-black above, fading into a lighter hue below. These animals feed on fish. Like the Dolphins, they are very playful, and their gambols often amuse voyagers at sea.

NARWHAL WITH BOTH TUSKS DEVELOPED.

The Killer-Whale or Grampus is found in almost every sea. The size ranges from 18 ft. to 30 ft.; the colour is black above and white below. From the great size of the dorsal fin, the Germans call these whales "Sword-fish," whence the true Sword-fish has been wrongly charged with attacking whales. Killers feed on seals and members of their own order, and they unite in herds to hunt down the larger whales, just as wolves and wild dogs hunt down their prey.

The Narwhal or Sea-Unicorn is only found in the seas round the North Pole, where it is met with in herds of from fifteen to twenty. It feeds on cuttles and small fish. There is no dorsal fin; the colour on the back is grey, and white below, both marked with shades of grey. The so-called horn of the males is really an enormously-developed tooth on the left side of the jaw; it often reaches a length of 7 ft. or 8 ft., and is valuable as ivory. The tooth corresponding to it on the right rarely pierces the jaw, but in some cases it is also developed.

The purpose they serve is not known. Sir William Flower compares them to the antlers of deer, which are both ornaments and weapons. These teeth do not pierce the jaw in the female.

The Beluga, called also, from its colour, the White Whale, is about 12 ft. long. Its home is in the Arctic Seas, but it comes down as far as the mouth of the St. Lawrence in the Atlantic, and has been seen off the Scotch coast. From its hide "porpoise" leather is made.

The Freshwater Dolphins, from the Ganges, Brahmapootra, and Indus, the Amazon and the Rio de la Plata, have the jaws produced into a long snout. There are only three of them—one from the Old World and two from the New; the largest is about 8 ft. long, and the smallest not more than 5 ft.

The Sperm-Whale is the largest and most important of this section. It is met with in most tropical and sub-tropical seas, but stragglers have been seen round our shores. An old male will reach from 56 ft. to 60 ft. in length, while a female is not much above half that size. The head forms about a third of the whole length, and the space between the skull and the upper jaw is filled with the oil from which spermaceti is procured. The long slender lower jaw bears about twenty teeth on each side, which fit into hollows above. These whales are hunted for sperm oil, procured from their blubber ; ambergris, obtained from their intestines, and spermaceti. The Sperm-Whale fishery is often attended with great danger. The author of "Stray Leaves from a Whaleman's Log" tells how, when the boats' crews were waiting for the whale to rise, "the immense glistening lower jaw, armed with two rows of polished teeth, flashed from the water, and the gigantic beast leaped into the air, carrying with it the head of the boat, which had been snapped asunder, and the unfortunate harpooner, whose head and arms were dangling from the corner of the monster's mouth, within which the body and legs were gripped as in a vice." The appearance of the slender lower jaw above the water before the ponderous upper jaw is due to the fact that the Sperm whale turns on its back to seize anything on or near the surface, though when below it swims with the lower jaw hanging down at an angle of about 45 degrees.

The Short-headed Whale, from Southern Seas, from 6 to 10 ft. long, also yields spermaceti, as does the Bottlenose Whale, from 20 ft. to 30 ft. long, from the North Atlantic, and the oil from the Bottlenose can scarcely be distinguished from sperm oil.

The Whalebone Whales are toothless, and the upper jaws are furnished with plates of whalebone, fringed at the lower ends, and forming a kind of sieve to prevent the escape of the small creatures on which these animals feed. The whale opens its mouth, and so fills it with water; then, closing it, allows the water to run out, and the small crustaceans and molluscs are entangled in the whalebone, from which they are swept off by the tongue.

The Greenland or Right Whale (Plate IV., No. 17) is a native of Polar seas. The head is longer in proportion to the body than that of the Sperm Whale, though not so deep and massive. The total length of a large male is about 60 ft. ; the skin is black, with some scattered white markings, and cream-coloured on the throat. The animals swim in small schools of from six to eight, and the females show great love for their young, and are very bold in defending them when attacked.

A whale between 40 and 50 ft. will yield from sixty to eighty barrels of oil and about half a ton of whalebone. The old method of whale-fishing by harpoons thrown by hand is, to some extent, superseded by harpoon-guns. As soon as the whale feels the wound, it dives, and remains under water till forced to come to the surface, when it is again attacked, and once more seeks refuge below. This is repeated till the whale succumbs to its wounds and floats on the surface dead. It is then towed to the ship; the blubber is cut up and melted down into oil, the whalebone removed and stowed away, and the skin and flesh abandoned to the Killer-Whales, Sharks, and sea-birds that are always found on the whaling grounds.

The Finners or Rorquals, of which there are several species, have a long slender body and a small dorsal fin. They are found in all seas, except those round the Poles.

The Humpback Whales, from the Atlantic and the Pacific, have very long fore-limbs. The upper surface is black, while the limbs and under surface are white, and the skin of the throat is so wrinkled that it has been compared to corrugated iron used for roofing. From their habit of lying on the surface and beating the water with their long fore-limbs, some of the stories of "sea-serpents" have perhaps arisen. The whalebone of the Finners and Humpbacks is of inferior quality, and their blubber yields little oil, but both are hunted.

DUGONGS AND MANATEES.

These animals, often called Sea-Cows, are fish-like in shape, aquatic in habit, and feed on water-plants and sea-weed. They have no hind limbs or fins; the tail is flattened, and the nostrils can be closed at will.

The Dugongs are confined to the Old World. There are three species : one from the Red Sea, another from the Indian seas, and the third from the coasts of Australia. The average length of an adult is about 8 ft. ; and the males have a pair of tusk-like teeth in the upper jaw. From their habit of raising the head out of the water, and carrying their young pressed to the body by means of the fore-limb, the stories of creatures half human, half fish-like, inhabiting the Indian seas, possibly took their rise, and certainly gave the scientific name to the order. The Australian Dugong is hunted for its oil, which is said to be as good for consumptive patients as cod-liver oil.

The Manatees are found in the estuaries of the rivers that flow into the Atlantic between the tropics. There are three species—two

MANATEE.

from America, and one from Africa. The best-known form, generally called the American Manatee, is about the same size as the Dugong, from which it differs in frequenting rivers rather than the sea, and in the incisor teeth being absent in adults. Both Manatees and Dugongs have molar teeth.

This animal has lived in captivity at the Zoological Gardens and at the Brighton Aquarium. At the Aquarium a pair lived for several months, and their daily allowance was 30 lbs. of lettuce and endive. Observations made upon these creatures show that the stories told of their coming on shore to feed have little foundation. When the water was drawn off they refused to feed, though vegetables were put into their tank; and though the male would sometimes make clumsy efforts at locomotion in the empty tank, his attempts were compared by an observer to those of a man lying on his face, with his feet fettered, and his elbows tied to his side.

CHAPTER XV.

SLOTHS, ANTEATERS, AND ARMADILLOS.

THIS group is sometimes called Edentates, or Toothless Animals. Many of them have teeth, but in all they are imperfect, and cutting and canine teeth are generally absent. All are natives of tropical countries, and they are most abundant in South America.

The Sloths live in the tropical forests of South America, east of

THREE-TOED SLOTH.

the Andes, feeding entirely on leaves, which they gather with the mouth, not with the paws. The body is covered with long, rough hair, which agrees with the colour of the mosses and lichens covering the trees among which they live, and a green vegetable growth on the hair itself disguises them still more effectually. The fore-limbs are longer than the hinder pair, and all four are armed with stout claws bent into a hook-like shape, by means of which these animals travel along the branches of trees with the body downward. They are nocturnal and rarely leave the trees for the ground.

There are several species, none much larger than a cat, known as Two-toed and Three-toed Sloths, according as there are two or three digits on the fore-limbs. Moseley bought a Three-toed Sloth (Plate III., No. 14) at Bahia, and kept it alive in his workroom on board the *Challenger* for some days, but as he could not get it to feed he had to kill it. He says : "The beast was the most inane-looking animal I ever saw, and never attempted to bite or scratch; none of us could look at its face without laughing. It merely hung tight on to anything within reach. It showed, however, one sign of intelligence. I hung it on a brass rod used for suspending a lamp beneath one of the skylights in our room.

It remained there half a day, hanging head downward, and constantly endeavouring to reach the bookshelves near by, but without success. At last it found out an arrangement of its limbs by which this was possible, and got away from the lamp-rod, and in future, whenever I hung it up on the rod, it climbed to the bookshelves within five minutes or so."

The Anteaters, like the Sloths, are found only in Central America. The long head tapers off to a small, narrow, toothless mouth, provided with a worm-like tongue. The tail is long, and in some species prehensile. The third digit on the fore-limbs is much larger than the rest, and is armed with a huge hooked claw. Anteaters are covered with long coarse hair.

The Great Anteater (Plate III., No. 16) is about 6 ft. long, of which the thick bushy tail amounts to a third. The general colour is grey, with a black band, bordered with white, passing backwards from the chest to the loins, where it ends in a point. These animals live on the ground, and feed on Termites, or White Ants, tearing open the ant-

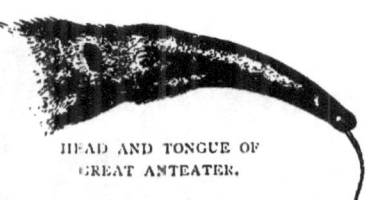

HEAD AND TONGUE OF GREAT ANTEATER.

hills with their claws, and sweeping up the insects with their extensile tongue, which is covered with a sticky secretion. They are harmless and inoffensive unless meddled with ; but if irritated can use their powerful claws with great effect, and they are difficult beasts to kill. A writer in the *Field* (November 12, 1892) was told by some Spanish planters that they had seen dogs killed on the spot with the claws of the Anteater actually interlocked in their bodies. The Tamandua, much smaller, and the Two-toed Anteater, about the size of a rat, live among the branches of trees. In both the tail is prehensile.

The Armadillos derive their name from the armour-like shields and bands with which they are covered. There are about twenty species from temperate and tropical South America, one ranging as far north as Texas. They are generally nocturnal in habit, and feed on anything that comes in their way, not excepting carrion. The body is long, the limbs are short, and armed with strong claws, well adapted for burrowing, in which all of them are very expert. Most of them walk on the soles and palms. The Great Armadillo, the largest living form, is nearly 5 ft. long, of which the tail is about one-third. It is never found in open country, but keeps near the great forests. This animal is said to dig up and devour the bodies of the dead, but the story is not supported by trustworthy evidence.

The Poyou or Six-banded Armadillo (Plate III., No. 15) is about 2 ft. long, including the tail, and between the bands on the body are long bristly hairs. The Hairy Armadillo, or Peludo, is smaller, and is nocturnal in habit. The edges of the shield and those of the bands end in sharp points. Hudson had a tame one, which caught and ate mice. A friend of his saw one kill a venomous snake by throwing itself upon it and cutting up the body of the reptile with the sharp edges of its bony armour. The Apar and some other of the same genus differ from the rest in walking on the tips of the front claws; they have the power of rolling themselves into a ball, like some of the wood-lice. "In this state," says Darwin, "it is safe from the attacks of dogs,

PICHICHIAGO.

for the dog, not being able to take the whole in its mouth, tries to bite one side, and the ball slips away." The flesh of most of these animals is eaten, and is said to be well-flavoured. The armour-like skins of some, lined with silk, are used as work-baskets.

The Pichichiago is a small animal. Its body is covered with long silky hair, and the long, banded shell covering the upper surface is loosely attached to the back along the spine.

The Scaly Anteaters from Africa and Asia, like the Hairy Ant-eaters of America, are quite toothless, and have a long extensile tongue. There are four African species, of which the Long-tailed Pangolin (Plate III., No. 17) is a good example, and three from Asia. They vary in length from 2 to 5 ft.; all of them are covered with large scales overlapping each other, ending in a sharp point behind, and have the power of rolling themselves into a ball. The tail is long, and is used as a support in climbing trees. They walk with the long claws closed, so that the back of the digits touch the ground.

Mr. Hornaday kept one of the Asiatic species, and says that in walking "the animal carried his back highly arched in the middle, and the long curved claws of his forefeet he bent until they pointed backwards, and literally walked on them. Whenever he found a colony of ants he would begin to dig most industriously, and after digging a little distance into the hill and exposing the interior, he would thrust his slender gelatinous tongue into the passage-ways and draw it out thickly covered with ants."

When the Pangolin rolled itself up, Mr. Hornaday and his servant tried in vain to unroll it; and though they called the dog to help, they failed to effect their purpose. The Cingalese say that the Pangolin often coils itself round the elephant's trunk, and squeezes it so tightly that the huge beast cannot breathe, and very soon drops dead from suffocation.

The Cape Ant-bear, or Aardvark, from Cape Colony, is somewhat pig-like in form, with a long snout, erect ears, short limbs armed with stout claws for digging, and long round tail. The light-coloured skin is scantily covered with bristles. These animals, which feed entirely on termites, are nocturnal, passing the day in their burrows. They have the strange habit of rising to the surface in unexpected places. Mrs. Martin, in her "Home Life on an Ostrich Farm," says that "during the war in Zululand a sentry was on guard at midnight, when suddenly, close to him, the ground opened, and out of it rose a ghastly living Jack-in-the-box. The moonbeams shone full on the horrid form, long head, and deadly pale, calf-like face; and the man— small blame to him—dropped his gun, deserted his post, and fled in horror, shouting to his astonished comrades the awful news that he had seen Old Nick himself." The Ethiopian Aardvark, from North-eastern Africa, ranges into Egypt.

AARDVARK.

CHAPTER XVI.

POUCHED MAMMALS.

THE Mammals of this Order are distinguished by the possession of a pouch, in which the young are nourished for a considerable time after birth, for the little creatures come into the world in a very

imperfect condition. Most of them are nocturnal animals, from Australia and the neighbouring region, and live on vegetable food.

At the head of the Order stand the Kangaroos, of which there are many kinds, one nearly as tall as a man, while others are no larger than a rabbit.

The Great Kangaroo (Plate III., No. 7), discovered by Captain Cook in 1770, is found in the open grassy plains of Eastern Australia and Tasmania. The fore limbs are short, the hinder pair are greatly developed, and the long powerful tail serves as a weapon of defence, and as a means of support when the animal raises itself nearly up-

KANGAROOS.

right on its hind-quarters. When pursued it gets over the ground by a series of jumps, and is said to be able to clear 15 feet at a bound. In walking, the fore-feet are placed on the ground to support the body, while the long hind-legs are brought forward outside them. These animals do great damage to the pastures of the colonists, and are hunted and shot in great numbers. When brought to bay, this Kangaroo will often rip up a dog with a stroke of its long claws, or kill it with a blow from its tail. The flesh is excellent eating, and the leather finer than calf-skin; and since these animals breed freely in confinement, they might possibly be acclimatised in this country, and become a source of profit to kangaroo farmers, though opinion is divided on this point. There are several other species closely allied.

The Rock Kangaroos range over the mainland of Australia, and derive their name from the situations they frequent. Gould describes them as leaping from rock to rock, often alighting on ledges so narrow that it appeared almost impossible for them to retain their footing. The Tree Kangaroos are found chiefly in New Guinea, only one of the four occurring in Queensland. They are about the size of small hares. Dr. Guillemard, who kept two on board the *Marchesa*, says that in the Tree Kangaroo we have

WOMBAT.

" an instance of a ground animal which is gradually becoming arboreal in its mode of life. But although a tree-haunting animal, it is as yet a tiro in the art of climbing, performing this operation in the slowest and most awkward manner. Our pets, for instance, would take a full minute or more in ascending the back of a chair, but their hold was most secure, and if we wished to pull them off we had no little difficulty in doing so." The Hare Kangaroo, found over all Australia, resembles the hare in appearance, and in forming "seats," like that animal. These animals run well, and can take marvellous leaps. One that Gould hunted leaped right over his head.

The Kangaroo Rats, none of which is larger than a rabbit, are found in Australia and Tasmania.

The Wombats are small bear-like animals, living on the ground or in burrows, and feeding on grass, roots, and the like. There are three species, one from Tasmania and the islands of Bass's Straits, and

M 2

two from South Australia.   The general colour is brownish-grey, and the length about 3 feet.   The incisor teeth resemble those of the Gnawing Animals.

The Phalangers are small woolly-coated nocturnal animals, from Australia and the Papuan islands.   The tail is often prehensile, and the great toes can be used like thumbs.   Some of them possess a flying-membrane, joining the fore and hind limbs, and by its aid they can take long leaps from above downwards.

The Woolly Phalangers range from Australia to Celebes.   Wallace tells us that they are opossum-like animals, with a long prehensile tail, of which the end is generally bare.   They have small heads, large

eyes, and a dense covering of woolly fur, often pure white, with irregular black spots or blotches. They move about slowly, and are difficult to kill, owing to the thickness of their fur and their tenacity of life.   A heavy charge of shot will often lodge in the skin and do them no harm, and even breaking the spine or piercing the brain will not kill them for hours.

KOALA.

Of the Flying Phalangers there are several species.   Of the Short-headed Flying Phalanger Dr. Guillemard says: "These creatures are common to New Guinea and Northern Australia, and are chiefly nocturnal in their habits, haunting the thick foliage at the crowns of palms.   The tail is not prehensile, as in the true Phalangers, but the stout sharp claws are well adapted for clinging to the smoothest bark. The fur is exquisitely soft and of a delicate shade of grey, against which the black dorsal stripe and white under-surface show to advantage.   We quite failed to tame this little animal, whose loud note of alarm and anger when its cage was disturbed made the sailors give it the nickname of 'the clockwork mouse.'   A whole page of description would not more accurately convey the nature of the sound."   The Pigmy Flying Phalanger, from East Australia, is not so large as a mouse.

The Koala, or Native Bear, from South-eastern Australia, is about 2 feet long, with ashy-grey fur.   It lives among the branches of gum-trees, on the buds and shoots of which it feeds.   The mother carries her young on her shoulders.

The Bandicoots, or Pouched Badgers, are found in Australia, Tasmania, and New Guinea. They are burrowing animals, of small size, and feed chiefly on insects.

The Banded Ant-eater, from South and West Australia, is about the size of a squirrel, and has a long bushy tail. The chestnut-red fur is marked on the back with broad white bands. It feeds on insects.

BANDED ANT-EATER.

In the so-called Pouched Weasels the pouch is reduced to a few folds of skin. These animals, which feed on insects, differ in habit: some take their prey on the ground, others live among the branches of trees. They are of small size, none of them larger than a rat, with dark fur, and range from Australia to New Guinea. In 1894 the members of the Horn expedition to Central Australia were fortunate enough to obtain several specimens of a rare form of Pouched Weasel, which lives in holes among rocks and stones. Up to this time but a single example of this animal was known.

The Dasyures are small in size, and something like a Civet-Cat in form. The fur is some shade of brown or grey, spotted with white. They occur in Australia, Tasmania, and New Guinea, and feed on birds and small mammals.

The Bear-like Dasyure, or "Tasmanian Devil," owes its name to its fierce temper and destructive habits. It is about the size of a Badger, with black fur, marked with white on the chest. It lives in burrows, and preys on sheep and poultry.

The Thylacine is a native of Tasmania. It is of dog-like form, whence the colonists call it "Wolf." The greyish-brown fur is marked with dark stripes on the back, flanks, and root of the tail, whence it is also called "Tiger." Its ravages among the sheep-folds have led to its extermination, except in the more mountainous part of the island.

The Mole-like Marsupial, recently discovered, is a small burrowing animal from the sand-plains of South Australia. Professor Stirling, who described it, says: "In penetrating the soil free use as a borer is made of the conical snout, with its

TASMANIAN DEVIL.

horny protecting shield, and the more powerful, scoop-like fore-claws are also early brought into play. As it disappears from sight, the hind-limbs, as well, are used to throw the sand backwards, which falls in again behind it as it goes, so that no permanent tunnel is left to mark its course. Again emerging, at some distance, it travels for a few feet along, and then descends as before. I could hear nothing of its making or occupying at any time permanent burrows."

The Opossums are found only in America. Most of them live among the branches of trees and feed on birds, birds' eggs, and insects. In some the pouch for the young is well developed ; in others it is represented by folds of skin. The tail is prehensile, and the great toe acts as a thumb, as in the Phalangers.

The largest species and one of the commonest is the Virginian

OPOSSUM.

Opossum (Plate III., No. 6). Its home is in temperate North America, and in the Southern States a variety is known as the Crab-eating Opossum. Azara's Opossum is also from the Southern States. Like most of its fellows, it feigns death when hard pressed. A correspondent of the *Field* says : " I rolled it over and over, and as it did not alter its position in the least, I thought I had killed it ; and as a mulatto, whom I knew, came up, I asked him to keep it for me until the morning, when I purposed skinning it and then burying the body, with a view of obtaining a perfect skeleton. But the man was far more conversant with the beast's ways than I, and said, 'It is not dead, sir, but only shamming.' I then realised what was meant by the saying, ' Playing 'possum.' "

In the pouchless Opossums the young are borne about on the back of the mother, who carries her tail erect or arched over her back, and round it the young ones twine their tails to steady themselves as she roams about.

The Yapock, or Water-Opossum, ranging from Guatemala to Southern Brazil, is aquatic. Its feet are webbed, and it lives on fish, crustaceans, and insect larvæ.

THE WATER-MOLE AND SPINY ANTEATERS.

These are the lowest of the Mammals, and possess some of the characters of Reptiles and Amphibians. The young are produced from eggs.

The Water-Mole (Plate III., No. 18) is found near the banks of rivers in the south and east of Australia and Tasmania, and feeds on insects and molluscs. The bill is duck-like, the body is covered with soft brown

WATER-MOLES AND SPINY ANTEATER.

fur, and the feet are webbed. There is a horny spur on the hind legs of the male, capable of inflicting a wound dangerous to man, owing to the secretion from a gland in connection with it. The young have teeth, but these are shed and replaced in the adults by horny plates.

The Spiny Anteaters are small, nocturnal, burrowing animals, feeding on ants. The fur is dark, and mixed with strong sharp spines. The snout is produced and the long tongue is thrust forth from a hole at its extremity. The claws are long and powerful; the males have a small spur. There are two species from Australia, Tasmania, and New Guinea.

CHAPTER XVII.

BIRDS. PASSERINE BIRDS.

THE chief character by which birds may be distinguished from all other backboned animals is their clothing of feathers. Like the mammals, they are warm-blooded ; the fore-limbs are organised for flight, though in some the power of flying has been lost, probably through disuse. The jaws are covered with a horny sheath, forming a beak or bill, and the feet are modified for walking, climbing, seizing prey, and scratching and digging ; and the toes may be greatly lengthened for walking on floating vegetation, or webbed for swimming. No bird has more

JAWS OF BIRD WITH HORNY SHEATH.

than four toes—that corresponding to our fifth, or "little" toe, being absent. The "great" toe is at the back ; and the inside toe on each foot when there are three toes in front, corresponds to our second toe. The foot of the Lark shows the arrangement of the toes in the Passerine Birds Parrots and Woodpeckers have their toes arranged in pairs—two in front, and two behind. In regard to Parrots, this fact may be easily verified.

FOOT OF LARK.   FOOT OF CASSOWARY.   FOOT OF OSTRICH.

All birds lay eggs, and the vast majority build nests and sit upon the eggs they have laid. Some few, as the Cuckoo, leave the duty of incubation to others ; and a few others, like the Mound Birds, deposit their eggs in the earth or in heaps of decaying vegetation, and leave them to be hatched out by natural heat. The backbone in Man may be roughly indicated by an upright stroke **|**, that of most of the lower mammals by a horizontal stroke **—**, but in birds it occupies a slanting position ✎. This will not seem strange if we remember that a bird is a "glorified reptile"; and though no one can say how "the slow, cold-blooded, scaly beast ever became transformed into the quick, hot-blooded, feathered bird," it is clear that when the fore-limbs left the earth, the backbone must have moved upwards with them.

From the annexed figure one may get a general idea of the bony framework of a bird. The different parts will be fixed in our minds if we compare them with the corresponding parts of our own body (p. 8), and with those of the Camel (p. 5).

Of Living Birds there are two sub-classes : (1) the Flying Birds, with a keel, or ridge, on the breast-bone, for the attachment thereto of the muscles that move the wings ; and (2) the Running Birds, with flat raft-like breast-bone.

The Flying Birds are divided into several orders.

PASSERINE BIRDS.

This is the largest order of the class, and contains nearly 6,000 species of birds, arranged in about fifty families. Dr. A. R. Wallace describes it as "comprising the most perfect, the most beautiful, and most familiar of our birds. The feathered inhabitants of our fields, gardens, hedgerows, and houses belong to it. They cheer us with their song, and delight us with their varied colours."

The Thrushes are well represented in Britain. All boys know the Common Thrush (Plate V., No. 16), and very many have taken its eggs. It is noted for its powers of song, and is a very favourite cage-bird. Nor does it seem to be unhappy in confinement: it sings as loudly and cheerily in its small cage as it does in the open. A writer in the *Field* recently told of one that his daughter took from the nest fifteen years ago, and says that "it still sings as much

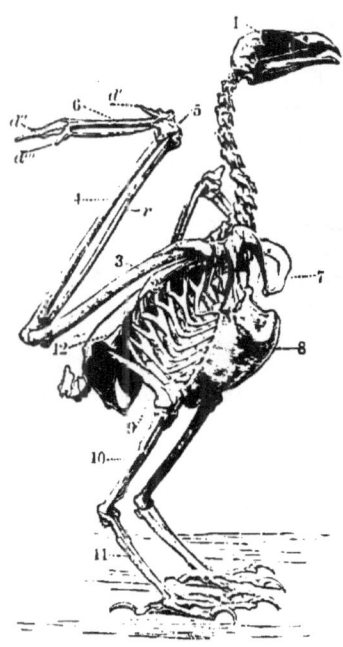

SKELETON OF PIGEON.

1, Skull ; 2, Shoulder blade (*scapula*) ; 3, Arm (*humerus*) ; 4, Fore-arm (*ulna*, r radius) ; 5, wrist (*carpus*) ; 6, Metacarpus (corresponding to the human palm). d', d", d''', correspond respectively to the thumb, second, and third fingers ; 7, Merrythought (*furcula*, composed of the two collar-bones united) ; 8, Breastbone (*sternum*, showing keel) ; 9, Thigh-bone (*femur*) ; 10, Leg (*tibia*) ; 11, Tarso-metatarsus (corresponding to the ankle and sole) ; 12, Pelvis. (For the toes, see text.)

as ever, and is apparently as contented as possible. Some years ago I allowed it to go free, but it came back to its cage." The Thrush feeds on worms, slugs, snails, and in some places on shell-fish. It breaks the shells by hammering or dropping them on a stone. In autumn it takes toll of fruit; but its services to the farmer and gardener in destroying snails and slugs should be allowed to count against its misdeeds. The length of the male is a little less than 9 inches. The basin-shaped nest usually contains five bluish-green eggs, spotted with brown, and there are generally two broods in the year.

There is a popular notion that if young Thrushes be taken from the nest and put in a cage in the open, the old birds will bring them poisonous food. In 1894 some young Thrushes which had been taken from the nest died in their cage soon after the parent birds had visited them; and their owner was told by his gardener that the old birds poisoned their young when they found they could not get them out of the cage. Their bodies were sent to the *Field* office for examination, so as to discover what the poison was, and their stomachs were found to be quite empty. They had died from starvation.

The Missel-Thrush is larger than the Common Thrush, but has not so sweet a song. Its food is pretty much the same as that of the last species, except that it has a greater fondness for berries, especially for those of the mistletoe, from which it derives its name. This bird sings in storm as well as in sunshine, whence it is known as the Stormcock.

The Fieldfare visits us in winter, returning northward in spring. It is about 10 in. long, chestnut on the back, bluish-grey on the head and neck; the reddish-yellow breast is marked with black, and the rest of the under surface is white.

The Redwing comes earlier and stays longer than the Fieldfare. Of these two visitors, Bishop Stanley says: "It is generally supposed that they are hardy birds; but the very reverse is the case, for in severe weather, should there be a dearth of food, they are the first to suffer."

The Blackbird (Plate V., No. 17), with its yellow bill, is familiar to every one who has walked in the country with observant eyes; though if disturbed, it quickly betakes itself to the shelter of a hedge or bush, darting in with a sharp, sudden turn. Its note is loud and clear, and is continued from early in the spring till the breeding season is over. It is a good mimic, and has been known to imitate the song of the Nightingale pretty closely. Bishop Stanley tells of one that was in the habit of crowing like a cock, and nearly as loud. "Perched upon the top-bough of an ash-tree, it might be seen crowing away, occasionally indulging in its natural song, but this only for a second or two, for it soon began again to crow, and when the cocks from a neighbouring poultry yard answered, it seemed delighted, as if it was trying to rival them in the shrilliness of its note." The bird was probably bred near the spot, and learned to crow from imitating the cocks.

Gardeners wage war upon the Blackbird, thinking that it robs them of their fruit. Professor Newton, however, says that when the enormous numbers of insects, slugs, and snails injurious to vegetation, and eaten by Blackbirds throughout a great portion of the year, are duly considered, it is pretty plain that the value of the fruit, consumed during a few weeks only, is counter-balanced by the services performed. In 1893, a white Blackbird—not an albino--was shot at Hinton St. George, Dorsetshire, and sent to Mr. Rowland Ward for preservation. The bird had been known in the district for two years, during which period many attempts had been made to shoot or trap it.

RING-OUZEL.

The Ring-Ouzel, a little larger than the Blackbird, with blackish plumage and a white half-moon on the chest visits Britain in the summer Its food is much the same as that of the Blackbird ; but it shuns cultivated or inhabited districts, preferring wild, hilly country. When the young are hatched, the parents fly around with anxious cries, and will venture to attack a dog or other animal.

The Mocking-bird is a North American Thrush about 10 inches long. The plumage above is grey marked with black and white, and brownish white below. Its own notes are full, strong, and varied. It also has the power of imitating the note of every bird it hears, the cries of domestic animals, and other sounds. A caged Mocking-bird has been known to call the hens together by screaming like a wounded chicken, and to draw the dog from the fireside by imitating its master's whistle.

The Warblers, some of which are resident in Britain, while others visit us regularly, have two moults in the year—one in spring, and one in autumn. Insects form the chief part of their food.

At the head of the list stands the Nightingale (Plate V., No. 12)—that is, the Night-singer—though it is by no means the only British bird that sings in the night-time. Its length is about 7 inches, and its plumage is of various shades of brown. The male birds arrive about the middle of April, and the females some ten days or a fortnight later. The males taken during this interval often do well as cage-birds, but those snared after they have mated generally pine and die. The song lasts from pairing time to the hatching of the young, and is so sweet that all who have written about it have been enthusiastic in its praise. Izaak Walton wrote : "He that at midnight, when the very labourer sleeps securely, should hear, as I have very often, the clear airs, the sweet descants, the natural rising and falling, the doubling and redoubling of her voice, might well be lifted up above earth, and say, Lord, what musick hast Thou provided for the saints in Heaven, when Thou affordest bad men such musick on Earth ! " A few instances are known of this bird having bred in confinement.

The Robin (Plate V., No. 13) lives with us—that is, some stay all the year, while others leave in autumn for warmer climes, and the following spring sees fresh arrivals on our shores, while even in this country there are periodical movements from one part to another. The Robin seems to be particularly fearless in approaching man, and in severe weather will often venture into houses in search of food. Robins frequently choose strange places in which to breed—an old tin teapot, that had been thrown up in the branches of a poplar. at Westgate-on-Sea, a desk in a study at Tunbridge, and the eaves of a house at Balham were recorded during one year as nesting-places of Robins.

The Wheatear, clad in grey and black plumage above, and white below and on the tail, visits this country in March, and leaves by the end of September. When these birds assemble on the South Downs for the southern journey, vast numbers are snared. They are then in excellent condition and highly valued for the table. The Stonechat, which is resident, and the Whinchat, are closely allied.

The Redstart (Plate V., No. 14), or Firetail. visits us in the summer. This bird feeds almost entirely on insects, which it takes on the wing almost as readily as if it were a Flycatcher or a Swallow. It has fair powers of imitation, and some have been taught to repeat tunes.

The Hedge Sparrow is a native bird, feeding on insects, worms, and seeds. Its nest is one of the earliest, and it is probably from the little care used to conceal it that it is so often chosen by the Cuckoo as a good place in which to deposit an egg.

The Reed-Warbler and the Sedge-Warbler are summer visitors, the former chiefly to the Eastern Counties. The Sedge-Warbler is well known to all anglers. A writer in the *New Review* says :—" He seems to take a kindly interest in you, sits on a reed near and watches, and begins his song when you come up ; if you don't see him or hear him, it is only necessary to call his attention by pitching something into the reed-bed, and he will address his song to you at once, and a very curious medley it is, being made up of the notes and bits of the song of other birds."

The Greater Whitethroat, or Nettle-creeper, the Lesser Whitethroat, the Garden Warbler, the Blackcap, the Wood Wren, the Willow-Wren, and some others, are summer visitors.

The Golden-crested Wren is a resident, though some leave this country in the spring, and there is an arrival in the autumn. In some places it is known as the Woodcock-pilot, from its arrival a few days before that bird. The Golden-crested Wren is about 3½ inches long. The Fire-crested Wren, a rare visitor, has the crest of a much redder hue. The general plumage of both is olive-green.

The Tailor-bird is Indian. The author of "The Tribes on my Frontier" describes it as a "most plain-looking little greenish thing, but a skilful workman. Aided by its industrious spouse it will spin cotton, or steal thread, and sew together two broad leaves of the laurel in the pot on your very doorstep ; and when it has warmly lined the bag so formed, it will bring up therein a large family of little tailors. At present it is burdened with no such cares, but still it is always busy, hopping from bush to

NEST OF TAILOR-BIRD.

bush, and prying with its sharp eyes for spiders and little green caterpillars."

The Babbling Thrushes, are small, lively birds, generally of sober plumage, abounding in Asia, and found less plentifully in Africa and Australia. The breeding habits of one species, observed by Forbes in Java, are curious. He says that the fledgelings were being tended by three parents, and his "boy," who on most subjects was well informed, told him that the female had always two husbands.

The Dippers inhabit Europe and the alpine districts of North and South America, with some close relatives of the True Dippers in Asia. The Dipper, or Water-Ouzel, is British. In form and carriage the bird is somewhat like a big Wren. The plumage is brown above and on the flanks, and white on the breast. Rocky streams are the Dipper's home, and there it finds its favourite food—water-snails, aquatic insects and their larvæ. Its diving powers have been compared to those of the Cormorant ; and Montagu says, "Where we have been able to perceive it under water, it appeared to tumble about in a very extraordinary manner with its head downwards, as if picking up some-

DIPPER.

thing; and at the same time great exertion was used, both by wings and legs."

The Wrens are small birds widely distributed. Our Common Wren (Plate V., No. 15) is nearly as well known as the Robin, and seems to have similar confidence in approaching the dwellings of man. Near the nest, which has a dome-like top and a hole in the side, are generally some nests partially finished or carelessly built. They are called "cocks' nests," and country boys will tell you the cock-bird builds them for sleeping-places for himself. There are several other explanations — no one knows which is the true one.

The Creepers are small, soberly-clad birds, widely distributed. In habits and in their stiff tails they resemble the Woodpeckers, but their toes are differently arranged— three in front, and one behind. The Common Creeper, or Tree Creeper, about 5 inches long, is clad in brown, marked with white and black, and has a curved bill. It feeds chiefly on spiders and insects, which it picks out from between the crevices in the bark of trees. Beginning at the bottom, it will work round and round a tree in a spiral, and then fly to the next and go over it in the same manner, generally confining itself to the trunk, but some-times searching the larger branches, under which it will occasionally travel back downwards. There are allied species on the Continent, in Asia, and America. The Wall Creeper, from the south of Europe, is a much larger bird.

The Nuthatches have a short, wedge-shaped bill, but the tail-feathers

CREEPER.

are not stiff like those of the Creepers and Woodpeckers. They are chiefly found in the temperate regions of both hemispheres. The Common Nuthatch is a little larger than the Tree Creeper, with the plumage slaty-blue above and pale cinnamon below. It searches trees for insects, but its mode of hunting differs from that of the Creeper, for it begins at the top and works downwards, often coming down head first. It will also prize up the bark to search for larvæ, which the Creeper cannot do, owing to its weak bill. Nuts form a large part of the food of these birds; and from their habit of wedging nuts or beech-mast into a chink and splitting them open with their bill they are also called Nut-hacks and Nutjobbers. If disturbed while sitting, the hen bird will hiss like a snake, and any attempt to remove the eggs will probably convince the intruder that the Nuthatch's bill can be effectively used for other purposes than splitting nuts.

NUTHATCH.

The Tits are noisy, lively little birds, with loose, fluffy plumage, often of gay tints, and short conical bill set at the base with hairs. They are more abundant in temperate than in tropical regions, and most are insect-eaters. The name Titmouse is sometimes given to these birds: but "mouse" is really the Old English *mase*, which means a small bird, and has nothing to do with the thieving little rodent.

The Great Tit, or Ox-eye (Plate V., No. 11), widely distributed over Europe, lives in Britain all the year round. Its length is a little less than 6 inches. Its notes are varied, and some of them are more loud than sweet, resembling the noise made in setting a saw, whence its local name of "Sawsharper." It is accused of killing smaller birds and feeding on their brains; but it rarely indulges in this bad practice when at liberty. The Blue Tit, or Bluecap, about 4½ inches in length, is unjustly persecuted by gardeners as a fruit-destroyer, when its only object is picking open fruit-buds in the quest of insects. Like most Tits, it is fond of flesh and fat, and if one wishes to attract Tits to the garden in the winter, it is only necessary to tempt them with food of this kind. A writer in the *Field* says that for many years each winter he suspended pieces of fat by a string to a stick, and on this numerous Tits used to feed: but that winter he put an upright stick, with a cross piece of wood, hanging the fat by strings at each end. The Tits, both Blue and Ox-eye, no longer took their meals hanging head downwards on the string, but, perched on the cross-pieces, hauled up the string "hand over hand," and so fed at their ease. The Coal Tit, which may be distinguished by the white patches on the head, feeds largely on vegetable diet, but has been seen devouring earthworms. The Marsh Tit, black on the head and brown on the upper surface, is locally distributed in Britain, generally near swampy places. The Crested Tit is also local, preferring woods in which fir and oak occur, feeding

COAL TIT.

on the insects that haunt the fir-trees, and when these fail, on the seeds of the fir. Holes in the oaks are utilised for nests. The Long-tailed Tit, sometimes called the Bottle-Tit from the shape of its nest, is the least common of the British Tits. Its length is about 5½ inches, but its body is very small; its upper surface is mostly black, with some white on the crown. All these Tits are native British birds.

The so-called Bearded Titmouse belongs to another family. There is little to be added to the first description given of it: "A little bird of a tawny colour on the back, and a blue head, yellow bill and black legs." Snails and insects form its chief food. It has its home in Central and Southern Europe; but it bred, and perhaps still breeds, in some parts of the fen-country.

The Hill-Tits are small birds, of delicately-tinted plumage, ranging from the Himalayas to China.

The Bulbuls, or Fruit-thrushes, are natives of Asia and Africa. The Madras Bulbul is trained in Ceylon as a fighting bird. It is taken young from the nest, and secured by a string to its perch. When properly trained it is matched with another bird similarly fastened, and the conflict would end in the death of one or both of the two if their masters did not separate them when likely to do each other serious mischief.

The Orioles, or Golden Thrushes, are found chiefly in Africa and Asia. The Golden Oriole (Plate VI., No. 17), with its brilliant yellow-and-black plumage (which is not attained till the third year), is well known

PLATE V.

1. Golden Eagle.　2. Lammergeier.　3. Iceland Falcon.　4. Egyptian Vulture.　5. Condor.　6. Horned Owl.　7. Goatsucker. 8. House Martin.　9. Swallow.　10. Great Grey Shrike.　11. Great Tit.　12. Nightingale.　13. Redbreast.　14. Redstart.　15. Wren. 16. Thrush.　17. Blackbird.

on the Continent, but is the only species found in Europe. These birds live in small flocks, fly well, and frequent high trees, amid the foliage of which they seek for caterpillars, soft insects, and fruits. The Golden Oriole sometimes strays to Britain, and in some few instances it seems to have bred here.

The Cuckoo Shrikes are insect-eating birds, from Africa, Asia, and the Malay Archipelago. Many of them are soberly clad, but some Indian species are gay with scarlet and black plumage. The Drongo Shrikes have nearly the same range as the Cuckoo Shrikes. Their feet are formed for grasping, and they rarely come down to the ground. They take their insect-food on the wing, and then return to their perches to watch for more. The Black Drongo Shrike is a handsome jet-black bird, with long tail-feathers, which, when it flies, open and shut like the blades of a pair of scissors.

The Flycatchers have the base of the bill well set with bristles. The species are numerous and widely distributed. They rarely feed on the ground, but watch from some perch for insects, which they take on the wing, and then return to their post. When insects cannot be obtained these birds subsist on berries.

The Spotted Flycatcher, plentiful all over Europe, visits Britain in the summer, beginning to build as soon as it arrives, and it is generally believed that the same pair return year after year to the same spot to breed. The length of this bird is about 5½ inches, and its plumage is brown above, and dull white, with brown markings, below. It has been accused of eating fruit, and in Kent is known as the "Cherry-sucker," but there is no evidence to support the charge of robbing gardens and orchards. Facts point the other way, for the stomachs of Flycatchers shot on fruit-trees have yielded nothing but insects. The hard cases of the insects are rejected in the form of pellets. The Pied Flycatcher, a rarer and rather smaller bird, with black and white plumage, also visits us in summer. The country round the lakes of Cumberland and West-morland is its favourite resort.

The Red-breasted Flycatcher, from Central and Eastern Europe, has strayed to England. Of the first specimen recorded, Mr. Copeland says : "We first observed it on a dead holly tree; this tree and the ground around the house were its favourite resort. It was particularly active, skimming the grass to within about a foot, then, perching itself, darted occasionally with a toss, resting either on a shrub or the wire fencing."

The Thick-headed Shrikes are natives of Australia and New Guinea. Their habits differ from those of most other insect-eating birds, par-ticularly in their quiet mode of hopping about the branches of trees in search of insects and their larvæ.

N

The Shrikes are most abundant in Africa, though many are distributed over the warmer and temperate regions of the globe. The feet are strong and the bill hooked, and sometimes furnished with a tooth.

The Great Grey Shrike (Plate V., No. 10), about 10 inches long, common on the Continent and in America, visits Britain in the winter. Yarrell says that it "feeds upon mice, shrews, small birds, frogs, lizards, and large insects. After having killed its prey, it fixes the body in a

forked branch, or upon a sharp thorn, the more readily, as is supposed, to pull off small pieces from it. It is from this habit of killing and hanging up their meat, which is observed also in other Shrikes, that they have been generally called Butcher-birds." The Lesser Grey Shrike, which adds fruit to its flesh diet, is an occasional visitor ; the Red-backed Shrike comes regularly every summer, and the smaller Woodchat, a rare visitor, has bred here.

The Crows are a very large family, with representatives in every quarter of the globe. The bill is strong, and broad at the base, and the nostrils are covered with stiff feathers, directed forwards. They are clever, cunning birds, many of them possessing powers of mimicry, and their habit of storing food in a wild state leads many of them, when tamed, to carry off and hide anything that attracts their attention.

RED-BACKED SHRIKE.

The Piping Crows, from Australia, Tasmania, and the Malay Archipelago, have long, conical beaks. They spend much time on the ground, and eat nearly everything that comes in their way. Some are distinguished by the flexibility of their voice, and all have great powers of imitation.

The Jays are widely distributed. The Common Jay (Plate VI., No. 16), a little over a foot long, is one of the handsomest of our native birds. Its home is in the woodlands, and it is rarely seen in open country, and scarcely ever on the ground. Insects, worms, and slugs form its chief food, but there is no doubt that it takes toll of fruit crops, and there is some truth in the charge that it destroys eggs and young birds, though the extent of this bad habit is probably exaggerated.

Unfortunately, gamekeepers believe it, and consequently shoot Jays whenever they have a chance. They are also killed for the blue feathers of the wing, which are used in dressing artificial flies. Jays have great powers of mimicry, and make amusing pets. The Blue Jays represent in America the Jays of the Old World, which they resemble in habits.

The Tree Crows are natives of Africa and Asia. In appearance they are like the Magpie, but in some the central tail-feathers are much longer than the rest.

The True Crows are found everywhere except in South America. There are several British species. The Raven, the largest of the family, is a little over 2 feet long, and has blue-black plumage, with a green metallic gloss. It is widely distributed in the northern hemisphere, but is becoming rare in Britain, especially inland, though there are still many breeding-places on the rocky parts of the coasts, and some land-owners protect these birds. The Raven will eat anything of an animal nature that comes in its way, and attacks

RAVEN.

sickly sheep and lambs. Professor Newton says : " No sooner does an animal betray any sign of weakness than the Raven is on the watch for the opportunity, and begins the attack on the eye, especially if the creature be large and still alive." It will hunt and devour the smaller mammals, as moles, mice, etc., and does not disdain to feed on carrion, and, failing animal food, will put up with grain. This bird is often kept as a pet, and it soon learns to " talk." Dickens was fond of Ravens as pets, and everybody will remember " Grip " in " Barnaby Rudge."

The Carrion Crow may be described as a small Raven, and is more detested than its larger relation ; for while it has all the bad habits of that bird, its numbers are so much greater in this country that the damage it does to farmers and gamekeepers, and breeders of poultry, is very considerable.

The Grey, Hooded, or Royston Crow (Plate VI., No. 14), called in Scotland the Hoodie, differs from the Carrion Crow only in its grey

N 2

back and under surface. It has all the Carrion Crow's love of animal food. Mr. Edward, the Scottish naturalist, says: "I observed a half-grown rabbit emerge from some whins, and begin to frolic about close by. Presently down pops a Hoodie and approaches the rabbit, whisking, prancing, and jumping. He seemed to be most friendly, courteous, and humorsome to the little rabbit. All of a sudden, however, as if he meant to finish the joke with a ride, he mounts the back of the rabbit. Up springs the latter, and away he runs. But short was his race. A few sturdy blows about the head from the bill of the Crow laid him dead in a few seconds."

The Rook differs little from the Crow in size or colour, but adult birds have the skin of the forehead and at the base of the bill bare of feathers. Rooks are, also, much better-mannered birds, and to a

great extent farmers' friends, for they devour immense quantities of worms, insects, and insect larvæ, though they take pretty heavy toll of corn and fruit. They are social birds, congregating in large flocks, and breeding in company, generally near the dwellings of man, sometimes in towns and cities. Rooks, though less easily tamed than Ravens, are sometimes made pets of, and may be taught amusing tricks, and even to repeat words and sentences.

HEAD OF ROOK.

The Jackdaw is a little over a foot long, and, with the exception of a collar of smoky grey, has black plumage, with metallic reflections. It builds in towers, steeples, and similar places in towns, and frequents cliffs rather than the wooded districts which Rooks love. In food and habits it resembles the Rook, and is much more often kept as a pet.

The Chough is a beautiful bird, confined to some of the bolder cliffs of our southern and western counties. Its length is about 17 inches, and its glossy-black plumage contrasts well with the curved bill and legs, which are coral red.

The Magpie (Plate VI., No. 15) is more plentiful in the wild and open parts of Britain than it is in the more cultivated districts. This is accounted for by the persecution to which it is subjected by game-keepers and farmers, both of them being mindful that the Magpie destroys leverets. young rabbits, and chickens, but forgetful of the fact that it also kills large numbers of rats and mice. The Magpie is a beautiful bird about 18 inches long, boldly marked with white and black, the latter glossed with metallic reflections of violet and green. It is often kept as a cage bird—generally in a cage far too small for it—

and will soon learn to "talk" and play amusing tricks. Its fondness for picking up and hiding glittering articles makes it a somewhat dangerous pet, and its misdeeds in this way, and the sufferings of innocent persons unjustly suspected of the theft, have been made the subject of a play, an opera, and one of Canon Schmidt's "Tales."

The Nutcracker, about a foot long or rather more, clad in brown spotted with white, is a rare British visitor from Southern Europe.

The Birds of Paradise, with which are included the Bower Birds, are natives of New Guinea and the neighbouring islands, some few living in the north and east of Australia. The dress of the male birds is exceedingly beautiful, and large and brilliant ornamental plumes are developed, chiefly from the wings and tail. But despite their fine feathers, they are very near relations of the crows. Old authors believed that they fed on dew, but fruit and insects constitute their food. Very curious legends are connected with these birds. Arabic writers of the Middle Ages gravely declared that there were islands in the East where grew trees bearing fruit resembling men's heads, that cried, "Wawk! wawk!" at sunrise and sunset. No one could find any

CHOUGH.

explanation of such a story till Mr. A. R. Wallace visited the Malay Archipelago, and found that the Birds of Paradise settled on the trees in flocks about sunrise, uttering this cry. It was also believed that they had no feet, and some thought they had no wings. This erroneous notion arose from the fact that the native traders who prepared the skins always cut off the feet and wings of the birds before selling them to Europeans. This seems to have been the general plan down to the time of Wallace's visit, though in some cases he says the feet and wings were left attached to the skins.

The Great Bird of Paradise, the largest species known, has the plumage of a rich coffee colour, while the long plumes, of a rich golden orange, on each side of the body, can be erected at pleasure. The King Bird of Paradise is a small species frequenting the thickest parts of the forests. Wallace describes it as frequenting the less lofty trees in places where the forest is not dense, flying strongly with a whirring sound, and continually hopping or flying from branch to branch. It eats hard stone-bearing fruits as large as a gooseberry, and often flutters its wings, at the same time elevating and expanding the beautiful fans with which its breast is adorned.

The Red Bird of Paradise (Plate VI., No. 18) is confined to the island of Waigiou. Wallace obtained several specimens, in the hope of bringing them alive to Europe. He says he had a large bamboo cage made, and fed the birds on fruit and grasshoppers, but as most of

SUN-BIRDS.

his birds died within three days of their capture, he gave up the attempt, and brought home preserved specimens.

The Plume Birds differ from the True Birds of Paradise chiefly in their sickle-shaped bills, whence they are sometimes termed Long-billed Birds of Paradise.

The Bower Birds owe their popular name to the fact that besides their nest they make a kind of run or gallery in which to play, and decorate it with shells, feathers, and any glittering objects they can find and carry off.

The Honeysuckers, from Australia and the Malay Archipelago, have an extensile tongue with a brush-like tip, with which they extract the juices from flowers. The Sun-birds, with similar habits, range from Africa, through Palestine, over the south of Asia to the Malay Archipelago, and thence to Australia. From their small size and brilliant plumage, they are often called the Humming-birds of the Old World.

A number of other foreign birds can only be mentioned. The Chatterers, from the northern parts of both hemispheres, have a single European representative—the Bohemian Waxwing, so called because some of the wing-feathers are tipped with what looks like pieces of red sealing-wax. The Hook-billed Creepers are confined to the Sandwich Islands. From the feathers of some of these birds the magnificent war cloaks of the kings of the Sandwich Islands were made. The Flower-peckers are small birds of gay plumage from Asia, Africa, and Australia. In habits they are said to resemble the Tits. The Sugar-birds, from tropical America, are small birds that live on fruits, seeds, and insects. The Wood Warblers, which range from Panama to the Arctic region, are allied to the Sugar-birds and to the Tits. The Greenlets are small American birds, allied to the Shrikes. The plumage is of various shades of green, and the food consists of insects, seeds, and berries.

The Swallows are distributed over the world, and few birds are better known. To us in Britain, the Common Swallow (Plate V., No. 9) is always welcome, as its arrival betokens the coming of spring. An old writer says of this bird :—"He lives a life of enjoyment among the loveliest forms of nature; winter is unknown to him, and he leaves the green meadows of England in autumn for the myrtle and orange groves of Italy, and for the palms of Africa." It is an insect-eating bird, and everyone must have seen it hawking over water for its winged prey.

A near relation is the House Martin (Plate V., No. 8), which arrives in this country a few days later than the Swallow. The difference in the plumage of these birds may be seen from the plate ; and the legs and feet are feathered. Professor Newton says that the Laplanders invite the Martin to breed near their houses by fixing narrow planks to the walls, with just room enough between them for the nests, which may be seen row upon row. They give it this accommodation because it keeps down the gnats with which the country is infested.

The Sand Martin, the smallest of our three Swallows, is generally the first to arrive in Britain from its winter quarters. The general plumage is brown above and white below. These birds nest in company, usually in banks, forming a tunnel or gallery, at the end of which the eggs are deposited.

The Hang-nests are American, and most abundant in the tropical regions. Though many of them feed largely on insects, yet they are so fond of fruit that they inflict considerable damage on the farmers' crops. The author of "Pepacton" says of the Baltimore Oriole, with its sharp, dagger-like bill, that "he has come to be about the worst cherry-bird we have. He takes the worm first, and then he takes the cherry the worm was after, or rather he bleeds it. . . He is welcome to all the

SAND MARTIN.

fruit he can eat, but why should he murder every cherry on the tree, or every grape in the cluster? He is as wanton as a sheep-killing dog, that will not stop with enough, but slaughters every ewe in the flock." The nests resemble those of the Weaver-birds.

The Tanagers, beautiful fruit- and insect-eating birds, are found in the warmer parts of America and in some of the West Indian islands. Some of them have considerable powers of song.

The Finches, with which the Buntings are here included, are small birds, differing chiefly in the shape of their bills, and feeding principally on seeds. They are widely distributed, and many build beautiful nests.

The Chaffinch (Plate VI., No. 8) is a native bird, common also in Europe, where it is more valued as a cage-bird than it is with us. Its song is heard from spring to midsummer, and it is a pretty general favourite. Probably the Chaffinch more than repays the damage it does to gardens by the numbers of insects it destroys.

The Goldfinch (Plate VI., No. 10) is a little smaller than the Chaffinch, with more brilliant plumage. Its chief food is the seeds of thistles, dandelion, groundsel, and plantain. Owing to the better cultivation of waste lands, and still more to the ravages of bird-catchers, Goldfinches are less numerous with us than formerly. They do well as cage-birds, and are often taught to draw water for themselves, to open a box in which their seed is kept, and other amusing tricks. The Siskin, a near relation, with plumage of a greenish hue, marked with black and yellow, comes from

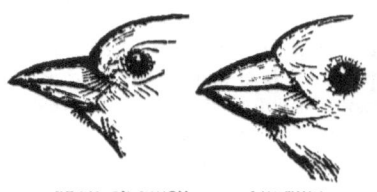

BEAK OF FINCH.   BUNTING.

the north to spend the winter with us, and has remained to breed.

The Hawfinch is resident in Britain, and its numbers are on the increase, but owing to its shy habits the bird is not often seen. The counties round London seem to be its favourite abode. The Greenfinch, or Green Linnet, soon becomes tame in confinement, and will readily pick up the notes of any bird caged near it. The Linnet is a very common British bird, congregating in flocks on waste lands, except at the breeding season. The general plumage is brownish, and from the presence or absence of red on the head or breast, according to the season, Red Linnets and Grey or Brown Linnets were distinguished, though they are the same species.

The Bullfinch (Plate VI., No. 11) frequents wooded districts, and does great damage in gardens by picking the blossoms and buds off fruit-trees and bushes. The Bullfinch is valued as a cagebird, because it can repeat a tune which has been played to it repeatedly, the bird being kept in the dark, without food, till it begins to imitate the tune. It is capable of great attachment, and makes an affectionate pet.

CROSSBILL.

The Canary, brought originally from the Canary Islands, is greenish-yellow in its wild state, the yellow plumage being due to careful breeding in domestication, as are also the strange forms of many of the breeds which are utterly unlike a wild Finch. Everyone knows what a charming songster the Canary is.

The House Sparrow (Plate VI., No. 12) is a well-known, pert, lively bird, that would be reckoned handsome if it were not so common. Opinions are divided as to its utility, some maintaining that the bird is a farmer's friend; others that its depredations far outweigh the good it does by destroying insects. Farmers seem to be of the latter opinion, for Sparrow clubs, which pay so much a dozen for the heads of these birds, exist in many parts of the country. The Tree Sparrow is much less common in Britain, and, unlike its relative, is never found in towns.

The Java Sparrow, or Rice-bird (Plate VI., No. 9), is a large Finch, with a wide range in Southern Asia. It is often kept as a cage-bird, but its notes are feeble.

The Crossbills owe their name to the way in which the two halves of the bill cross at the tip, forming an admirable tool for breaking up pine cones, the seeds of which are extracted by the spoon-shaped tongue. The Common Crossbill is a British visitor, and some remain to breed. There is a legend that the red plumage (of the males) and the crossed bill are due to the fact that when Christ was crucified, a Crossbill tugged at the nails in the hope of releasing Him.

The True Buntings generally have a knob on the palate, used for crushing seeds. The Yellow Hammer, or Yellow Bunting (Plate VI., No. 7), is an extremely handsome bird, but its song, "Little bit o' bread and no-o-o cheese," is monotonous. The young are fed on insects, and the old birds eat the seeds of many noxious weeds. In Italy this bird is fattened like the Ortolan. The Reed Bunting, which must be distinguished from the Sedge-bird, frequents marshy places, and feeds largely on molluscs and crustaceans. When the nest and young are threatened, the old birds endeavour to draw intruders away from the nest by feigning lameness. The Ortolan sometimes strays to Britain in its summer visits to Europe. As it comes northwards, and again on its return, vast numbers are netted and fattened for the table. The Snow Bunting is an autumn visitor to Britain, and some remain to breed. Saxby says: "Seen against a dark hillside, or a lowering sky, a flock of these birds present an exceedingly beautiful appearance, and it may then be seen how aptly the term 'Snow-flake' has been applied to them."

The Weaver Birds are mostly African, though some range to Asia and Australia. Many of them nest in company, building their flask or bottle-like nests side by side. The Sociable Weaver Bird goes far beyond this, and Le Vaillant called these birds "republicans," from their nesting habits. A company of them will select a large tree, and, using the branches as supports, construct a compact sloping roof of grass, underneath which the separate nests are built.

Some of the Asiatic species build nests in shape like a chemist's retort, the long tubular part serving as the entrance. When Dr. Forbes was travelling in Java he met with a colony of Yellow Weaver Birds. "Each nest was artfully suspended between the interlacing leaf-stems of one or two reeds. . . . These nests were not made fast to, but strung lightly on the leaves, sometimes passed through the fork of another leaf to form a pulley, so as to permit of their retaining their upright position, which they must do, weighted as they are by a layer of clay in the bottom of the nests."

The Starlings are widely distributed over the Old World, Australia being the only region without them. The Common Starling (Plate VI., No. 13) is a beautiful bird, and as useful to the farmer as it is beautiful, its services in destroying insects far outweighing the harm it does to gardens and orchards by eating fruit. Waterton was so convinced of this that he fitted up in his gateway at Walton Hall a number of holes, which were afterwards used as nesting-places for these birds. Like the Beefeaters and Buffalo Birds, they will remove ticks from the backs of cattle and sheep. The Starling has considerable power of song. Professor Newton compares it to that of the Mocking

NESTS OF WEAVER BIRDS.

Bird, and it will readily learn to "talk." The brutal and too common practice of slitting its tongue can have no possible effect in enabling it to "speak."

The Rose-coloured Starling is an irregular summer visitor.

The Beefeaters from Africa, and the Buffalo Birds from Asia, owe their popular names to their habit of picking out grubs from the backs of cattle. Mr. H. O. Forbes says : "I was never tired of watching the friendly relations between the Buffalo Birds and their bovine hosts. They used to collect in impatient flocks about the hour of the return of the herd to their feeding-grounds, and as soon as the cattle arrived they would alight on their backs in crowds, to the evident satisfaction of the buffaloes, which they relieved of troublesome parasites. Although the herd boys commonly lay dozing at full length on the buffaloes' backs, the birds seemed to know that they were quite safe, and would

BUFFALO BIRDS.

even alight on the bare back of the sleeper, and from that hop on to the haunches of the quadruped; and when the herds were driven away at nightfall the birds flew off to the forest."

The Larks have their home chiefly in the Old World, and are most abundant in Africa. The hind claw is generally long. The Skylark (Plate VI., No. 6) is common all over Europe, and is well known to everyone, in its wild state and in captivity. A good deal of false sentiment has been indulged in about caging Larks. Captivity, however, has some advantages—for birds, at any rate. In his cage the Lark has no Sparrow-hawk to fear, and is sure of good treatment. Cases are known of Skylarks living nineteen or twenty years in captivity, and many instances of their breeding in that condition are recorded. The Woodlark, a smaller bird, is also native, and some other species visit us.

The Wagtails and Pipits are found all over the world, but they are most numerous in the Old World. The Pied Wagtail, with its black and white plumage, is a very common British bird of elegant form and great activity. Insects and their larvæ form its chief food, though in confinement it has been seen to take minnows from the basin of a fountain in the aviary, only to be robbed by Thrushes and Blackbirds that had developed a taste for fish. Pied Wagtails have bred in the Zoological Gardens. The Grey Wagtail also lives in Britain, but has a partial migration from north to south in winter. It loves the water as much as the Dipper does, and water-beetles and pond snails make up a large part of its food. The Yellow Wagtail is a regular summer visitor. It has some of the habits of a Starling, for it follows sheep and cattle for the sake of the insects hovering round these animals.

The Tree Pipit spends the summer with us. It is closely allied to the Meadow Pipit, which lives with us all the year, but the former frequents woodlands and is fond of perching on trees, while the latter prefers waste and marsh lands. The Shore Lark, with a tuft of feathers on each side of the head, visits us in winter from the north of Europe.

GREY WAGTAIL.

The Tyrant Shrikes range over America, from Patagonia to the Arctic regions, and, in the shape of the bill and the bristles at its base, they resemble the Flycatchers of the Eastern hemisphere. The Kingbird is one of the best-known species. Its name refers to its crest, and to the boldness with which the male, during the breeding season, will attack birds much larger than himself, though his pugnacity is said to be exaggerated. The Fork-tailed Tyrant, or Scissor-bird, derives its popular name from the fact that during flight the long tail-feathers open and shut like the blades of a pair of scissors.

The Manakins are South American, chiefly from the forest regions near the Equator. They are small, shy birds, and little is known of their habits in a wild state.

Among the South American Chatterers are some very remarkable birds, mostly of brilliant plumage. The Cock of the Rock, about the size of a small pigeon, is a shy, solitary bird, found in the districts bordering the rivers of Surinam, Cayenne, and Guiana. The plumage is orange-yellow, and the head bears a semi-circular crest. The Umbrella-bird, with black plumage, a large crest, and a pendent tuft of feathers hanging down in front from the neck, is a native of the forests of Brazil. The Bell-birds, so named from their note—like the clang of a church bell—have white plumage. In one species there is a caruncle on the forehead, which, when extended, hangs down on one side of the beak. The Broad-bills, with brilliant plumage, allied to the South American Chatterers, range eastward from the Himalayas to Java. The Plant-cutters, from the temperate regions of South America, are said to associate with some of the Tanagers, to which family they are perhaps allied.

The American Creepers live in the tropics of the New World.

UMBRELLA BIRD.

They are small birds, generally with brown plumage, and the stiff tail-feathers are some help to them in climbing. Here belong the Oven-birds, so called from the shape of the nests which some of them build. One species, which the Spaniards call the Little Housebuilder, makes its nest at the end of a long gallery, which sometimes runs 6 feet underground, and a sandy bank is generally chosen for the purpose. Sometimes, however, these birds bore into mud walls, mistaking them for banks, though they were constantly flitting over them. It is clear they have no notion of thickness, and Mr. Darwin says, "I do not doubt that each bird, as often as it came to daylight on the opposite side, was greatly surprised at the marvellous fact."

The Bush-Wrens, from South America, have very large feet, and carry the tail upturned like our common Wren. One species is known to English settlers as the Barking Bird, from the fact that its cry resembles the yelp of a small dog.

The Pittas, or Old World Ant-thrushes, are birds of brilliant plumage, chiefly from the Malay Archipelago. Wallace describes the giant Pitta as "one of the most beautiful birds of the East. It has very long and strong legs, and hops about with such activity in the dense tangled forest, bristling with rocks, as to make it very difficult to shoot."

The Lyre-birds are Australian. The common Lyre-bird (Plate V., No. 18) is somewhat larger than a pigeon, and feeds chiefly on insects and their larvæ. The tail-feathers of the male are exceedingly beautiful. They generally droop, like a Peacock's train; but when the birds are on their "dancing-beds" or playing-grounds, they are erected and expanded. "The Lyre-bird is a most wonderful mocker, not only of other birds, for he will imitate to the life the bullock-driver with his whip, the step of the teamster's horses, the rasping of the cross-cut saw, and the blows of the axe and the tomahawk."

The Scrub-birds, of which but two are known, are Australian. They are like large Wrens, with a long tail. They are extremely shy, and a naturalist who after long watching succeeded in catching sight of one, says:—"All of a sudden it would begin to squeak and imitate first one bird and then another, now throwing its voice over my head, then on one side, and then again apparently from the log on which I was standing. This it will continue to do for hours together; and you may remain all day without catching sight of it."

LYRE BIRD.

CHAPTER XVIII.

PICARIAN BIRDS AND PARROTS.

THE Picarian birds, grouped round the Woodpeckers, whence they derive their title, are sometimes classed with the Parrots as "Climbing Birds," from their habit and the arrangement of the toes—two in front and two behind—in most of them.

The Woodpeckers are widely distributed, and owe their popular name to their habit of making holes in trees in which to deposit their eggs, and seeking their insect food on and under the bark, which is

GREEN WOODPECKER.

not infrequently stripped off, the strong wedge-shaped bill being used as a lever. The extensile tongue is furnished at the end with barbs, and covered with a sticky secretion, which serves to secure their prey. Many of them feed largely on ants, laying open the ant-hills with their powerful beak, and gathering up the insects with their tongue. The tail-feathers are stiff and pointed, and are of great assistance to these birds in climbing, supporting them on the trunk of a tree, the legs and tail thus forming a natural tripod, as do the hind legs and tail of the kangaroo when it sits upright.

There are three British species, of which the most abundant is the Great Spotted Woodpecker, about the size of a Thrush, with black and white plumage; the Lesser Spotted Woodpecker, about half the size, has a very similar dress. The Green

PLATE VI.

1. Grey Parrot. 2. Yellow-crested Cockatoo. 3. Rose-crested Cockatoo. 4. Scarlet Macaw. 5. Blue and Yellow Macaw. 6. Lark. 7. Yellowhammer. 8. Chaffinch. 9. Java Sparrow. 10. Goldfinch. 11. Bullfinch. 12. House Sparrow. 13. Starling. 14. Hooded Crow. 15. Magpie. 16. Jay. 17. Golden Oriole. 18. Red Bird of Paradise. 19. Humming Bird. 20. Hoopoe. 21. Green Woodpecker. 22. Great Hornbill. 23. Kingfisher. 24. Cuckoo. 25. Toucan. 26. Woodpigeon. 27. Passenger Pigeon. 28. Crowned Pigeon.

Woodpecker (Plate VI., No. 21), about 1 foot long, with green and yellow plumage and the head marked with crimson, is perhaps the most often seen, for it is by no means a shy bird. Mr. Dixon says: " I often watch him fly from tree to tree in drooping flight, and either settle on the bark at once, or perch among the slender twigs, usually uttering his loud, laughing cry as soon as his wings are at rest." This bird becomes very noisy before rain.

The Wrynecks are small tree-creeping birds from Europe, Asia, and the north of Africa. The tongue resembles that of the Woodpeckers, and serves the same purpose of procuring food. The Common Wryneck, about the size of a Lark, with mottled plumage, is a spring visitor, generally arriving a little before the Cuckoo, whence it is popularly known as the Cuckoo's Mate or Cuckoo's Leader. The name Snake-bird, also applied to it, refers to the wavy motion of the neck and the loud hiss uttered by the female when disturbed while sitting. Wrynecks do not make a hole for a nesting-place, but utilise one already made either in a tree or a bank. The Wryneck, when taken young, is easily tamed, and soon becomes extremely attached to its master, creeping about his person and nestling in his pockets and sleeves.

WRYNECK.

The Honey-Guides are nearly all African, and derive their name from their habit of pointing out or leading the way to bees' nests containing honey. There is a wide spread belief among the natives that these birds, from malice, will lead a man to the lair of savage creatures ; but for this there is no evidence. They seem to guide men to bees' nests, with the very natural object of getting some of the honey, or the grubs, which they greedily devour. These birds are allied to the Cuckoos, and have the same bad habit of depositing their eggs in the nests of other birds.

The Barbets are tropical fruit-eating birds. living in the forest, and generally keeping to the tops of trees. The plumage. especially on the head and neck, is brilliant. The bill is short but strong, and with it they make holes in trees, in which the eggs are laid, generally on a few chips at the bottom.

o

The Toucans are South American birds, with enormous bills, generally richly coloured. Although of such great size the bill is very light, being full of air-cells; the edges are toothed, and the long, thin, bony tongue is barbed at the sides like a feather. The body-plumage is dark, with brilliant markings, and the naked parts gaily coloured. They live principally on fruit, which they seize with the beak and throw into the air, so as to catch it readily, and it is swallowed whole. There are many species. The flesh is valued for food, and the feathers are used for decorations and ornaments.

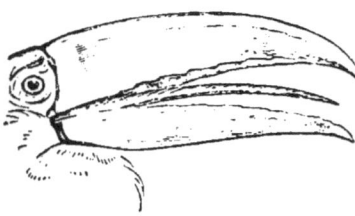

BEAK AND TONGUE OF TOUCAN.

The Toko Toucan (Plate VI., No. 25), from Brazil, nearly 2 feet long, is dressed in glossy black, with some white on the cheeks and throat, and red on the rump. The Prince von Wied says that these birds near Bahia were very shy, from their having been fired at by the inhabitants to drive them away from the fruit-trees; but their fondness for oranges and guavas induced them to approach the town when these fruit were getting ripe, and many of them were killed for the table.

Toucans are often kept as pets. Bates met with one, near Ega, that had escaped from its owner, and as no one claimed it, he made a pet of it. He describes it as a most amusing companion, very intelligent and confiding. It had the run of the house, and always made its appearance at meal times. At last it took to rambling about the streets, and was stolen.

TOUCANS AND NEST.

" But two days afterwards he stepped through the open doorway at dinner hour, with his old gait and sly magpie-like expression, having escaped from the house where he had been guarded by the person who had stolen him.

The Araçaris have a smaller beak and long, conical tail. Some smaller species are called Toucanets.

The Plantain-eaters, or Touracoes, are African. In general appearance they resemble the Game Birds, but have a fine crest, which can be raised or lowered at will. Most of them have some of the wing-feathers brilliant red, and a heavy shower of rain will wash out the colour and leave the feathers of a pale pink. In a few days, however, the colour becomes as bright as it was before ! At one time this was thought to be a traveller's tale, but experiments made on the feathers of birds kept in confinement have shown it to be true.

In the Colies, or Mouse-birds, also African, all the toes turn forward. Captain Shelley says that they are fruit-eaters, living in small bands and frequenting thick bushes, flying to some covert when disturbed. " They place themselves in the most extraordinary attitudes when they rest or scramble among the boughs, and they roost at night in thickly-packed companies, generally, if not always, with their feet above their heads. The general plumage is dull."

The Cuckoos, of which there are nearly two hundred species, are natives of the warmer regions. Many deposit their eggs in the nests of other birds, and some are strangely like birds of other families. The Common Cuckoo is very much like a Sparrow-hawk, and this probably gave rise to the old belief that cuckoos turn to hawks in winter, as it certainly does to some of the stories of the appearance of the Cuckoo in this country long before the usual time. One of the Bush-cuckoos from Borneo resembles a pheasant in gait and appearance. The Common Cuckoo (Plate VI., No. 24) has its summer quarters in Europe, the north of Africa, and in Asia, as far south as the Himalayas ; in winter it reaches Natal, some of the islands of the Eastern Archi-pelago, Burmah and Ceylon. The length is about 1 foot, plumage ash-grey above, white barred with black below. In Britain the males arrive first, generally about the middle of April, and the females follow a few days after. At the end of July or the beginning of August the old birds go to their winter quarters ; but the birds of the year stay later, sometimes till October. The note *cuck-oo* is heard as soon as the birds arrive, and the following lines sum up the history of the male bird's stay with us :—

> In April, come he will ;
> In flowery May he sings all day ;
> In leafy June he changes his tune ;
> In August go he must.

O 2

The hen-bird returns to the same locality year after year. She lays her egg on the ground, and taking it in her mouth flies with it to the nest of the foster-parent. The young bird has a hollow in its back; when it is about ten days old it gets the other nestlings, one after another, into this hollow, and so literally heaves them out of the nest. This cavity is filled up when it has served its purpose. These birds feed on insects and their larvæ.

, The Puff Birds, of small size, with thick plumage of sober hues, are natives of Central and Southern America, and feed on insects, for which they watch patiently, perched on the lower branches of a tree, darting on their prey and returning to the same branch to devour it. Their habit of raising their feathers has given them their popular name.

The Jacamars, from Central and South America, have the beak long, the tail wedge-shaped, and the plumage generally rich metallic

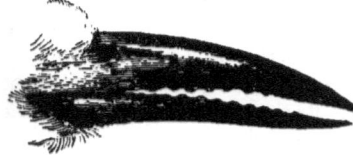

BILL OF MOTMOT.

green. They feed on insects, and sit motionless on a branch, often over water, and dart upon their prey, somewhat in the fashion of King-fishers, for which they were formerly taken. Waterton describes them as indolent, and shunning the company of other birds.

The Rollers are bright-plumaged birds confined to the Old World, and owe their popular name to the habit of the males of turning somersaults in the air at certain seasons. Most of the species are African; the Common Roller is abundant in Central Europe, and sometimes visits Britain and Ireland. The male is about 1 foot long; the sexes are alike, with plumage of shades of blue and a chestnut-brown mantle.

The Bee-Eaters are found over nearly the same countries as the Rollers, and, like them, have one European species. The popular name is rather misleading, for, though these birds do damage to bee-keepers, they destroy a number of noxious insects, wasps, locusts, and beetles. They nest in colonies, in holes in banks, like the Sand Martin. Mr. Howard Saunders says: "Sackfuls of birds are taken in Spain by spreading a net over the face of an occupied bank and pouring water into a parallel trench, cut at some distance back; for the Bee-eater is hated by the peasants, owing to the ravages inflicted upon their numerous hives."

The Todies are delicate, bright-coloured insect-feeding birds from the West India islands, some of which have species not found in the rest. The Green Tody, from Jamaica, is, Mr. Gosse tells us, " conspicuous from its bright grass-green coat and crimson velvet gorget." He never saw this bird on the ground, but hopping about on the twigs of low trees, searching for small insects, and occasionally uttering a low, hissing note.

The Motmots are natives of the New World, ranging from Mexico southwards, and are most abundant in Central America. They are forest-loving birds, and feed on insects and berries. Green is the prevailing colour of their plumage, and the central tail-feathers have a spatule or racket at the end, formed by the birds themselves. Waterton says : "This bird seems to suppose that its beauty can be increased by trimming the tail, which undergoes the same operation as one's hair in a barber's shop, only with this difference, that it uses its own beak (see the Figures) in lieu of a pair of scissors. As soon as the tail is full grown, he begins about an inch from the extremity of the two longest feathers in it, and cuts away the web on both sides of the shaft, making a gap about an inch long.' This was confirmed by Mr. Bartlett, the Superintendent of the Zoological Gardens, who says : "I have seen the bird in the act of picking off the webs of the central feathers of its tail, and have taken from the bottom of the cage the fragments of web that fell from the bird's bill."

TAIL OF MOTMOT.
(Partly trimmed.)

The Trogons are forest-dwelling birds, with thick loose plumage of pink, crimson, orange, brown, or metallic green, often banded with white. The True Trogons are South American, and feed mostly on fruit ; but some other members of the family feed on insects. One of the best known is the Resplendent Trogon, or Quesal, from Guatemala, clad in golden green, except on the breast, which is scarlet. The male bird has the tail-feathers produced to a length of about 3 feet, and its head is crested. These birds are met with in forests about 6,000 feet above sea-level, generally resting on the lower branches. They are shot for their tail-feathers, which are valued as ornaments.

The Kingfishers are distributed all over the world, in a scattered kind of way. The sole is flat, and the three toes in front are joined for the greater part of their length, thus giving these birds great grasping power. There are two groups : in the first, of which the Common Kingfisher is a good type, the bill is long, compressed, and ridged on the top, and the birds feed chiefly on fish ; in the second, represented by the Laughing Jackass of Australia, the bill is stout and flat, and the birds indulge in a mixed diet, of which insects form a good part. The Common Kingfisher (Plate VI., No. 23) is about 7 inches long, with brilliant plumage of shades of blue, marked with black and white. It is found nearly all over Britain, and frequents the ponds and gently-running streams, perching on a branch over the water, whence it

COMMON KINGFISHER.

plunges on its prey, which, however, is sometimes taken by hawking. The nest is a hole in a bank near water, and the eggs are laid on the ground or on a mass of fishbones. Minnows, sticklebacks, and fish-fry form the chief food, but snails, water-beetles, and dragon-flies are not despised. Instances are known of these birds being choked by fish which they have taken but could not swallow.

The Hornbills range from Africa through India eastward to New Guinea. They are large, heavy birds, in appearance not unlike Toucans but in nearly every one the bill has a casque or helmet. The feet are like those of the Kingfishers. The diet is chiefly fruit, but lizards and fish are eaten by some of the species. The male plasters up the hen-bird while sitting, and till the young are hatched, feeds her through a small hole left for the purpose. The Great Hornbill (Plate VI., No. 22), about 4 feet long, ranges from India eastward to the Malay peninsula. The African Ground Hornbills kill snakes.

The Hoopoes. from Asia, Africa, and the South of Europe, are insect-eating birds, allied to the Hornbills. The Common Hoopoe (Plate VI., No. 20) are about 10 inches long, visits Britain in the spring, and some few have bred in this country. It is a beautiful and graceful bird, especially when the crest is raised, but is extremely dirty in its nest-building, and its favourite food consists of the insects found in cattle-droppings.

The Wood Hoopoes are African birds with the habits of Wood-peckers.

The Goatsuckers are insect-eating birds with soft, mottled plumage, and the gape fringed with bristles. They come out at dusk, and, with swift, silent flight, take their prey on the wing. The popular name refers to the erroneous belief that they suck goats and cattle. Their presence near the udders of those creatures is really to pick off the insects with which those parts are infested. These birds have a curious way of resting lengthways on, and not across, a branch. The Common Goatsucker (Plate V., No. 7), about 10 inches long, arrives in England in May and generally leaves in September. The Whip-poor-Will of America, named from its cry, belongs to

SWIFT.

the same family. Closely allied are the Oil-bird of Trinidad—killed for the sake of its oil, which equals that obtained from the olive—and the Frog-mouths of the Australian region ; of the latter the best known is the " More-pork."

The Swifts are Swallow-like in appearance and habits, but all the toes are directed forwards. The Common Swift, about 6 inches long, is blackish-brown with a greyish patch under the chin. It arrives in the south of England about the end of April, and departs southwards about the end of August. To this family belong the birds that form, from the mucus they secrete, the nests prized as delicacies in the East.

The Humming-Birds are confined to America, from Cape Sitka to Tierra del Fuego, and the West Indian islands. These birds rarely alight on the ground. The plumage is extremely beautiful, generally more brilliant in the male, though in some species that of the females has a metallic gloss. There are over four hundred species, all of small size, some no larger than a Hawk-moth. Mr. Gould, to whom we owe the magnificent collection of Humming-birds in the British Museum, describes them as restless, irritable, and quarrelsome, and says that "they not only fight persistently among themselves, but they will even venture to attack much larger birds, and it is said that several of them will combine and attack a Hawk and drive it away. People are often attacked by them when they approach too near their nests. It is stated

HUMMING-BIRD.

that they have also a great dislike to the large Hawk-moths, which they themselves somewhat resemble in their flight, the vibration of the wings producing in both a similar humming sound." It is from this humming that the birds get their popular name. They feed on the juices of flowers, and insects.

The Ruby-throated Humming-bird (Plate VI., No. 19) is found in North America during the summer, and migrates southwards and to the West Indian islands in winter. Gould brought some specimens with him on his homeward voyage, feeding them with honey, or sugar-and-water, and the yolk of egg. Only one reached London alive, and it died on the second day after its arrival.

PARROTS.

In these birds the bill is large and strong, the upper half turning down over the lower half, and it is often utilised as a climbing organ; the tongue is usually large and fleshy, and the toes are arranged in pairs, two in front and two behind. They are most abundant in the tropics, but one ranges far north in America, and there are many species in the Australian region. In most the plumage is brilliant, in some glaring, and in a few dull and sober. Parrots differ greatly in size, some being quite 3 feet long, while the Love-birds are not more than a few inches. Fruits, seeds, and leaves form their chief diet. The voice is naturally harsh, but many have great power of imitation, and soon learn to repeat words and sentences. This power varies in individuals, and the skill and patience of the trainer ought to be taken into account when a parrot is a good talker.

The Cockatoos, from Australia and some islands of the Malay Archipelago, are large birds, having a crest that can be raised or lowered at will. They nest in hollow trees and in holes in cliffs and rocks. As cage-birds they soon become tame and affectionate, but their cry is harsh and their imitative power small, so that they rarely learn more than a few words. The name is said to be derived from their cry.

The Yellow-crested Cockatoo (Plate VI., No. 2), spread over the Australian region, has white plumage, with a lemon or sulphur-coloured crest. It is probably the best known of the Cockatoos, for it is the species most often kept as a cage-bird in this country. Wallace met with these birds in his travels in the Malay Peninsula, and says that "their loud screams, conspicuous white colour, and pretty yellow crests rendered them a very important feature in the landscape."

The Rose-crested Cockatoo (Plate VI., No. 3) is a native of the Moluccas, has a red crest, and the plumage with a blush of rose-colour, which is most apparent when the feathers are ruffled.

The Inca, or Leadbeater's Cockatoo, from Australia, has similar plumage, but a much more showy crest, the feathers of which are red, spotted with yellow and tipped with white.

The Black Cockatoo, from New Guinea, is described by Wallace as having "an enormously developed head, a magnificent crest, and a sharp-pointed hooked bill of immense size and strength. The plumage is entirely black, and the bare cheeks are of an intense blood-red." Its note is a plaintive whistle. The birds feed on the kernel of a nut "the shell of which is so hard that only a heavy hammer will crack it," breaking a hole with their powerful bill, and extracting the kernel with the tongue, which has been compared to a spoon.

The Large-tailed Parrots, principally from Australia and Tasmania, are generally gaily clad, except some few that are ground-feeders, whose plumage corresponds with their surroundings. The Grass Parrakeets belong to this group, and the Waved Grass Parrakeet, or Budgerigar, is well known in England as a cage-bird. The general hue is green, with black markings. Its note is soft and agreeable, and it soon becomes tame in captivity.

The Parrakeets are found in India and the Eastern Archipelago. The Alexandrine Parrakeet was long supposed to be the bird brought from India to Greece by Alexander the Great; but the Rose-ringed Parrakeet answers the description better, and the home of the first-named species is now known to be Java. The Rose-ringed Parrakeet is about 16 inches long, and has green plumage with a black band extending from the chin nearly to the nape, and a collar rose-coloured in the male, but emerald-green and narrower in the female. These birds appear to have been common pets in Ancient Rome, and there were schools in which these birds were taught to talk. The discipline was of a stern kind : Pliny says that they were corrected with a rod of iron, though, of course, it must have been very slender. Probably it was no thicker than wire; otherwise it would have killed the bird, rather than have taught it to mend its evil ways. In India the common practice when teaching a Parrakeet, is to keep the bird in the dark, which is often effected by lowering the cage into a well. These birds are taught to perform many tricks by the native showmen. In Mr. J. L. Kipling's "Beast and Man in India" is a picture showing a Parrakeet with a lighted match in his claw, and about to fire off a tiny cannon. The bird is on a T-shaped stand, the base of which the showman holds between his naked toes (see p. 12).

The Lories, or Brush-tongued Parrakeets, have the tongue divided at the tip into a bunch of horny fibres, with which they extract from flowers the nectar on which they feed. The plumage is brilliant, and the flight swift and powerful. Broderip says that one of these birds, which was kept as a pet, but supplied with unsuitable food, "when a coloured drawing of a flower was presented to it, applied its parched tongue to the paint and pasteboard, and even did this, in the extremity of its distress, to the ruder image on a piece of flowered chintz."

The Macaws are American, and distinguished by their long tails and bare cheeks. They feed principally on fruit, and many of them have very brilliant plumage. The Scarlet Macaw (Plate VI., No. 4), from South America, is nearly 3 feet long. It frequents flat well-watered forests, but is not found in mountainous districts. The Prince von Wied says that, except at pairing time, they fly in small companies in search of different kinds of fruits; and that, notwithstanding the noise they usually make on the wing, as soon as they have found a tree with suitable food and have settled down on it, their presence is only indicated by the fall of the husks which they bite off and throw down. The Blue-and-Yellow Macaw (Plate VI., No. 5), from the banks of the Amazon, resembles the Scarlet Macaw in habit. It feeds on the hard nuts of palms, which are crushed by the powerful bill. The Caroline Macaw ranges into North America, and was formerly abundant along the banks of the Mississippi.

The True Parrots have a short, square tail, and are crestless: species occur in Asia, Africa and America. The Grey Parrot (Plate VI., No. 1), a native of Africa, is a common cage-bird, and its power of learning and repeating words and phrases is well known. It is perhaps the most intelligent of the Parrots: at any rate its powers have been more highly cultivated than those of any other species. Sometimes its remarks are so appropriate to the circumstances that the bird has been credited—of course wrongly—with a greater share of intelligence than it possesses, though there can be little doubt that it does attach a meaning to some of the words it repeats.

OWL PARROT, OR KAKAPO.

The Amazon Parrots from South America also learn to talk, but less readily than the Grey Parrots. Their plumage is chiefly brilliant green, and although conspicuous against a dry branch, conceals the birds admirably when settled on a tree in full foliage. The sportsman may watch a flock descend upon a tree, and hear their screams, though unable to distinguish a single bird.

The Owl Parrot, or Kakapo, from New Zealand, generally lives on the ground in holes or burrows, and is to some extent nocturnal. The face is owl-like and the green plumage marked with black and yellow. It is intelligent and good-tempered, and makes an amusing pet. A gentleman who kept one says: " It will run from a corner of the room, seize my hand with claws and beak, and tumble over with it, exactly like a kitten, and then rush back to be invited to a fresh attack."

The Kaka Parrot comes from New Zealand and Norfolk Island. Its plumage is olive-brown above, crimson below, and light grey on the crown. This bird is remarkable for having developed flesh-eating habits, probably owing to the abundance of scraps round the slaughter-houses on sheep-stations, and the scarcity of their natural food. From picking up offal, they have come to attack live sheep, and do such damage to the flocks that sheep-farmers shoot them without scruple.

KAKA PARROT.

DOVES and pigeons are widely distributed, but are most numerous in warm countries. The bill, which is somewhat like that of a Plover in shape, is covered at the base with a soft skin, which in many of the domestic breeds of the Common Pigeon develops into wattles. They live much among the branches of trees, but most of them take their food—chiefly grain and seeds—on the ground, and they are strong on the wing. The crop is double, and from it food is forced into the mouth of the young by the old birds. As this half-digested food is curd-like in appearance, it probably gave rise to the old joke about "pigeons' milk." Pigeons drink, while all other birds scoop up water, and then, raising the bill, swallow it. They pair for life, and the eggs in a sitting are two in number, but there is more than one brood in a season. The young are born helpless. There are three native British species, and the Turtle-dove is a summer visitor.

The Ring-dove, or Woodpigeon (Plate VI., No. 26), owes the first name to the white feathers of the neck. The length is about 17 inches: the plumage of the upper surface is grey of different shade, the breast is rich purple, and the sides of head and neck have a metallic gloss. These birds are becoming more numerous, and extending their range in Britain, probably owing to the destruction of their natural enemies, the birds of prey, by keepers for the sake of game. They build in some of the London parks, and in 1894 a pair nested in Finsbury Square. These birds do great damage to the farmer, paying him the compliment of preferring the produce of his fields to that of the woods and hedgerows.

The Stock-dove, so named from its favourite nesting-place in the "stocks" or stumps of trees, is smaller than the Woodpigeon, which it resembles in habits, being only about 14 inches long. There is no white on the sides of the neck, and the purple patch on the breast is smaller. This bird is generally found in woodland districts, and prefers forests where the timber is old and decayed. It is not, however, confined to such places, for it nests in crags and cliffs on the South Coast, and keeps company with sea birds at Flamborough Head, while in the Eastern Counties and on the moors it breeds in rabbit burrows. It visits the farmer's fields, and in winter ventures into his rick-yards, but, as Mr. Dixon tells us, "it makes ample amends for its depreda-tions in consuming millions of seeds of the most troublesome weeds, such as charlock and dock, which if not kept in check, would soon change fertile fields into unproductive wastes."

The Rock-dove justifies its name. In a wild state it is only found on the rocky parts of our coast, preferring those places where the cliffs are weather-worn into holes and caves, which afford convenient nesting-places. The Rock-doves that breed in the inland cliffs in some counties are probably not really wild, but have escaped from domestication, or are descended from birds which did. The bird greatly resembles the Stock-dove in size and appearance, but has two black bars on the wings and some white on the rump. It repays the farmer to some extent for the grain it devours, by eating the roots of couch-grass, the seeds of noxious weeds, and snails. From this species are descended all the breeds of the Domestic Pigeon, of which our fanciers are so proud, as affording proofs of their skill. How greatly these differ from the parent stock may be seen from the picture. In the Central Hall of the Natural History Museum, Cromwell Road, is a large case in which the most remarkable of these breeds are arranged round the Rock-dove, so as to show what may be effected by judicious selection in breeding, so as to develop any variation.

Of the domestic breeds the only one we can mention is the Homer or Homing Pigeon, remarkable for the fact that it can be trained to find its way home from great distances, and for swift flight. The "pigeon-post" organised by the French during the siege of Paris was carried on by these birds. Balloons carrying Homers left Paris, and though no doubt a good many were shot by the Prussians, and still more lost their way or perished, about one bird out of every four found its way back, and of these three out of four brought messages rolled in quills and attached to the tail-feathers. These birds, however, could not have taken a message out of Paris. The reports of matches for pigeon-flying show similar results. In no case do all the birds return, and, as a rule, the longer the distance the fewer the number that find their way home. Mr. Tegetmeier says : "Pigeons must be regularly trained by stages, or they will be lost if flown one hundred or two hundred miles from home."

The Turtle-dove visits Britain in the summer in its northward migration, probably breeding in the south-west of Scotland, as it has been known to do in the north of England. Its length is rather less than a foot ; the upper plumage is bluish-ash, there is some black and white on the neck, the tail-feathers have white tips, the throat and breast are pale red and the belly white. The Collared Turtle-dove, so

DOMESTIC PIGEONS.

1. Homing Pigeon.   2. Tumbler.   3. Carrier.   4. Barb.   5. Pouter.   6. Fantail.   7. Satinette.
8. Turbiteen.   9. Jacobin.   10. Trumpeter.

often kept as a cage-bird, comes no farther westward into Europe than Turkey.

The American Passenger Pigeon (Plate VI., No. 27) may be taken as the representative of a group in which the tail is long and tapers to a point. Its length is about 16 inches ; with slaty-blue plumage above and reddish-grey below, and violet metallic gloss on the neck. These birds make partial migrations in immense flocks, from one part of the country to another, in search of food. On these occasions the pigeons are destroyed in immense numbers, and their abundance over large tracts of the country has been greatly reduced.

The Fruit Pigeons are found in the tropical parts of the Old World, and their plumage is generally green. Jerdon describes them as gliding about the branches, like Squirrels, hunting for fruit. Some that Wallace described, from the Malay Archipelago, were bluish-white in colour, with the back, wings, and tail intense metallic green. " These pigeons have a very narrow beak, yet their jaws and throat are so extensible that they can swallow fruits of very large size."

The Ground Pigeons are named from their habit of feeding on the ground, nesting in shrubs or in trees at no great height. Grain forms their principal food. Here belong the Bronze-wing Pigeons of Australia, so named from the brilliant metallic gloss on the wing-feathers. Of the Crested Bronze-wing, Gould says that the elegance of its form and the graceful crest render this bird one of the most lovely members of the family. It frequently assembles in very large flocks, and when these visit the lagoons or river sides for water, they generally select a single tree, or a particular branch, on which to congregate before coming down to drink. Its flight is extremely rapid, and after a few flaps it goes sailing on, apparently without any further exertion. When one of these Pigeons settles on a branch, it raises its tail and throws back its head, so as to bring them nearly together, at the same time erecting its crest so as to show itself off to the greatest advantage.

The Common Bronze-wing is a plump, heavy bird. In the early days of the colony the settlers, when travelling with their waggons, used to watch for the Bronze-wing's evening flight, which was always in the direction of water, the habit of the bird being to drink before going to roost.

The Wonga-Wonga Pigeon, also Australian, frequents the shades of the forests, feeding upon seeds and the stones of the fallen fruit. When disturbed these birds rise with a "whirr" like pheasants, but the flight is not strong, and is only continued till a place of safety is reached.

The Nicobar Pigeon, a native of the islands of the Malay Archipelago, is a remarkable looking bird, the long feathers of the neck and breast forming a sort of collar. The general plumage is green with metallic lustre, and the tail-feathers are white. Wallace, from whom we have the best account of its habits, describes it as a very heavy, fleshy bird, feeding on the ground, and only going upon trees to roost. Its wings, however, are very large, and the muscles of the breast, by which the wings are moved, of immense size. Consequently it is not surprising to find that the bird can fly for a long distance. From a small coral island about one hundred miles north of New Guinea, with no land between, "a bird was seen flying from seaward which fell into the water exhausted before it could reach the shore. A boat was sent to pick it up, and it proved to be a Nicobar Pigeon, which must have come from New Guinea, and flown a hundred miles."

HOOK-BILLED PIGEON.

The Hook-billed or Toothed Pigeon, from the Samoan Islands, has the lower part of the bill with three tooth-like projections. This bird is about 14 inches long, and the plumage is for the most part glossy greenish-black, and on the hinder parts chestnut-brown. A few specimens have lived in the Zoological Gardens, Regent's Park. Mr. Whitmee, who sent one of these birds, brought it up from the nest, feeding it for some time with bread-fruit till it was able to peck for itself. He describes it as exceedingly savage, and capable of giving one's finger a severe nip with its toothed bill. He believes that these birds are increasing in numbers, owing to a change of habit brought about by altered circumstances. Formerly they fed upon the ground and roosted low, having no enemies to fear. But the introduction of cats, and probably rats, by European vessels caused such diminution in their numbers that they were in danger of becoming extinct. They appear to have learnt wisdom by experience, and now build in lofty trees where no four-footed foe can reach them.

The Crowned Pigeons, from New Guinea and the neighbouring islands, are the largest of the group, being about the size of a domestic fowl. The plumage is slaty-blue, and the head is adorned with a fan-like crest. In habit they resemble pheasants, living upon the ground

P

and wandering about the woods in small parties in search of fallen fruit. The Common Crowned Pigeon (Plate VI., No. 28) has the shoulders chestnut-red and a white stripe on the wings. The Victoria Crowned Pigeon has the under-parts reddish. Crowned Pigeons do well in confinement, and have frequently bred in the Zoological Gardens. They are fine, handsome birds, and walk their spacious cage with stately dignity, which is immediately put to flight if a visitor approaches and throws them some crumbs. Like the Passenger Pigeon and some of the Fruit Pigeons, these birds lay but one egg.

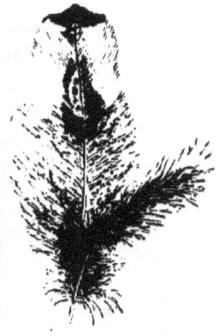

FEATHER OF FOWL.
(*Showing Aftershaft.*)

FOWLS AND GAME BIRDS.

The Domestic Fowl is a well-known type of this order, which is widely distributed. The Turkey is one of the largest forms, and some of the Quails, little bigger than a Sparrow, are the smallest. The plumage is close, and of gorgeous hue and exquisite pattern in some of the male birds, and the feathers bear an aftershaft. This accessory plume is also present in many other birds, but is most highly developed in the Cassowaries, where it is nearly as long as the main shaft. These birds obtain their food chiefly by scratching up the soil for grain, worms, and insects. They generally nest on the ground; in most cases the males have several mates, and the young birds can run about as soon as they come out of the egg. The hen-birds shelter their young under their wings, and are very brave in their defence.

The Sand Grouse have long, pointed tails and sandy-coloured plumage marked with black, which corresponds well with their surroundings in the deserts of Africa and steppes of Asia. From

SAND GROUSE.

the latter continent there have been occasional migrations of these birds in great numbers into Europe. In 1888 some reached Britain and bred here.

At the head of the True Grouse family stands the Capercaillie (Plate VII., No. 3), or Wood Grouse. This handsome bird was formerly common over the British Isles. It is not known when it became extinct in England and Wales, but it disappeared from Scotland and Ireland rather more than a hundred years ago. It was introduced into Scotland from Scandinavia in 1837, and is now once more a native bird. The length is about 2 feet, but individuals vary greatly in size. Capercaillie feed on the buds of trees, and berries, and in winter on the "needles" of pine, fir, etc. The young are reared on insects. In this family the legs are feathered to the toes.

The Black Grouse (Plate VII., No. 4) is spread over England as far south as Sherwood Forest, Scotland, and the north of Europe, and is found in Cornwall, Devon, Somerset, and Norfolk, and in Wales. It is smaller than the Capercaillie, and the male is known as a Black-cock, while his mate is called a Grey-hen. Mr. St. John says that "where these birds take well to a place, they increase rapidly, and from their habits of taking long flights, soon find out the cornfields, and are very destructive, more so probably than any other winged game."

PTARMIGAN IN WINTER DRESS.

The Grouse, or Red Grouse, is peculiar to Britain and Ireland: in England the Trent seems to form its southern boundary. The male bird is about 16 in. long, with rufous-brown plumage marked with black. This is the "Grouse" of the sportsman, and is represented in Northern Europe and Asia by the Willow Grouse, and there are allied species in America.

The Ptarmigan, slightly smaller than the Grouse, is a native of Scotland, and the colder parts of the northern hemisphere. Mr. Dixon says: "In spring and summer it dons a dress of mottled grey and brown, which absolutely shields the bird from its enemies, and as soon as the wild mountain tops begin to get covered with the wintry snows the Ptarmigan assumes a snow-white garb which renders it invisible among the eternal whiteness of its dreary haunt."

The Partridges have their legs bare of feathers. The Common

Partridge (Plate VII., No. 5) is a good type of the family. It is native in Britain, and widely spread over Europe. The male is about 1 foot long, and may be distinguished from his mate by the dark horseshoe mark on his breast. These birds frequent the open fields, and consume great quantities of slugs and harmful insects. They are excellent parents. Mr. Howard Saunders says that he saw "old birds show a bold front to a Hen Harrier for several minutes while covering the retreat of their brood to the shelter of a hedge." Partridges very rarely breed in confinement, but two instances were recorded in 1894. The Red-legged or French Partridge has been introduced into England, and is increasing in some parts. Its flesh is not very highly valued.

PEACOCK.

The Quail is very much like a small Partridge, but has several mates, and fights fiercely with his rivals at the breeding season. Quails arrive in Britain in the spring, and most of them go south again in October, but some pass the winter in this country.

To the Pheasant family belong the Peacocks, the Pheasants, the Fowls, Turkeys, and Guinea-fowls. The legs bear spurs, in some cases two on each leg. No Grouse or Partridge has spurs, though these are represented by knobs in the Red-legged Partridge.

The Peacock is an Asiatic bird, domesticated in this country, and well known to every one. Its flesh was formerly eaten at state banquets, the bird being sent to table with its train spread, but Peacocks are now kept only as ornamental birds. The long tail-coverts, with which the male makes such a brilliant display, do not appear till the third year. In India Peacocks are plentiful; in some places they are venerated by the natives, and shooting them is forbidden by law. A writer in the *Field* says: "I have known the peacock shot by English soldiers in the following way. The men first of all black their faces and hands, and then don the large white cloth of a native. Thus disguised, they mount a camel, driven by a *real* native, and enter the low scrub and jungle in which the peacock delights to dwell. The birds, deceived by the clothes and blackened face, allow a near approach, when they are easily shot by the wily 'Tommy.'"

The Argus Pheasant from Malacca has marvellously beautiful

plumage, of sober hue, with eye-like spots. Mr. H. O. Forbes says that "the closer they are examined, the greater is seen to be the extreme chasteness of their markings and their rich, varied, and harmonious colouring. When alarmed, the Argus escapes by running through the thick underscrub, when the brilliancy of the plumage, by being gathered close about its body, is quite concealed."

ARGUS PHEASANT.

He saw the love-display of these birds, and says that the male makes a large ring, 10 feet to 12 feet in diameter, in the forest. On the edge there is invariably a projecting branch or root a few feet above the ground, on which the female perches, while in the ring the male shows off all his magnificence. The male is often trapped, but the female invariably returns with a new mate, even if two or three times in succession her lord should be caught. She avoids the snares and traps by flying to her perch ; the males walk into the ring, which is barricaded by the natives, except at the spot where the trap is set.

The Peacock Pheasant, from Burmah, has brownish plumage, with green eye-spots on the tail-feathers.

The Monaul, or Impeyan Pheasant, from the Himalayas is a large and beautiful bird, with brilliant metallic plumage and the male is crested.

The Eared Pheasants, from China and Tibet, have a tuft of feathers on each side of the head. These bear some resemblance to the plumicorns of the Horned Owls, but are directed backwards.

The Horned Pheasants, or Tragopans, from India and China, owe their popular name to the fleshy appendages that project behind the head. The naked skin on the throat hangs down like a lappet on each side.

The Common Pheasant is native in Britain. It was introduced before the Norman Conquest, though how long before is uncertain. By the end of the sixteenth century it was introduced into Ireland, and then into Scotland. This bird came originally from the Asiatic shores of the Black Sea, near the river Phasis, now called the Rioni, and is spread over Europe. It is one of the most highly valued, as it is the most beautiful of our game birds. The general plumage is

EARED PHEASANT.

HORNED PHEASANT.

brown marked with black, the head and neck green, glossed with blue, and the bare space round the eye is red. But no description can do justice to the beauty of the plumage, with which, however, most people are well acquainted. Pheasants are woodland birds, and like damp ground and the neighbourhood of water. They roost low, thus offering an easy prey to poachers. They feed on grain, berries, acorns, insects and their larvæ, and have been known to pick up a field mouse. They generally try to escape from danger by running rapidly; when pressed they rise suddenly, whence they are often called "rocketers." They are very strong on the wing, and their flight is swift and long sustained.

The Ring-necked Pheasant, from China, with a white collar, is now also native in our woods, and breeds freely with the common species. The Golden Pheasant (Plate VII., No. 2) is a gorgeous bird from China. The Plate will give a far better idea of its form and brilliant colouring than the most elaborate description. The feathers of the frill, present only in the males, can be raised, and the bird has been seen to do this when showing himself off before the females. The Amherst Pheasant, also Chinese, has a black and red plume on the head ; the feathers of the frill are silvery with dark edges. The Silver Pheasant, with white plumage, pencilled with black on the upper surface, and black below, has no frill, but the head bears a long black crest.

The Jungle Fowl (Plate VII., No. 1) is widely distributed in India, and is supposed to be the original stock whence our breeds of the domestic fowl have descended. It is smaller than the Domestic Cock, which it closely resembles in plumage. Among Eastern nations it was

WILD TURKEYS AT HOME.

domesticated at an early date, and was known to the Greeks and Romans, who trained it for fighting. Dr. Jerdon says that in a wild state these birds are partial to bamboo jungle, but also inhabit lofty forests and dense thickets. When they are put up by dogs, they fly at once to the nearest trees. When cultivated land is near their haunts, they may be seen morning and evening in the fields.

The Guinea Fowls, from Africa, have the head adorned with a crest or helmet. The general plumage is bluish-grey marked with white spots. The Common Guinea Fowl is a well-known domestic species. Its cry of alarm is *"Come back!"*

The Domestic Turkey was introduced into Britain from America about 1541, and received its name from the mistaken notion that it came from Turkey. The general plumage is brown, marked with black above and darker below; the head and neck are wattled, a tuft of bristle-like hair hangs down from the breast, and the tail can be expanded like a fan. Most people have seen the display of the Turkey, which may be in love or anger. " They spread out and erect the tail, draw the head back on the shoulders, depress the wings with a quivering

BRUSH-TURKEYS AND THEIR EGG-MOUNDS.

motion, and strut pompously about, emitting at the same time a succession of puffs from the lungs."

The American Turkey is larger than the Domestic breed, which, contrary to the general rule, has degenerated under domestication. The wild species was formerly very abundant, but its numbers are lessening as land is brought under cultivation. The Honduras Turkey has the tail-feathers marked with eye-spots.

The Curassows are confined to South America, where they are found from Mexico to the southern districts of Brazil. They do well in confinement in this country, but they do not breed freely.

The Mound-birds and Brush-Turkeys are confined to the Australian region, ranging from the island continent to the islands of the Malay Archipelago. These birds do not sit on their eggs, but deposit them in the sand, leaving them to be hatched by the heat of the sun, or on a mound of decaying vegetable matter, large quantities of which are heaped up over them, the heat of the decomposing mass acting as a natural incubator. Several birds use the same hole or heap.

These birds constitute an order by themselves. They live among the branches of trees on the banks of lagoons in South America. There is but one species, which, owing to its loose feathers, appears to be about the size of a Pheasant, but is really not so large. The plumage is dark brown, variegated with reddish, and the head is adorned with a long crest. The hind toe is on a level with the rest, making the foot a good grasping organ. The bird is never seen on the ground. Bates says that its voice is a harsh grating hiss, and that its flesh has a mingled smell of musk and wet hides, which accounts for its not being eaten. The young have little hooks at the end of the digits of the fore limb, and with these they can, unaided, climb up out of the nest when threatened with flood.

BIRDS OF PREY.

This order contains more than five hundred species, widely distributed all over the world. They have strong bills, sharply curved at the point, and large, strong feet, armed with powerful claws. Most of them feed almost entirely on flesh ; some taking living prey while others prefer carrion. The former have stronger beaks and claws than the latter. In most cases the female is larger than her mate.

The Vultures are carrion-feeders ; the head and neck are naked, or clothed with down instead of feathers, and the claws are blunt. Waterton says, " Providence has conferred a blessing on hot countries in giving them the Vulture, and in ordering it to consume that which, if left to dissolve in putrefaction, would infect the air and produce a pestilence."

The Griffon Vulture is pretty common all over the south of Europe, spreading into Africa and Asia. It is about 4 feet long, tawny yellow in colour, with a white ruff round the neck. This bird was made the subject of an experiment by Colonel Drayson, to test whether Vultures discovered their prey by sight or by smell. He had shot an antelope early in the morning and concealed it in a hole, and covered the mouth of the hole with grass. Several vultures were flying at a great height ; and, thinking they would come down and attempt to drag out the body, he rode off and kept watch. But they failed to follow the scent to the spot where the body was hidden, and in the evening, when Colonel Drayson came back from shooting, the antelope was found undisturbed.

The Black Vulture, a rather smaller bird, with the head and neck flesh-coloured, is found on both sides of the Mediterranean, ranging eastward to China. The Eared Vulture, from Africa, and its relative the Indian Vulture, have folds of skin on the neck, standing up by the side of the head somewhat like large ears. Like the King Vulture of the New World, they do not allow other carrion-feeding birds to approach a carcass till they themselves have eaten their fill. The Egyptian Vulture (Plate V., No. 4), or Pharaoh's Chicken, from Southern Europe and the neighbouring parts of Africa and Asia, is about 30 inches in length, with whitish plumage and dark brown quill-feathers. The

CONDORS ON THE WING.

bare parts of the face, the bill, and legs are yellow. This bird is an excellent scavenger, and for its services in this respect is protected by law, a heavy fine being levied on any person who wilfully kills one. Besides clearing the streets of offal and garbage, it consumes vast numbers of rats, mice, and lizards.

The American Vultures have no bony partition between the nostrils, which are open from side to side, as shown in the Figure.

The Condor (Plate V., No. 5), the largest of these Vultures, has its chief home in the Andes, where it breeds at great elevations, depositing two white eggs on a bare ledge of rock. The general plumage is black, with a soft, downy white ruff round the neck, the naked parts of which are flesh-coloured. There is a comb on the head of the male bird. A Condor that Darwin shot measured 4 feet from beak to tail, and 8½ feet between the tips of the outstretched wings. These birds live principally on carrion, but they frequently attack young goats and lambs. Condors are very strong on the wing, and in describing their flight Darwin says: "Near Lima I watched several for nearly half an hour, without once taking off my eyes: they moved in large curves, sweeping in circles, descending and ascending without giving a single flap."

NOSTRIL OF AMERICAN VULTURE.

The King Vulture, from Central and Southern America, is a very handsome bird. Its body is about as large as that of a Goose, and the plumage of the adult is satiny white, tinged with fawn on the back, pure white below. The long feathers of the wings and tail are black, and the bare head and neck are brightly coloured. The young birds are clothed in snow-white down. One that was sent to the Zoological Gardens in 1893 was sketched by Mr. Frohawk for the *Field*, and he thus describes the behaviour of the bird whilst sitting for its portrait :— "The tameness, or rather the fearlessness, of the youngster is remarkable. It is up to all sorts of playful tricks with me—snatching my pencil away while I was sitting on the ground so as to be on a level with it, sitting on my legs and feet trying to unlace my boots, tugging at my breeches, and picking my pockets. It seemed perfectly happy when I tickled it under the chin. It delighted in nestling down by my side, and resting its head on my legs. It certainly is a delightful pet."

The Turkey Vulture is so called from its resemblance to the Wild Turkey, and it is said that it is frequently shot by inexperienced sportsmen in mistake for the latter bird. It is a carrion feeder, and is protected by law for its services as a scavenger.

The Secretary Bird, from South Africa, about 3 feet long, with long, Crane-like legs, and slate-grey plumage marked with black, derives its name from its erectile crest, which the early Dutch settlers compared to pens stuck behind the ear of a clerk. It is extremely serviceable in destroying snakes, which constitute its principal food. It is often tamed at the Cape and kept in poultry yards, but it has a bad habit of snapping up young chickens ; and there is a story that the whereabouts of a missing kitten was discovered by hearing a faint mew as the pet Secretary Bird stalked to and fro, looking as innocent as if it knew nothing at all about the matter.

SECRETARY BIRD.

The Caracaras of South America, are often placed here. They are much smaller birds, clad in brown, striped and marked with white.

One of them has the habit, when uttering its harsh cry, of throwing back its head till the crown almost touches the back, as in the Figure.

The Bearded Eagle, or Lämmergeier (Plate V., No. 2) has some of the habits of a Vulture. It is fairly common in the mountainous parts of Central Asia, but in Europe is almost entirely confined to Spain. It is about 40 inches long, with a wing-spread of nearly three times as much. The plumage is greyish-brown above, dashed with white, the under surface is light, and there are tufts of bristles round the nostrils and at the base of the bill. Strange stories are told of its carrying off lambs, kids, and even children. Brehm, however, considers it " a weak, cowardly bird of prey, gifted neither in mind nor body to any great extent, and one that but rarely carries away small mammals. Its food usually consists of bones and other carrion."

CARACARA.

The Golden Eagle (Plate V., No. 1) is found in both hemispheres, and though extinct in England, still breeds in Scotland, where it is preserved. An adult male is about 3 ft. long, and his mate some 6 in. more. The plumage is dark tawny brown, and the golden appearance of the lighter feathers on the head and neck has given these birds their popular name. These Eagles frequent rocky places, and prey on birds and small mammals. Occasionally they drive roe deer over the rocks, and they have been seen to beat the bushes for ground game. Sometimes they will stoop to feed on carrion. The Golden Eagle

GOLDEN EAGLE AND NEST.

is called the King of Birds. Macgillivray describes it as "powerful, independent, proud, and ferocious, regardless of the weal of others, and intent solely on the gratification of its own appetites; without generosity, without honour, bold against the defenceless, but ever ready to sneak from danger." There are allied species, two of which, the Spotted Eagle and the Imperial Eagle, occur in Europe.

The Bateleur Eagle is confined to Africa. It is about 2 feet long, and the head is crested. The general plumage is black, with maroon patches on the shoulders and back, and the tail is of this same hue. The bare skin of the face is red. Its name is French, and refers to its habits of turning somersaults in the air like a tumbler pigeon.

The Erne or White-tailed Sea Eagle is found in Northern Europe and Asia, and though becoming rare in Britain, is more common than the Golden Eagle. It is somewhat smaller than the former species, and frequents woody places near water, for fish forms a large part of its food. There are several species, which differ from the Golden Eagles and its allies in the leg-feathers not extending to the toes.

KITE AND NEST.

The Bald Eagle, from Northern Europe and America, a little smaller than the British species, has the head and neck white as well as the tail. This bird is the emblem of the United States; and owing to its bad habits in robbing the Fishing Eagle, Benjamin Franklin said, "For my part I wish the Bald Eagle had not been chosen as the representative of our country. He is a bird of bad moral character; he does not get his living honestly. . . . Besides, he is a rank coward: the little King-bird, not bigger than a sparrow, attacks him boldly and drives him out of the district."

The Buzzards have one representative in Britain—the Common Buzzard, about 20 inches long, clothed in loose brown plumage, the shade and markings of which differ greatly in different birds. They feed on mice, rats, snakes and insects.

Kites are small Birds of Prey, distinguished by their forked tail and the absence of a notch on the bill. The Common Kite was formerly abundant in England, and till the sixteenth century seems to have acted as a scavenger in the streets of London. It is about 2 feet long, with reddish-brown plumage. The Honey-Buzzard, found all over Europe, is really a Kite that feeds principally on insect larvæ, especially those of bees and wasps. It is subject to great changes in colour, frequently assuming a blackish hue.

GOSHAWK.

At the head of the True Falcons, distinguished by their strong curved beak, with a tooth in the top fitting into a notch in the lower half, stands the Gyr Falcon, of which there are species or varieties from Norway, Greenland (Plate V., No. 3), Iceland, and Labrador. The plumage is white, more or less marked with black. These birds are larger than the Peregrine Falcon, but not so highly valued for sporting. They prey chiefly on other birds.

The Peregrine Falcon is about 15 inches long, bluish-grey barred with black above, and reddish-white barred with black below. Falconers call the female a *falcon*, the male a *tiercel*, and a bird taken wild in full plumage a *haggard*. The Falcon is very strong on the wing, and its speed has been calculated at more than a hundred miles an hour.

The Goshawk, nearly 2 feet long, blackish brown above and reddish below, feeds on birds and small quadrupeds. It is becoming rare in England, but is sometimes employed to hunt hares, rabbits, and rats. A correspondent of the *Field* recently told how, when two rats sprang out of a rick that was being threshed, his Goshawk "instantly collared the first one, and holding it in one foot, flew after and caught the second one with the other foot, and sat on the ground with a large rat in each foot, to the great admiration of the machine men."

The Sparrowhawk is one of the commonest British birds of prey. It is about 1 foot long, and its plumage brownish-grey above, and white marked with brown below. It flies with ease and rapidity. Brehm tells of one that followed its prey into a railway carriage in rapid motion. In some parts of Asia Sparrowhawks are trained to hunt small birds.

MERLIN.

The Hobby and the Merlin are small British falcons: the former feeds largely on beetles: the latter prefers flesh-food, and is very bold and daring, fearlessly attacking birds much larger than itself.

The Kestrel is fairly common in Britain, and is often called the Windhover, from its remaining poised in the air, head to wind. Kestrels feed principally on mice, lizards, and beetles, but do not disdain small birds. Two were recently shot—right and left—in Essex: one was carrying a blackbird, the other a starling.

The Fishing Eagle, or Fishing Hawk, is about 2 feet long, with brown plumage, varied with black, grey, and white. It is still found in Scotland on the coast, or by the bank of some rivers. Mr. St. John says : "This very beautiful bird drops like a stone on any unlucky fish that her sharp eye may detect, and I believe she seldom pounces in vain Having caught a trout or a small salmon, she flies with it to land, or to some rock, and there tears it up." When fish is not attainable, no dead carcass comes amiss.

FISH HAWK.

The Harriers, of moderate size, have the feathers of the face forming a disk, though not so perfect as in the Owls. They take their prey on the ground. The Hen Harrier and Marsh Harrier are British; they feed on reptiles and amphibians, and are destructive to game. Mr. St. John saw a Hen Harrier strike a Grey Hen, and the head of the victim "was cut as clean off by the single stroke as if done with a knife."

The Owls are distinguished by their soft, fluffy plumage, which enables them to fly noiselessly, the position of the eyes in front of the face, and the disk of feathers which surrounds them. The body-feathers bear no aftershaft, as do those of other Birds of Prey, except the New World Vulture and the Fishing Eagle. In many the legs are feathered to the toes.

The Barn Owl is rather over a foot long, and has buff plumage marked with grey, white, and black. It is common in Britain, and nests in old buildings, barns, and hollow trees. From the nature of its food—rats and mice—this bird is one of the farmer's best friends, but it is ruthlessly shot by gamekeepers in the erroneous belief that it is destructive to feathered game. It is often called the Screech Owl, but its cry is rather a hoot.

The Brown, Wood, or Tawny Owl, with deep grey or reddish-brown plumage, is common in the wooded parts of Britain. It feeds on small birds and field-mice, and does not disdain caterpillars. It is easily tamed, and makes an interesting pet. In defence of its young it is very courageous, and fights fiercely with beak and claws. The *Field* records an instance in which a pair of Brown Owls killed a cat which attacked their nest.

The Snowy Owl, from Arctic and sub-arctic regions, is a day-flier. It is over 2 feet long, and when adult has the plumage quite white. It is an occasional visitor to Britain.

The "Horned Owls" owe their name to the plumicorn or tuft of feather on each side of the head. The Great Eagle Owl (Plate V., No. 6) is a native of Northern and Central Europe, but a rare visitor to Britain. It is about 2 feet long, and has russet plumage, marked with black. The Virginian Horned Owl is closely allied. The Long-eared Owl and the Short-eared Owl are British. The latter was a rare visitor, but since the plague of field-voles in Scotland (p. 161), numbers of these birds have remained to breed.

The Fish Owls are natives of India and Africa. When fish is not procurable they feed on small mammals, birds, and reptiles, and in some cases on carrion.

The Burrowing Owls are American, and live in the holes of the Prairie Dog and the Viscacha. Hudson calls them the most Darby-and-Joan-like of birds, for the male is never seen without his mate.

BARN OWL.

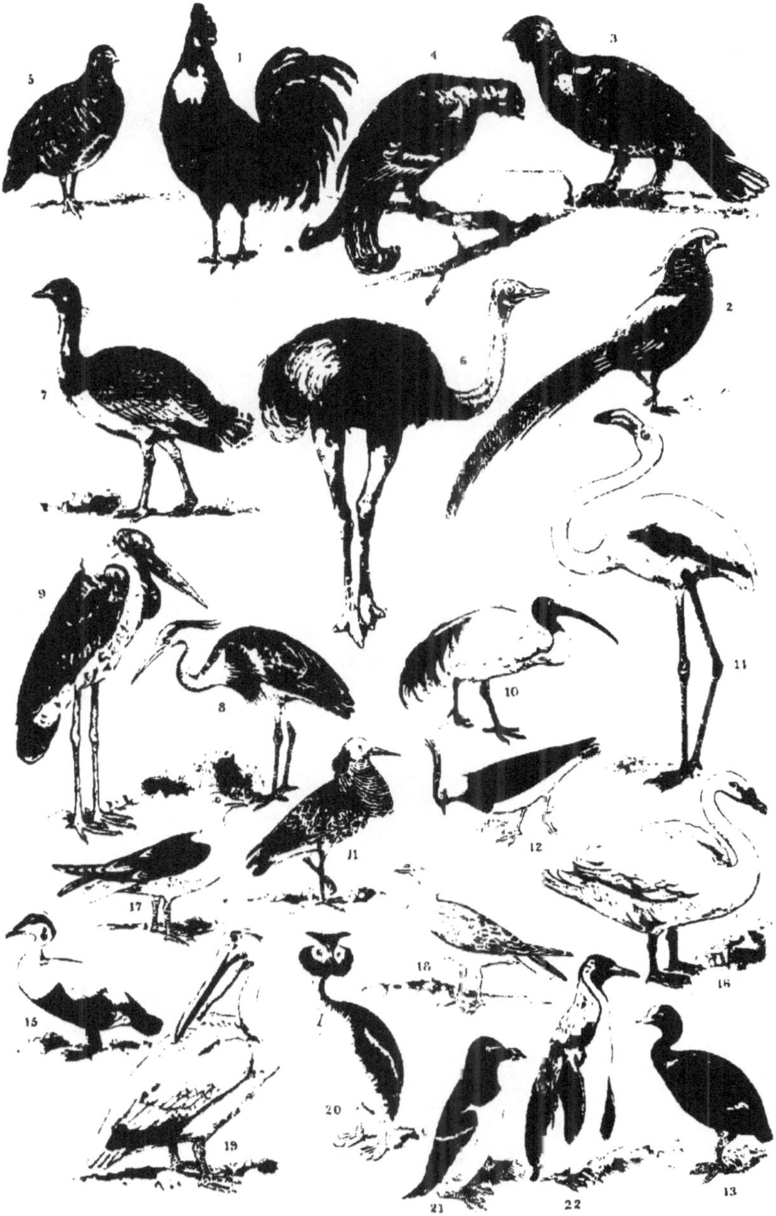
5
1
4
3
7
6
2
9
8
10
11
12
17
11
15
18
16
19
20
21
22
13

CHAPTER XX.

WADING AND SWIMMING BIRDS.

 HE distinguishing marks of the Wading birds are their long legs and toes. The latter are four in number, three in front and one behind, rarely webbed. They frequent the shore or thick vegetation near the banks of rivers and lakes, and small aquatic animals constitute the bulk of the food of many of them. The young shift for themselves as soon as they leave the egg.

The Rails have short wings that fit close to the body, and can thread their way easily and swiftly through reeds and long grass.

The Jacanas, from the warmer countries of both hemispheres, have very long claws, which spread their weight over such a large surface that they can walk with ease over floating vegetation, a lily pad affording them ample support. The Pheasant-tailed Jacana, or Water-Pheasant, owes its name to its long tail.

The True Rails have short claws, though the toes are long. These birds are very numerous and widely distributed. In Britain we have several species. The Water-Rail, rather less

LAND-RAIL.

than a foot long, is brown, marked with black above, lead grey on the neck and breast, and has the black flanks barred with white. It is a marsh-loving bird, frequenting by choice water thickly fringed with reeds, rushes and rank grass, in which it can skulk, or through which it can slip into the stream and swim to a place of safety. The Land-rail, or Corn-crake, is rather smaller, with yellowish-brown plumage marked with black above, and buff fading to white beneath; the flanks are barred with brown. These birds reach our southern coasts towards the end of April, and most of them have left again

Q

by the end of September for their winter-quarters in Africa. The Corn-crake is very shy, and much more often heard than seen. The note of the male, whence the name "crake" is derived, resembles the

MOORHEN.

noise made by drawing the nail along the teeth of a comb. Like some other Rails, the Corn-crake will feign death when taken or hard pressed. The Spotted Crake, the Little Crake, and Baillon's Crake are also visitors, but the last two are rare.

The Moorhen, a little more than a foot long, with brownish plumage above and dark grey below, has the base of the bill carried up on to the forehead, there forming a "frontal plate." It frequents ponds rather than running streams, resorting to the latter chiefly when the standing water is frozen over. There are two and sometimes three broods in the season, and the birds of the first will help the parents in nest-building and in caring for the second brood. Allied to the Moorhens are the Purple Gallinules, showy-looking birds with brilliant blue metallic plumage, contrasting strongly with their red legs and frontal shields.

The Coot (Plate VII., No. 13) lives with us all the year round, and differs from the Moorhen by its larger size, lobed feet, and the white patch on the forehead, which has given rise to the proverb "As bald as a coot." In winter these birds may be found on the coast. The Finfoots, from Asia, Africa, and America, have lobed feet like the Coots, but the bill is long.

In the Snipe family the bill is long, slender, and flexible, and with it the birds probe mud and damp earth for worms, insects, and molluscs. Many are highly valued for the table.

FOOT OF COOT.

The Woodcock is a winter visitor, generally leaving in spring, though some remain to breed, and where there are plantations of pine and fir. and larch, the number increases year by year. The length is about 14 inches, and plumage reddish-brown with small black markings. These birds lie low during the day, the plumage harmonising well with their surroundings of dead and dying leaves, and come out at dusk to feed. There is but one brood in the season, and the old birds carry the young from place to place. The call-note of the male is a loud "whirr."

The Common Snipe is a native bird, and large numbers also arrive in October and November, leaving again in the spring. It is a little more than half the size of the Wood-cock, and has mottled plumage. The "drumming" or "bleating" of the Snipe, which serves as a call-note, is produced by the wings and the out-spread tail-feathers. The Great Snipe and the Jack-Snipe are visitors.

The Curlew, about 2 feet long, with mottled plumage of brown and white, is a shore bird, with a long, curved bill. The Whimbrel, much smaller, is a visitor, sometimes called the May-bird, from the month in which it arrives.

WOODCOCK.

The Sandpipers are shore-birds with straight bill, and the toes joined at the base by a fold of skin. Here belong the Common Sandpiper, the Knot, the Dunlin, the Greenshank, the Redshank, the Curlew Sandpiper, and some others. Perhaps the most interesting is the Ruff (Plate VII., No. 11), which owes its name to the frill of feathers developed on the head and neck in the breeding-season. It serves as an ornament to attract the females (called Reeves, and having no such adornment) and as a defence in his battles with his rivals. These birds formerly bred in England, but now they are known only as visitors, chiefly to the eastern counties. The length of the male is about 1 foot, and the "ruff" is variously marked with black,

SNIPE.

shades of brown, grey, and white, no two individuals being alike in this respect. How great the difference is in the ornaments of these

Q 2

birds may be seen in the specimens exhibited in the Great Hall of the Natural History Museum.

The Phalaropes, from the northern parts of both hemispheres, are Sandpipers, with feet like those of the Coot.

The Stilt Plovers are distinguished by their long legs. The True Stilts have the bill straight; in the Avocets it has an upward curve. The Avocet, formerly numerous in England, has not bred here for many years. The plumage is black and white; the feet are webbed, but the bird does not take to water unless compelled. It feeds on small aquatic animals, which it scoops up from the mud with a sidelong motion of its bill.

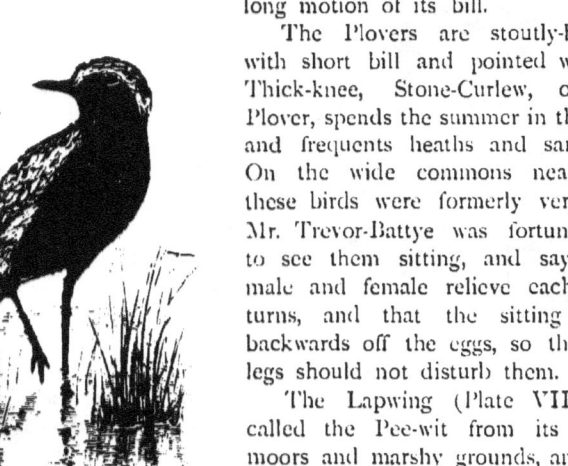

GOLDEN PLOVER.

The Plovers are stoutly-built birds, with short bill and pointed wings. The Thick-knee, Stone-Curlew, or Norfolk Plover, spends the summer in this country, and frequents heaths and sandy places. On the wide commons near Thetford these birds were formerly very plentiful. Mr. Trevor-Battye was fortunate enough to see them sitting, and says that the male and female relieve each other by turns, and that the sitting bird rises backwards off the eggs, so that its long legs should not disturb them.

The Lapwing (Plate VII., No. 12), called the Pee-wit from its cry, loves moors and marshy grounds, and is widely distributed in the British Islands. Its eggs are collected and sold as "Plovers'" eggs. When disturbed while sitting the hen-bird steals away, and the male bird does his best to attract the attention of the intruder; but when the eggs are hatched she joins him in his endeavours, and both will often feign lameness—a trick generally successful, for one is sure to be led far from the nest if the birds be pursued. The Grey Plover, the Golden Plover, the Ringed Plover, and several more, are well known, and the first two are highly valued for the table.

The Turnstone, with varied plumage of black and white, is a summer visitor, and frequents the sea-shore and sometimes the banks of streams. These birds derive their name from the fact that they turn over stones with their bill in search of insects, which form their chief food. When the stones are too heavy to be moved with the bill, they push with their breast, and Mr. Edward says "it would seem that the birds are willing to assist each other, just as masons or porters will do in turning over a stone or a bale of goods." He once saw two Turnstones

TURNSTONE.

busily endeavouring to turn over a dead fish, fully six times their size. They were boldly pushing at the fish with their bills, and then with their breasts. Their endeavours, however, were in vain; the object remained immovable. On this they went round to the opposite side, and scraped away the sand from beneath the fish, and they carried on their mining operations till they were able to turn the fish over.

The Oyster-Catcher lives with us all the year round. Its boldly marked black-and-white plumage and long, orange bill make it a very conspicuous object on the seashore, where it picks up its living by probing among the stones, or in the pools, and scooping out limpets and mussels from their shells.

The Bustards are confined to the Old World. The Great Bustard (Plate VII., No. 7) was formerly abundant in Britain, but since about 1840 it has only occurred as a visitor. The *breks* or sandy commons of Norfolk and Suffolk were the last places from which it was driven, and even now the appearance of a straggler from the Continent will generally bring out all the guns in the neighbourhood. The length is about 45 inches, and an old cock-bird will weigh 30 lbs. The Little Bustard, a much smaller bird, sometimes strays to this country. The

OYSTER-CATCHERS.

Australian Bustard is protected in Victoria as an insect-destroying bird.

Cranes are natives of warm countries, and make long migrations.

The legs are long and slender, and in many the feathers are plume-like. They are noted for their extraordinary caperings in their love-display, always a great source of amusement to lookers-on. The Common Crane, about 4 feet in length, with slate-grey plumage, is a rare visitor to Britain. In the sixteenth century it bred in the fen-lands, but the drainage of those parts deprived it of food and so put an end to its regular visits. It feeds on grass, grain, seeds, and insects. The Numidian Crane, or Demoiselle, and the Balearic Crane, with a fan-like crest, are generally to be seen in zoological gardens.

The Sun Bittern, or Peacock Heron, from South America, with brilliant plumage, is kept as a pet in Brazil. Bates says: "It soon

becomes tame, and walks about the floors of houses, picking up scraps of food, or spearing insects with its long, slender beak."

The Kagu, from New Caledonia, with loose bluish ash-coloured plumage, is described by Wallace as partaking something of the appearance of Rail, Plover, and Heron.

The Trumpeters, from the Amazon Valley, are named from their cry.

The Çariamas, or Crested Screamers, from the mountains and

NUMIDIAN AND BALEARIC CRANE.

open plains of Brazil and La Plata, in some respects resemble the Birds of Prey. There are two spurs on the wings, and one of the two species has a horn-like protuberance on the head. They are domesticated, and run with the poultry, which they defend from rapacious birds.

HERONS.

The birds of this order have the bill hard and horny. They generally nest in trees, and the young need the care of the old birds for some time after they are hatched. The True Herons have the hind toe large and resting on the ground.

The Common Heron (Plate VII., No. 8) is found throughout the British Islands. These birds breed in colonies, called heronries, more common in England than in Scotland or Ireland. Herons feed on fish, reptiles, frogs, water-voles, mice, and young water-fowl ; and, failing these, on snails, slugs, and insects. They may be often seen standing in the water watching for any prey that may come within reach of the sharp spear-like bill. Some allied species are visitors to Britain ; among these is the Night Heron, which would probably breed here if undisturbed.

The Egrets are distinguished by their white plumage, and in the breeding season long crests and feathery plumes are developed. The great White Egret, a rare visitor to this country, was formerly common in the valley of the Danube, but its numbers have been greatly reduced, these birds being taken for the sake of the plumes.

The Bittern is now a visitor which has no chance of breeding, being generally shot as soon as it makes its appearance. Its length is from 27 to 30 inches, and the buff plumage with small black markings matches well with the reed-beds which these birds frequent, and in which they remain during the day, coming out at night to feed. It is dangerous to approach a wounded bird, for

BITTERN.

it will throw itself on its back and fight fiercely ; and the long, sharp bill and strong claws are capable of inflicting serious wounds.

The Little Bittern is also a visitor, but occurs more rarely. Like the larger species, its plumage is highly protective, as is that of the American species. An American bittern, started by Mr. Hudson, flew into a reed-bed. He searched for nearly an hour, but did not observe that the bird was sitting on a reed 2 feet from him, having postured itself so that it exactly resembled the branch of a reed. Mr. Hudson tested the instinct of the bird by walking round and round it, and found that it in like manner turned round on its perch, always keeping its reed-coloured breast to the enemy, and its long bill shot up like a spike in the air. He actually forced the bill down with his hand : the bird did not fly away ; but resumed its original position when his hand was removed. Finally it took an opportunity to slip off, like a flash of lightning, to another reed-bed, and it was a quarter of an hour before it was discovered.

The White Stork was formerly a much more frequent visitor than it is in these days. It is common on the Continent, whither it comes from its winter quarters in Asia and Africa to breed; and in Central Europe the people protect it and encourage it to nest by setting up cart-wheels or rough platforms, on which it may place the huge structure of twigs and sticks on which the eggs are deposited. It also builds in trees and towers, and on the ledges of cliffs. Canon Tristram says that year after year—indeed, generation after generation—a

STORK'S NEST.

pair of birds return every spring to the same place, and rebuild or repair the old nest. Insects, worms, snakes, frogs, mice, and birds, form its diet, and the old birds insert their bills in the mouth of the young ones, and feed them with half-digested food, in somewhat similar fashion to that adopted by the Pigeons. In some parts of Europe the Stork is looked upon as a public scavenger, and protected accordingly, as some of the larger species are in India.

The Black Stork is a much rarer bird in this country than the White Stork: only about a dozen specimens have been met with since the first was noticed, in 1814. The plumage of the upper surface and neck is black, with brilliant metallic reflections, and the under parts white. It feeds chiefly on fish, and nests in high trees, generally near water. One of these birds lived in the Zoological Gardens, Regent's Park, for thirty years.

The Adjutant (Plate VII., No. 9) is a gigantic Indian Stork, 5 feet or more in height. It has a long pouch on the neck, and another, which is inflated when the bird flies, on the back of the neck. Dr. Jerdon says that in Calcutta and other large towns these birds are protected by law, and become so tame as to walk about fearlessly, lurking near the slaughter-houses and burning-grounds of the Hindoos, or examining the refuse heaped up in the streets, ready to be carted away by the scavengers. Fish, frogs, and small mammals form part of their diet. In the crop of one which he shot, Mr. Hornaday found a dog, the weight of which he estimated at about 5 lbs. There is an African species, and the tail-feathers of these birds are sold as Marabou plumes for ladies' head-dresses.

The Jabirus, or Giant Storks, are found in Africa, America, and Australia; the Wood Ibises in Asia, Africa, and America; the Boat-Bill lives in tropical America, and the Shoe-billed Stork on the Upper Nile.

The Spoonbill used to breed in Norfolk, "on the tops of high trees," which are favourite nesting-places of the birds in India and Ceylon. In Holland it breeds among the rushes. Its length is about 32 inches, the plumage is white, and the crest tipped with

SPOONBILL.

yellow. A South American species has the plumage tinged with rose-colour.

The Sacred Ibis (Plate VII., No. 10) about 30 inches long, with an arched beak like that of the Curlew, is a native of Africa. It is now rare in Egypt, where it was formerly a sacred bird, worshipped as the herald of spring. Figures of it occur on the ancient monuments of that country, and a pyramid is dedicated to it. The Scarlet Ibis, from America, is a very beautiful bird. The Glossy Ibis, with dark metallic plumage, is an occasional visitor to Britain.

The Flamingo (Plate VII., No. 14) is linked to the Waders by its long legs, and to the Geese by its webbed feet and beak furnished with straining-plates. These birds are natives of Asia and Africa, and some parts of the south of Europe. In feeding, the bill, with the top part undermost, is plunged into the mud, which is stirred up with the feet. The small aquatic animals are retained, and the refuse allowed to flow away through the straining plates on the edges of the bill. Flamingoes build nests of mud, high enough to avoid danger of flooding, and sit on their eggs just like any other bird.

GEESE AND WATERFOWL.

These birds are web-footed, and the legs are set far back, which, while it increases their power as swimming organs, makes them of less use for walking; hence the gait of Geese and their relations on land is an unsteady waddle. The bill is furnished at the sides with a fringe of plates, serving to sift the mud in which these birds seek a great part of their food worms, insect larvæ, crustaceans, and small molluscs. There is an under garment of down, and the upper feathers lie close and thick, and are almost waterproof. Everybody knows how easily water runs off a Duck's back.

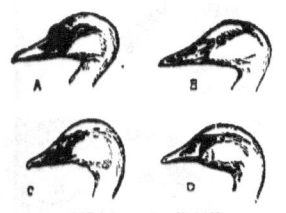

HEADS OF SWANS.
A. Mute Swan. B. Whooper Swan.
C. Bewick's Swan. D. Polish Swan.

Grey Geese and Black Geese, so called from the prevailing hue of their plumage, are among our winter visitors. The Grey Lag Goose, one of the rarest, is the stock whence the White Domestic Geese of our poultry-yards and commons is descended. Why they assumed a white coat no one seems able to tell us; but there would be no difficulty in keeping up the colour, and there is a motive for keeping it up, for white feathers fetch a higher price than grey ones. The Bernacle Goose and the Brent Goose both belong to the "Black" group; they are common winter visitors, often appearing in flocks in severe weather. Both these birds were formerly believed to be hatched from barnacles, attached to floating timber, and more than one writer has described the tiny birds which he had seen in the shells of these crustaceans. The Snow Goose, from North America, has strayed to Ireland some few times. One was trapped, and put with some tame geese. Mr. Harting says that "this bird, after slaying a rival in a fair fight, paired with one of the common geese, and afterwards assisted to rear a family of goslings."

The Swans are the largest birds of the order, and are known to everyone for their graceful movements in the water. The Mute Swan (Plate VII., No. 16), with which we are familiar as a half-domesticated bird adding beauty to our lakes and rivers, is found wild in many parts of Europe. At Abbotsbury is a swannery, with many hundred birds; they breed freely in the Backwater at Weymouth, on the Exe, on the Norfolk Broads, and in many other suitable places. The Whooper, Bewick's Swan, and the Polish Swan are winter visitors, the former by far the most plentiful. The plumage of all these birds is

MALLARD.

white, and for ages it was thought that no Swan could be dressed in any other colour. Australia, however, has a Black Swan, and the Southern American Swan, with white body-plumage, has a black head and neck.

The Mergansers are large fish-eating sea-birds, having the edges of the bill armed with saw-like teeth, the points of which are turned backwards, so as to prevent the escape of their slippery prey. They are British visitors, and some remain to breed. From the nature of their diet their flesh is rank. The stumps of trees near water are favourite nesting-places, and of the common species, the Goosander, we are told that "as soon as the eggs are hatched, the mother takes the chicks gently in her bill, and carries and lays them down at the foot of the tree, where she teaches them the way to the river, in which they instantly swim with an astonishing facility."

Of the Freshwater Ducks the most important is the Mallard or Common Wild Duck, from which are descended all our domestic breeds. Some of these birds still nest in this country, but the drainage of the fen lands has destroyed many of the old breeding-places. Large numbers, chiefly from Northern Europe, visit us in winter, when many are shot for the table and market.

Under the general name of Wild Duck are also to be reckoned the Gadwall, or Grey Duck, the Shoveller, with its spoon-shaped bill, the Pintail,

SHOVELLER.

the daintily-marked Teal, and its near relation the Garganey, and the Wigeon. All breed more or less freely in Britain or Ireland. Allied to these are the American Summer Duck and the Mandarin Duck, both of which are domesticated in this country as ornamental birds.

The Sheldrake, or Burrow Duck, is a large goose-like bird, of brilliant plumage, in which green, chestnut, white, and black are broadly marked. It breeds in sand-hills and rabbit-burrows on the East Coast. In Orkney it is called the "Sly Goose," because it feigns lameness to decoy intruders away from its nest, and when it has led them a safe distance it spreads its wings and takes to flight.

The Eider Duck (Plate VII., No. 15) is the most important of the Sea Ducks. It is a native of some parts of Britain, and breeds from Northumberland northwards. It breeds also in some parts of the north of Europe, where it is protected by law. Selby, in his account of the breeding of these birds at the Farne Islands, says : "The nest is composed of fine seaweed, and as incubation proceeds, a lining of down plucked by the bird from her own body is added; this increases from day to day, and at last becomes so considerable in quantity as to envelope and entirely conceal the eggs from view." In Iceland the yield of each nest is between two and three ounces of down, which is worth there from twelve to fifteen shillings a pound, and about a pound and a half is required, according to Mr. Howard Saunders, to make a single coverlet. The Icelanders pickle the eggs, and the Greenlanders eat the flesh of this bird. The King Eider and Steller's Eider have occasionally strayed to Britain from their Arctic home.

SHELDRAKE.

Other Sea-ducks that visit Britain are the Pochards, the Scaup-Duck, the Tufted Duck, the Golden Eye, the American Buffel-headed Duck, the Long-tailed Duck, the Harlequin Duck, and the Scoters.

PELICANS.

In the birds of this order all the four toes are connected by a web. True Pelicans are found in tropical and temperate regions in both hemispheres, and may be readily known by their long bill, from the lower part of which there hangs down a large pouch, not noticeable when the birds are at rest. This pouch is capable of great expansion, and serves as a bag to stow away the fish, which is not eaten when caught, but is brought on shore to be devoured. The Common Pelican (Plate VII., No. 19) from Asia, Africa, and the south and east of Europe, is common in zoological collections. Its habits are pretty regular: a fishing excursion in some shallow in the early morning, flight to land, where the catch is devoured, a long doze, another turn

EIDER DUCKS AT HOME.

at fishing in the evening, supper, then to roost, make up the ordinary day of a Pelican. There are several species, of which but one, the American Pelican, can dive for its prey.

The Darters, of which there are four from South America, Africa, India, and Australia, haunt the banks of rivers and lakes. They generally perch on trees overhanging the water, and dart down upon fish as they pass by. They are also excellent swimmers, as anyone will allow who has seen a Darter in its tank in the Zoological Gardens at feeding time.

Cormorants and Gannets, of which there are many species, are found on every coast. The Common Cormorant about 3 feet long with sombre plumage, is a familiar British sea-bird, frequenting by preference rocky parts of the coast. They feed chiefly on fish, of which they devour a great quantity; and though they have no pouch like the True Pelicans, the gullet is so elastic that they can swallow prey of considerable size. A fish 14 inches long, the body of which was 4 inches deep, was taken from the gullet of a Cormorant shot in Plymouth Sound. Their skill in fishing was formerly utilised in this country, as it still is in China, where the birds, usually with a collar round the neck to prevent their swallowing, are sent into the water to fish for their master, being afterward, as a reward, allowed to fish for themselves. Of two that bred in the Zoological Gardens, Mr. Howard Saunders says that the task

DARTER.

of feeding the young ones was undertaken entirely by the male bird. "After he had been fed and retained the fish about an hour, he mounted the side of the nest, and as each young bird came out from under the hen, he opened his great mouth, and in went the nestling as far as the outstretched wings would allow, and helped itself to the now macerated fish in the old one's crop."

The Shag is a smaller British species with green plumage, differing little in habit from the larger bird, excepting in rarely venturing inland or frequenting fresh water.

The Gannet or Solan Goose, breeds on Lundy, on the Scotch coast, where its chief station is the Bass Rock, and on some parts of the Irish coast. The plumage of the adults is white, that of the young till their fifth year is dusky, but more and more white is gained each year. These birds swoop down from aloft on their prey.

The Booby is a southern species from both hemispheres. Mr. Forbes saw them in the Keeling Islands, and remarks that they were shamelessly robbed by the Frigate-birds, who took advantage of their industry. When sailing landward with their fish-supper, "they were often seized by the tail by the Frigate-birds, and treated to a shake that rarely failed of satisfactory results."

The Frigate-birds are found between the tropics, and are noted for

GANNETS.

their powers of flight. The Cocos Islanders eat them, and train tame birds to act as decoys for their fellows. The sportsman takes out his decoy-bird, and throws some offal on the water, over which the tame bird plays, without touching it. Soon the attention

GULLS ON THE THAMES IN LONDON.

of the wild birds is drawn to the bait; and as they swoop down, away flies the decoy, leaving its master free to fire.

GULLS, TERNS, AND PETRELS.

These birds have only the three toes in front webbed, and the hind toe is small and raised above the ground. The wings are long, and the flight consequently powerful. The general plumage is grey above, and white below; but the clothing of some is dusky.

The Skimmers are tropical birds, about the size of a pigeon, with the under part of the bill longer than the upper. They procure their food by skimming along the top of the water, and so picking up the small marine animals that live near the surface. With the bill they also scoop out molluscs from their shells.

The Gulls are a numerous group, some of which are found in all seas and round all coasts. The Herring Gull (Plate VII., No. 17), about 2 feet long, is a very common British bird, especially on the South Coast. These Gulls feed on fish, which they take hovering over the water, and do not despise shell-fish, or even offal that may be picked

NOSTRILS OF VARIOUS PETRELS.

up on the shore. In winter and spring they travel in bands over the fields searching the pastures and ploughed land for worms, grubs, and insects. The habit of feeding on land is shared by the Black-headed Gull, especially in the breeding season, whence it is looked upon as a farmers' friend. There are many other species resident in, or visitors to Britain.

The Skuas, or Robber Gulls, well deserve their popular name, for though they can fish for themselves, and will eat birds and their eggs, and small mammals, they prefer to live upon the labour of other diving birds, by forcing them to drop what they have caught, swooping down upon it and snatching it up before it falls into the sea.

ALBATROSS.

The Terns, or Sea-swallows, are extremely elegant birds, owing their popular name to their forked tails and their powers and mode of flight. Everyone who has been at the seaside must have seen them skimming over the water or hunting the rock-pools on the shore.

The Petrels are marked off from all other birds by their tube-like nostrils above the bill. They are called Petrels, or Peter's Birds, because some of the smaller species appear to walk on the water.

At the head of this group is the Albatross, remarkable for its great size and its powers of flight. The Wandering Albatross, from the Southern Ocean, is nearly 4 feet long, and has white plumage with some black on the wings. Gould says that "although during calm or moderate weather it sometimes rests on the surface of the water, it is almost constantly on the wing, and is equally at ease during the stillest calm or the most furious gale."

GREAT NORTHERN DIVER.

Several Petrels visit Britain — the commonest is the Fulmar (Plate VII., No. 18) which breeds in St. Kilda, where its flesh and eggs are eaten, and the oil obtained from it burnt in lamps. Its home is in the Arctic regions, and it is always a constant attendant on whale-ships for the sake of the offal which falls to its share when a whale is cut up. Scoresby says that when carrion is scarce Fulmars follow the living whale, and sometimes, by their peculiar motions when hovering at the surface of the water, point out to the whaler the animal of which he is in pursuit. The Stormy Petrel breeds in a few places in Britain, and in Sark.

DIVING BIRDS.

In all these birds the feet are placed very far back, so that when standing they are nearly upright. They differ from the Penguins in being able to fly. Three species of Divers visit Britain in winter, and of these the Great Northern Diver, or Ember Goose, is the commonest. It is a very hand-some bird from 30 to 33 inches long, and the general plumage above is black, with white spots, and white below. In all the beak is long, straight, and pointed. They live mostly in the open sea feeding on fish, for which they dive, and rarely coming on land except in the breeding season. Montagu had a tame one, which would come at call and take food from his hand. Unlike the Guillemot and Auks, these birds do not use their wings in swimming and diving.

RAZOR-BILL.

R

The Razor-bill (Plate VII., No. 21), about 17 inches long, is a well-known British bird, and has many breeding stations round our coasts. A writer in the *Zoologist* describes how the mother teaches the young

bird to dive by taking it under with her. " Up comes the young one again, only to get another dose ; but the young one cannot remain so long under water as the mother, and it often dodges her by diving for an instant." The Guillemot, rather larger than the Razor Bill, is more abundant in Britain. Both birds have the same habits, and they breed in the same localities.

GUILLEMOT.

The Puffin, or Sea-parrot, is a droll-looking bird, about 1 foot long, black above and white below, moulting not only its feathers but its bill, which is larger and more gaily coloured in summer than in winter. Puffins breed in holes, frequently using rabbit-burrows, and driving out the rightful owners.

The Grebes are Diving Birds, principally frequenting fresh water. They can dive well, and their feet, though not webbed, are admirable swimming organs. The Great Crested Grebe (Plate VII., No. 20) is a British bird, and has a wide range over the Old World. From the silvery-white plumage of the under-surface, ladies' muffs are made. The Dabchick, or Little Grebe, is the smallest and commonest British species. It seems to be a greedy feeder, for specimens are often found choked by the Bullhead or Miller's Thumb. The like fate often befalls Kingfishers

PUFFIN.

PENGUINS.

These birds represent in the Southern hemisphere the Auks and Guillemots of the North. The wings are useless for flight, but are employed as fins in the water; and on land, when alarmed, Penguins will throw themselves down, and push themselves along with feet and wings so rapidly that a man would have difficulty in overtaking them. The wing-feathers are short and scale-like, and when the birds moult these flake off like the shedding of the skin of a serpent. The King Penguin (Plate VII., No. 20) is found on the coasts of the South Pacific. The Jackass Penguin of the Falkland Islands is described by Darwin as "crawling, it may be said, on four legs, so that it might be easily mistaken for a quadruped, or moving with such agility in the water as to resemble a fish leaping." Those who have seen these birds scrambling on the floor of the Fish-house of the Zoological Gardens, Regent's Park, and diving in their tank for fish, will have some idea of their mode of progression on land and in the water.

PENGUINS.
(*From a Photograph.*)

R 2

OSTRICHES ON A FARM.

CHAPTER XXI.

THE RUNNING BIRDS.

TO this order belong the largest living birds. The wings are small and quite useless for flight, though they may be extended and raised so as to form a kind of sail, and help their owners in their course, thus adding considerably to the speed with which these birds get over the ground. This habit is referred to in the Book of Job (xxxix. 18), where the Ostrich is described as "lifting up herself on high, and scorning the horse and his rider." The breast bone has no keel; the plumage is hair-like, and in the feathers of the Ostrich and Rhea, there is no plumule or aftershaft, as there is on those of the Emu and Cassowary.

The Ostrich (Plate VII., No. 6) is confined to the plains and deserts of Africa and to some parts of Western Asia. Its height is from 6 feet to

8 feet; the head, neck, and thighs are naked, and it has only two toes. The plumage of the body in the males is black, and in the females ash-grey; the wing and tail plumes in both sexes are white, and it is the plumes from the wings of the male that are most highly prized. For the sake of these feathers Ostriches have been partially domesticated, and are now bred on Ostrich farms, where the feathers are regularly cut from the time the birds are about eight months old; and the operation is repeated at intervals of eight months, except during the breeding season. In the wild state the Ostrich takes several mates, which have one nest between them, and the male sits during the night.

Ostriches are exclusively vegetable feeders, but swallow stones to assist the process of digestion. In captivity they have a bad habit of taking and swallowing whatever is offered. One that died in the Zoological Gardens had ninepence halfpenny in bronze money in its stomach, given it by visitors who certainly had more money than sense.

CASSOWARY.

These birds soon become tame and learn to recognise their keepers. If interfered with or irritated they defend themselves by kicking, and no one would care to be kicked twice by an Ostrich. On the African farms the people who tend the birds carry the branch of a native thorn, the dexterous use of which will keep the birds off. Mrs. Martin tells of an Ostrich farmer who believed all animals could be quelled by the human eye, and tried the experiment on one of his own birds. " He was presently found in a very pitiable predicament, lying flat on the ground, whilst the subject of the experiment jumped up and down on him, occasionally varying the treatment by sitting upon him."

The Rheas, or South American Ostriches, of which there are three species, have three toes, and the neck and head are feathered. They are found from Brazil to Patagonia, and are much smaller than the True Ostriches from Africa. The plumage is brown or grey above and lighter below. These birds are hunted for their plumage, and their stomachs, dried and powdered, are said to be a cure for indigestion. The Gauchos used this medicine long ago. Hudson says that " science has gone over to them, and the Ostrich hunter now makes one profit

from the feathers, and one from the dried stomachs, which he sells to
the chemists of Buenos Ayres. Yet he was once told that to take the
stomach of an Ostrich to improve his digestion was as wild an idea as
it would be to consume birds' feathers in order to fly." The Rheas
take readily to water and swim well, as do the Emus.

The Emus, of which there are two species, are confined to Australia.
Large specimens may attain a height of 6 feet. The head and neck
are feathered : there are three toes, all armed with claws, and the
general plumage is brown. These birds pair, and the male takes part
in sitting. The flesh of the hind-quarters resembles beef, and on this
account the Emu is hunted ; its
eggs are also valued for food.

KIWI OR APTERYX.

The Cassowaries, of which
there are eight species from New
Guinea and the South Sea Islands,
and another in North Australia,
may be known by the horny
helmet on the head and the
brightly coloured wattles. The
neck is short, and the average
height is about 5 feet. The plumage
is long and hair-like, and brownish-
black in colour. Like the Emus,
Cassowaries pair, and both sexes
sit on the eggs.

The Tinamous are small birds,
not much unlike Partridges in
appearance, and are confined to
South America. The flesh is valued for food, and shooting them is
a favourite form of sport.

The Kiwis, or Wingless Birds of New Zealand, have short legs and
neck, and a long bill with the nostrils at the tip. There are four toes,
three in front and one behind, armed with claws which can be used
with effect if the birds are hard pressed, though they generally seek safety
in flight. The Large Grey Kiwi is about 2 feet high, the other three
species being about the size of a domestic fowl. Of the habits of these
birds when at liberty little is known beyond the fact that they pass the
day in holes in the ground or among the roots of trees, coming out at
dusk to seek for the worms and insects on which they feed.

CHAPTER XXII

REPTILES and Birds, though so unlike in appearance, are closely allied and are generally classed together in one group. Reptiles are also connected with the Amphibians lower down and the Mammals higher up in the scale of life ; whence it has been thought that the upper part of the genealogical tree might be represented thus—Y—where the upright stroke stands for the class Reptiles, and the two arms for Birds and Mammals respectively.

The skin is covered with scales or bony plates ; there may be two pairs or one pair of limbs, or they may be absent, or represented by small bones. The blood is cold—that is, very little warmer than the air—and the young are produced from eggs ; in some few cases hatched with the body of the parent. Reptiles never breathe by means of gills.

TORTOISES AND TURTLES.

Some of the members of this order live on the land, others in fresh water, and others, again, are marine. All have four limbs, and the body is enclosed in a kind of box, the rounded upper part of which is called the carapace, and the flat lower part the plastron, or breastplate. There are no teeth, but the edges of the jaws are sheathed in horn, constituting a sharp beak, capable of inflicting a severe bite.

The Leathery Turtle is a native of the Atlantic Ocean and of the Mediterranean. It differs from all the rest of the order in having the carapace soft and leathery. The skin of the back is raised in ridges running from the head to the tail, and this has probably given rise to the story that from the empty shell of this animal the Greeks derived the idea of the lyre. The Leathery Turtle, when full grown, is from 6 to 8 feet long. Its flesh is said to be unfit for food, and, if eaten, to produce serious results. Specimens have been washed ashore on our coasts.

The Land Tortoises have some small representatives in Europe, of which the best known is the Common Tortoise, from the southern parts of the Continent. Leaves, grass, and roots form its food, and, like its fellows, it takes a long winter sleep. Like the rest of its family, it has club-shaped feet, with blunt claws.

Gilbert White says of the hibernation of his pet tortoise, which had gone to its hole about the middle of the previous November : " The moist and warm afternoon (April 21st) brought forth troops of shell-snails, and at the same juncture the tortoise heaved up the mould and put out its head, and the next morning came forth, as it were, raised from the dead, and walked about till four in the afternoon. This was a curious coincidence, to see such a similarity of feelings between the two house-bearing animals, for so the Greeks called the shell-snail and the tortoise."

Gigantic Land Tortoises are found in the Galapagos Islands and in the Seychelles, but in both localities they are becoming scarce. Dampier, writing of the former, states a fact which will account for the decrease in their numbers since his day: "They are so extraordinarily large and fat, and so sweet, that no pullet eats more pleasantly." Whalers used to call

GIGANTIC GALAPAGOS TORTOISE.

at these islands to take tortoises on board for provisions for their crews; and they are also killed for the sake of the oil obtained from their fat.

It is well to know, on the authority of Mr. Harold Baty, the agent for the lessee of the Aldabra Islands, that "the Administrator of Seychelles, when leasing the Aldabra Group, had a clause inserted by which the lessees are bound to protect the land tortoises. They still exist in fairly large numbers, but are seldom to be seen, as they have a preference for the most inaccessible spots. Large as are the tortoises that exist in the Seychelles, there is every reason to believe that there are some veritable monsters amongst the patriarchs in Aldabra. One of my overseers at Aldabra informs me that several years ago he saw in the thickets at the south end of the island two tortoises nearly seven feet long, and others not much shorter. A very interesting account of the Seychelles, or Aldabra, Tortoise was written by the late General Gordon, who was here for some weeks before his last visit to Khartoum. He mentions one very intelligent tortoise that literally "flattened out" a sucking pig by which it had been much worried. The tortoise raised its huge body as high as it could on its legs and watched. It was not long before the little pig was again running round his foe, but the instant he tried to run underneath, down came the tortoise like a sledge-hammer. Almost every bone in the pig's body was afterwards found to be broken. From Mr. Baty Admiral Kennedy received the two specimens which he presented to the Zoological Society.

The Marine Turtles, which have the limbs fashioned into paddle-like swimming organs, are found in warm seas, and only come to land to deposit their eggs in the sand.

The Green Turtle is the animal from which is made the turtle soup that figures so largely at State banquets. This animal feeds on small marine creatures, fish, seaweed, and zostera, the single flowering plant found in the ocean. It attains a length of from 5 to 6 feet. Mr. Baty says that since the British annexation, the preparation and export of calipee, or dried sea-turtle's fat, from Aldabra has very greatly increased.

The Hawksbill Turtle (Plate VIII., No. 1) yields the "tortoise-

GREEN TURTLE.

shell" of commerce. This animal is seldom more than 3 feet long, and its flesh is not good for food. Darwin says of the method of taking this turtle at Keeling Island : "The water was so clear and shallow that, although at first a turtle quickly dives out of sight, yet in a canoe or boat under sail, the pursuers, after no very long chase, come up to it. A man standing ready in the bow at this moment dashes through the water upon the turtle's back : then clinging with both hands by the shell of its neck, he is carried away till the animal becomes exhausted and secured." Tortoise-shell is the thin covering of the carapace, and to obtain it much cruelty is often employed. The natives of the Chagos Archipelago cover the animal with burning charcoal, which causes the tortoiseshell to curl upwards ; it is then forced off with a knife, and before it becomes cold, flattened between boards. The animal is then allowed to return to the sea, where a new shell is gradually formed, but too thin to be of any commercial value. In the Celebes, whence comes the finest tortoiseshell, the animals are killed, and the carapace soaked in boiling water to detach the plates.

The Loggerhead Turtles belong to this group. Their flesh is not used for food, except by some of the lower races, and their "tortoise-shell" is too thin to be commercially valuable.

The Freshwater Tortoises are amphibious rather than aquatic—that is, they can live as well on land as in the water. Most of them are natives of South America, but some live in Africa, and others in Australia. They feed principally on fish, and have the feet armed with sharp claws. Of this family one lives in the south of Europe, and here also belong the Terrapins of America, valued for food. The Bearded Matamata of South America is about 3 feet long, and has outgrowths of skin on the head and neck. These are supposed to act as decoys, or bait, to fish, which, mistaking them for worms, dart at them, and are themselves snapped up. Here also belong the Alligator Terrapin, with a long, crested tail, and Temminck's Snapper, a specimen of which has lived for some years in a tank in the Reptile House of the Zoological Gardens, Regent's Park.

Some of these are commercially important. Their flesh is valued, as food and oil is prepared from their eggs. Wallace says that of one species alone millions of eggs were used for this purpose every year, and that these reptiles were consequently becoming scarce.

The Soft Turtles, which have the feet webbed and armed with claws, live in or near the tropics of both hemispheres, and are carnivorous in habit. The snout is produced into a kind of tube, so that they can breathe without raising the head out of the water. The Snapping Turtle, from North America, feeds on young alligators, and old alligators return the compliment by devouring Snapping Turtles wherever they can find them. An allied species is said to be fond of young crocodiles.

CROCODILIANS.

These are large carnivorous reptiles, living in fresh water. The body is covered with bony plates; there are four limbs, and the long, compressed tail is an admirable swimming organ. On land their motion is slow and clumsy. The nostrils are at the end of the snout, and the eyes and ears near the top of the head, so that these reptiles can see and hear with the head just above the surface. Moreover, nostrils, eyes, and ears can be closed so as to keep out water when the head is below the surface. Young individuals feed chiefly on fish, while adults lurk near the edge of the water, and seize larger animals as they come to drink, dragging them into the water and drowning them, and then tearing them to pieces and devouring them. Nor do they despise carrion. In 1893 the Bombay officer of health protested against the decision of the authorities of that city to adopt means for the utter destruction of the crocodiles. He said they were the best and only scavengers possible of the water reservoirs in which they dwell, as they cleared away all obnoxious substances, which there were no human means of removing.

Between Crocodiles and Alligators there is not a great difference in appearance, but the head of the former is longer and narrower than that of the latter, and the crest on the tail is larger.

Alligators are found in America, and there is one Chinese species, specimens of which have been exhibited in the Zoological Gardens, Regent's Park. The fourth lower tooth on each side bites into a pit in the upper jaw, and the hind feet are never completely webbed.

The best-known species is the Mississippi Alligator. The fashion of using alligator hides for travelling bags and smaller articles, caused such destruction of these animals, which kept down the voles, that the farmers raised an outcry, and in at least one State of the Union these creatures enjoy a close time.

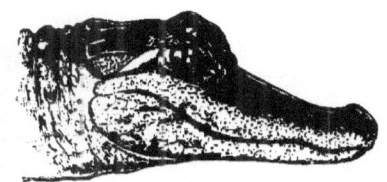

HEAD OF ALLIGATOR.

They are often called Pike-headed Alligators, from the resemblance of the head to that of a Pike. South America has several species.

The Alligator never attacks man when the intended victim is on his guard. According to Bates, it is cunning enough to know when it may attack with impunity. On one occasion, he tells us, when the river had sunk very low, a large one had been seen in the shallow water. Most of the people were very cautious when they went to bathe, but a tipsy Indian sailor, during the greatest heat of the day, when nearly everybody was enjoying his afternoon's nap, went down alone to bathe. He was seen by a feeble old man, who, as he lay in his hammock, shouted to him to beware of the alligator. Before the warning could be repeated the Indian stumbled, "and a pair of gaping jaws, appearing suddenly above the surface, seized him round the waist and drew him under the water. The village was aroused, and the young men seized their harpoons and hurried down to the bank; but, of course, it was too late. A winding track of blood on the surface was all that could be seen. They embarked, however, in their canoes, determined on vengeance; the monster was traced, and when, after a short lapse of time, he came up to breathe

—one leg of the man sticking out from his jaws — was speedily despatched."

The Crocodiles are found in the rivers of Africa, the south of Asia, Australia, Central America, and the West Indies. A few frequent estuaries, and occasionally go out to sea for a mile or two. The fourth lower tooth on each side bites into a groove in the upper jaw, and is visible when the mouth is shut. There are about a dozen species. The Nilotic Crocodile (Plate VIII., No. 2), which may attain a length of from 18 to 20 feet, has been known from very early times. To some of the ancient Egyptians it was a sacred animal, and Herodotus tells us how the priests kept and fed tame Crocodiles, just as in some parts of India the Muggars, or Marsh Crocodiles, are tamed and venerated by the Fakeers at the present day. It was of the Nilotic Crocodile that Herodotus told the story of a bird entering its mouth to rid the huge reptile of the leeches that infested it. The story was doubted for a long time, but now it is certain that the "Father of History" had some foundation for what he wrote. Mr. J. M. Cook, of the celebrated tourist agency, when in Egypt in 1876, "watched one of these birds, and saw it deliberately go up to a crocodile, apparently asleep, which opened its jaws. The bird hopped in, and the crocodile closed its jaws. In what appeared to be a very short time, probably not more than a minute or two, the crocodile opened its jaws, and we saw the bird go down to the water's edge." There were several of these birds about, and Mr. Cook shot two of them, which Dr. Sclater identified as Spur-winged Plovers; so that the question as to what bird enters the mouth of the crocodile is now set at rest.

In habit these reptiles resemble the alligators, and no kind of animal food comes amiss to them. Women and children often fall victims to them, and even a strong man seized by their terrible jaws would be powerless to escape.

Crocodiles are of some commercial value. The skin is tanned for leather, the glands of the lower jaw yield a musky substance, and the fat is rendered into oil.

The Gharials (or Gavials, as they were long called) are found in the Ganges, in some of the rivers of Borneo and of North Australia. They have the snout greatly elongated, with the nostrils in a knob at the end. They feed principally on fish. The Gharial of the Ganges is said to reach a length of 20 feet.

SPHENODON OR TUATÉRA.

This animal, the sole living representative of an order, is of lizard-like form, from 1 foot to 2 feet in length. The upper surface is dull olive-green, spotted with yellow, and the under surface whitish. It is found on some exposed rocky islets on the New Zealand coast, where it burrows among the rocks, and feeds on small animals. These animals may be seen in a small enclosure containing rockwork,

THE CROCODILE'S FRIEND.

just outside the Reptile House in the Zoological Gardens, and so protective is their colouring that one may watch for some time before detecting these creatures among the stones. The New Zealanders have a great horror of this little reptile, and will run away in abject fear if one is shown them. Yet it is quite harmless: it has no weapons of offence or defence, nor energy to use them if it had them. In this animal a third eye, on the top of the head, still represented in Man by the pineal gland, was first detected.

LIZARDS.

The animals of this order are most abundant in warm countries, and absent from the Arctic regions. The body is usually covered with scales, and though fore and hind limbs, bearing clawed digits, are generally present, yet either pair, or both pairs, may be absent. The tail is very brittle, and lost limbs may be regrown. Most of them take animal food, but some are vegetable feeders. The majority live on the ground, some among the branches of trees, a few are more or less aquatic, and one is marine.

The Chameleons are chiefly African, but the Common Chameleon (Plate VIII., No. 3) ranges eastward to Ceylon and northward to Spain. All are slow, arboreal animals, feeding on insects, which they take by protruding the tongue, which is covered with a sticky secretion. The eyes are large, and move quite independently of each other. The tail is prehensile, and the digits are divided into two sets, something after the fashion of the toes of a woodpecker or a parrot. They have the power of changing their colour.

Miss Marianne North says (in "Further Recollections of a Happy Life"): "Another curious inhabitant of Smyrna is the chameleon, which may be seen darting over any old heap of stones in search of flies; but it becomes very stupid and slow in captivity. One was given to me, which afterwards died at Athens, and was buried in the Acropolis. While at sea I used to catch flies for it. At the sight of one it would open its mouth very wide, and very slowly take it in. After a sufficient while I offered another, and again it opened its mouth slowly and wide, when the first fly would walk out again! It was too limp and depressed to swallow its food, poor thing, and died of starvation diet: but it had not lost the power of changing its colour—black as coal on the dirty floor, straw colour in its basket."

The Skinks are a large family, absent only from the Arctic and Antarctic regions. Their scales are smooth, and they frequent dry, sandy places. Some are snake-like in form; others have one pair, and the typical Skinks two pairs, of limbs. The Common Skink, from 6 to 8 inches long, reddish dun in colour with dark bands, is African. It was formerly used in medicine.

The True Lizards, which are confined to the Old World, have representatives in this country. There are five toes on each limb, the tail is generally long, and the tongue is cleft at the tip and extensile. The Land Lizard (Plate VIII., No. 4) is about 7 inches, though larger specimens are recorded, and varies in coloration. Professor Bell found plenty of them round Poole, and they seem to be common on the heaths near Bournemouth. They feed on insects, and are preyed on by snakes and adders. One now in the Natural History Museum was swallowed by an adder, which was killed and put into a bottle, its captor being ignorant of what was in its stomach. The next day, spirit was poured in to preserve the adder, when the lizard crawled out of its mouth, only to be drowned in the spirit. The Viviparous Lizard, which brings forth its young alive, is also British, and extends to Ireland. Professor Bell says that its chief diet consists of flies, though it will eat beetles, crickets, and grasshoppers.

The Green Lizard, common on the Continent, is found also in the Channel Islands.

Simony's Lizard, from the rock of Zalmo, at the east end of the island of Ferro, is interesting on account of its habitat. A specimen was exhibited for the first time in the Zoological Gardens in 1891. The natives said that these animals lived on crabs, but the lizard brought to London had no taste for such diet, and did very well on raw meat and fish, with grapes and bananas for dessert.

The Teguexins are natives of the New World. Some of them live in marshy places and take to water freely. The Common Teguexin is a large lizard, with a wide range in South America. It frequents the banks of rivers, and though it does not willingly go into the water, its long, slender tail makes a capital swimming organ. Fruit, insects, small reptiles, birds and their eggs constitute its diet ; and in captivity it is fed principally on meat. This animal was formerly supposed to give notice of the approach of wild beasts by a warning sound ; but there is as little truth in the story as in a similar one told of the Monitors. Bates says that the fat of these lizards " is much prized by the natives, who apply it as a poultice to draw palm spines, or even grains of shot, from the flesh."

The Iguanas are also natives of the New World. There is a crest along the back and tail, and under the throat a pouch that can be inflated at will. They live chiefly among the branches of trees, but take readily to water, and swim well. Many of them have the power of changing colour like the Chameleons. The Common Iguana (Plate VIII., No. 6) is widely spread over South America. It feeds on vegetable substances and insects, and is hunted for its flesh. It is of gentle disposition, but if hard pressed can bite severely, and will use its long, whip-like tail with considerable effect.

A writer in the *Field* says :—" On our way home the dogs came across an iguana, which showed fight in the most determined manner, looking for all the world like the pictures of the dragon done to death by the brave St. George. . . . It is said that the bite of these reptiles is severe enough to cut the tendons of a horse's leg and lame him for life ; but this is possibly only a Gaucho yarn."

The Basilisk has none of the evil qualities attributed to the fabulous monster which the ancients called by this name. It is a harmless tree-lizard, with the habits of the family.

Two strange members of this family live in the Galapagos Islands.

HORNED LIZARD.

One frequents rocky beaches, and takes readily to the sea, though it objects to be driven into the water. Its food consists of seaweed. The other species lives in burrows, which were so numerous on James' Island that Mr. Darwin's party, for some time, could not find a spot free from them on which to pitch their tents.

The Horned Lizard, often miscalled the Horned Toad, from Texas, has the power of ejecting a red fluid, like blood, from the region of the eyes. These reptiles lie buried in the sand, the spines round the head closely simulating the dry, thorny vegetation of the desert, while they are admirably protected against both beasts and birds of prey by their prickly armour.

The Moloch, or Thorn Devil, some 6 inches long, has the body covered with stout spines, and the two long ones on its head are miniature copies of those of the rhinoceros. It is a harmless little creature, confined to Australia.

The Heloderm is a Mexican lizard, about 2 feet long, of repulsive appearance. The skin is covered with tubercles, and the pale-salmon ground colour is dotted with black. Its teeth are grooved, and its bite is fatal to small mammals.

The Agamids are confined to the Old World, and to this family belong the Flying Dragons of the East. They live among the branches of trees, and leap from bough to bough by means of a

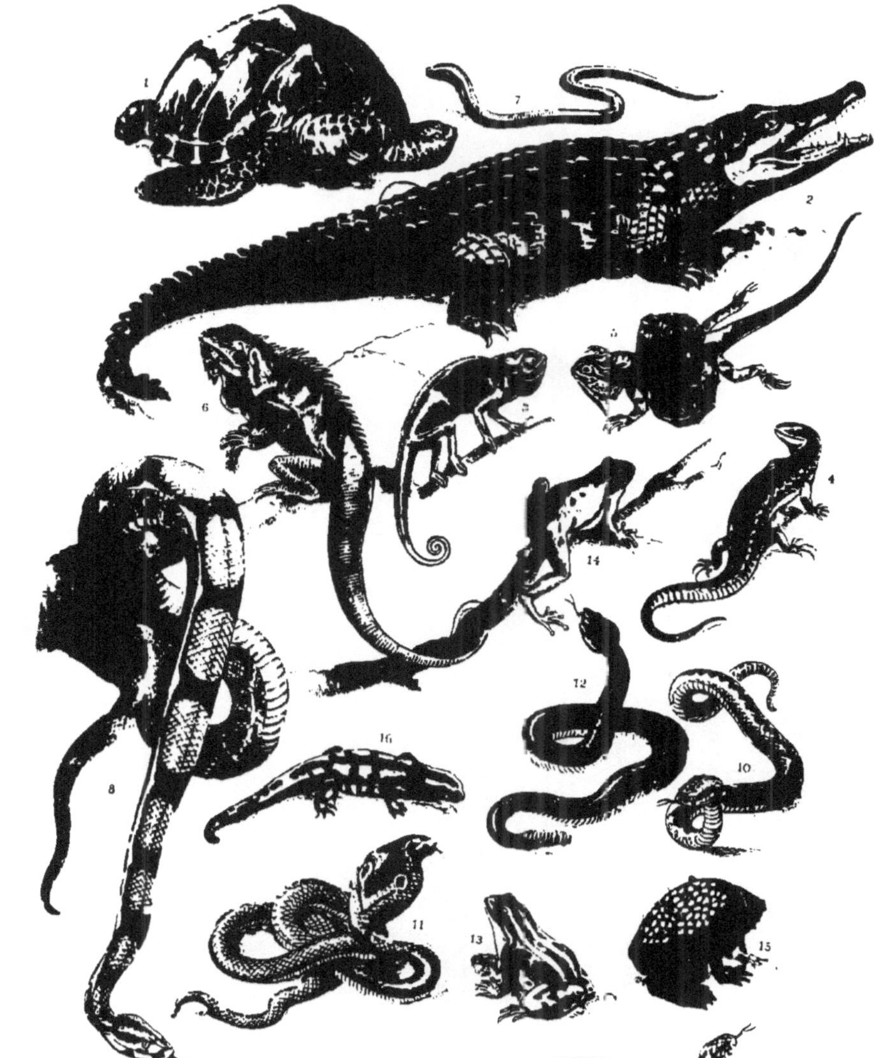
1
2
3
4
5
6
7
8
9
10
11
12
13
14
15
16
17
18

parachute, formed by the extension of the lower ribs and the erectile folds of skin about the head and throat. Many of them are brilliantly coloured. The Common Flying Dragon (Plate VIII., No. 5), from the Eastern Archipelago, is perhaps the best known. It lies motionless on a branch, waiting till some passing insect comes within reach, when it expands its "wings," and with a sudden spring seizes it, alighting on a branch below.

The Frilled Lizard of Australia has round the throat a large membranous ruff, which can be opened and erected at will.

The Monitors, or Water-Lizards, are confined to Africa, Asia, and

GECKO.

Australia. Their name refers to the popular notion that they warn man of the fact that a crocodile is near. The only foundation for this story seems to be that these animals are often found in the same streams and lakes, and that the Nile Monitor, about which the warning story was first told, devours the eggs of the crocodile when it can find them. This lizard often grows to a length of 6 feet, of which the tapering tail counts for a full half. It frequents the shores of rivers and lakes, and if disturbed takes to the water. The flesh and eggs of many of the species are eaten.

The Geckos are found in all the warm countries of the globe. The body is stout, and the leaf-like expansions of the toes enable them to cling to and walk on upright surfaces. They may often be seen running up the walls and on the ceilings of houses, hunting for

S

spiders and insects. The name refers to the noise made by some of the species. The Wall Gecko is common in the South of Europe. Of the Common East Indian species, Sir Emerson Tennent says : "In a boudoir where the ladies of my family spent their evenings, one of these familiar and useful little creatures had its hiding-place behind a gilt picture-frame, and punctually as the candles were lighted it made its appearance on the wall to be fed with its accustomed crumb ; and if neglected it reiterated its sharp, quick call of *chic-chic-chit !* till attended to. It was of a delicate grey colour tinted with pink, and having by accident fallen upon a work-table it fled, leaving its tail behind it, which, however, it reproduced in less than a month. This faculty of reproduction is doubtless designed to enable the creature to escape from its assailants ; the detaching of the limb is evidently its own act."

The Amphisbænas—the "Two-headed Snakes" of travelling show-men—are found in Asia Minor, Africa, and South America, and one of the African species lives in Spain. They are worm-like in appear-ance, and with scales arranged in rings round the body, and the largest is not more than 18 inches in length. Bates found several species at Pará. He says that their peculiar form, added to their habit of wriggling backwards as well as forwards, has given rise to the fable that they have a head at each extremity. The South American species live in the nests of the Saüba Ant, probably for the same reason that Rattlesnakes live in the burrows of the Prairie Dog —to be near an abundant food supply. The natives, however, say that the ants treat the reptile with affection, and that if it be taken away, they will forsake the nest.

The Slow-worms are also Snake-like Lizards, and of these our Common Slow-worm or Blind-worm (Plate VIII., No. 7) is a good example. It is familiar to almost everybody, and though the spread of cultivation is driving away our native reptiles, dwellers in the country need not search long in spring and summer before finding one. To my thinking, it is extremely beautiful ; and its usefulness equals its beauty. for it feeds on caterpillars, slugs, and insects that do harm to the farmer and gardener. Both its names are wrong : it is not a "slaying" worm, for that is the meaning of "Slow" worm : but a strange prejudice attaches to this little reptile, and in most places it is killed, under the erroneous impression that it is venomous. Nor is it blind : on the contrary, a single look at it will convince the observer that it has a pair of very bright eyes. Like many of the limbed Lizards, the Slow-worm will readily part with its tail to save itself from capture, and the detached part will jump and twist about, while its late owner is making off with all speed to a place of safety.

The so-called Glass Snakes, also limbless, are closely allied.

SNAKES.

The popular notion of a Snake—a long, limbless animal, covered with scales, is fairly correct, though in some there are internal traces of hind limbs, and even external hooks or "claws." The widely dilatable jaws, by means of which these creatures can swallow prey much larger than the normal size of the throat, is a much safer mark of distinction than the absence of limbs, which is found in some Lizards, Amphibians, and Fishes. No snake is slimy.

The Vipers constitute a group of venomous Snakes, of which most bring forth their young alive, the eggs being hatched within the body of the parent. The head is more or less triangular in shape, and has been compared, not inaptly, to an ace of clubs. In each jaw there is a large erectile hollow fang, having a little hole at the side,

JAWS OF RATTLESNAKE.
A, Poison gland.   B, Tubular fang.

through which the venom passes into the wound. The True Vipers are confined to the Old World, being most abundant in Africa, where the most deadly species are found.

The Common Viper (Plate VIII., No. 10) is widely spread over Europe, and is common in Great Britain, frequenting dry heaths and waste, sandy places. The general length is a little under 2 feet. The ground colour varies greatly, but in all there is a dark zigzag line, running from head to tail, by which this reptile can be readily distinguished from the two harmless British Snakes and from the Slow-worm. The story that in time of danger the young seek refuge in the throat of the parent seems probable, but is not yet proved. The rapid disappearance of the young when alarmed is beyond doubt, but they can easily take refuge in any crack in the ground, or under a stone, and so escape observation. The fat of the Viper is used as a remedy for its bite, though probably olive-oil would do just as well; and viper-broth, though no longer used, was formerly prescribed as a remedy for ulcers.

S 2

Some Vipers have horn-like projections on the head. Of these the most notable is the Cerastes, from Africa, which is said to hold on like a bull dog, its poison flowing into the wound all the while. The Puff-Adder, which has the power of inflating its body when irritated, is also African, as are some other species. There are some Asiatic Vipers, of which the Daboia, or Russell's Viper, is often seen in captivity.

The Pit-Vipers, so called from a depression on each side between the eye and the snout, are confined to Asia and America, and of these

CERASTES.

the Rattlesnakes are best known. The Common Rattlesnake (Plate VIII., No. 12) is widely spread over the United States, and there are other species, one of which ranges to Brazil. The rattle whence they derive their name is formed from the skin, a portion of which is left behind and hardens, every time the creatures cast their coats, which occurs at irregular intervals, not—as was formerly thought to be the case—once a year. Rattlesnakes are nocturnal, and rarely attack unless first provoked. Their poison is exceedingly powerful, and generally proves fatal to man.

Bates met with Rattlesnakes on the Lower Amazon. When he was collecting, a favourite dog of his rushed into the thicket and made a dead set at a large snake whose head was raised above the herbage. "The foolish little brute approached quite close, and then the serpent reared its tail slightly in a horizontal position and shook its terrible rattle. It was many minutes before I could get the dog away, and this incident shows how slow the reptile is to make the fatal spring."

The Yellow Viper of Martinique, also called Fer-de-lance, may attain a length of 6 feet. Of an allied species, the fierce Lance-Snake, from the mainland and Trinidad, Messrs. Mole and Urich record the fact that one, which was being irritated with a stick, ejected its poison, and it fell on the face of a woman some 12 feet off.

The Asiatic Pit-Vipers, though smaller than those of the New World, are equally venomous. Many of them live among the branches of trees, and when disturbed attack fiercely. The tail is prehensile, and aids them in their arboreal life.

The Poisonous Colubrines resemble our Common British Snake in form, and have not the peculiarly shaped head of the Vipers. The fangs are fixed in the jaw, and the poison passes down a groove, not through a canal. They are natives of the warmer parts of both hemispheres, but none is found in Europe.

The Snakes of the first family of this group have the power of expanding the skin of the back so as to form a kind of hood. In none is this better seen than in the Cobra de Capello, or Hooded Snake (Plate VIII., No. 11), widely distributed over India. These snakes may attain a length of 5 feet or rather more, and there are many varieties, some of which have a mark on the hood like spectacles, or rather, like the " eye " used as a fastening for ladies' dresses. In habit it is nocturnal, and does not attack man unless provoked or disturbed. The Cobra can lift its head and the fore part of the body, but when it does so the greater part still remains on the ground.

The poison of these snakes is very deadly, and if a person is bitten by a Cobra whose poison-glands are pretty full, medical treatment is of little use. The number of deaths every year in India from the bite of these reptiles is very large, but the efforts of Europeans to diminish the number of Cobras are rendered powerless by the natives, who regard these creatures with veneration. It was formerly believed that the natives bred them for the sake of the Government reward for their heads. This seems improbable, though Dr. Guillemard tells a story of a Dutchman in the Malay Archipelago who kept a kind of crocodile farm. He had staked off a small reach of the river, where these animals multiplied at a rapid rate, and their heads brought him two dollars apiece.

The Cobra is the serpent commonly carried about by the charmers in India, and they seem to have strange command over these creatures. With regard to these performers, Mr. E. C. Cotes, of the Indian Museum, Calcutta, says (*Maclure's Magazine*):—" It is a mistake to suppose that he invariably removes the poison fangs from his dangerous pets. Even if he did do so, it would not be a complete safeguard. It is easy enough to break off a snake's fangs, but as a general rule fresh ones soon grow again; and, even without this, the poison glands continue to pour their deadly secretion into the creature's mouth, where it is ready to be inoculated into the scratches which a snake is quite able to inflict by means of the numerous minute teeth which it possesses in addition to its poison fangs."

The Egyptian Cobra is an allied species, but has no " spectacle " markings, nor are these found in the Ring Hals Snake from South Africa. The Egyptian snake-charmers render the former species stiff as a stick by pressing the back of the neck between the finger and thumb.

The Hamadryad, or King Cobra, from India, is the largest of the family. Fortunately it is not common, for it is much fiercer than the Common Cobra, and the natives say that it will not only attack man if irritated, but will also pursue an aggressor. This animal feeds principally on other snakes, and is by no means particular whether they are poisonous or not. There are generally very fine specimens in the Zoological Gardens, Regent's Park.

India has other snakes of the same family equally deadly. Australia has some, but these vary greatly in the power of the venom; some are deadly, while the bite of one produces scarcely more effect than the sting of a bee. Others are also found in South America, and are remarkable because their colouring of black and red or yellow rings is mimicked by several harmless snakes. The brilliant colouring of the Coral Snakes probably acts as a warning, so that the harmless snakes similarly coloured derive advantage from the resemblance.

To the same group belong the Sea Snakes of tropical climates, which may attain a length of from 10 to 12 feet. In all, the tail is flattened from side to side, and serves as a swimming organ. The sea is their native element, and if thrown on shore by the force of the waves, most of them are helpless or nearly so. One, however, lives in marshy ground. The bite of all is intensely venomous.

The next group contains what are called "harmless" Snakes. Most of them have solid ungrooved teeth and are without poison glands, but some of the Tree Snakes and Sand Snakes, and the Freshwater Snakes of Asia, have grooved teeth of the same character as those of some deadly species, but placed farther back. From the position of these fangs, and the very small quantity of poison secreted by them, the bite of these Snakes is harmless to man; but it is certain that some, and probable that all, paralyse or kill the small mammals, birds, and lizards on which they feed.

The Whip-Snakes are so called from their long, slender shape, which has caused them to be compared to the thong of a coach-whip. They are generally green in colour, and are found in tropical America, West Africa, and Asia. Mr. Whitehead, in his account of the exploration of Mount Kina Balu, North Borneo, gives the following interesting description of the method in which these Snakes hunt their prey:—"I saw what I took to be a small red bird settle on a tree-stump. After waiting a few minutes, one of those long green

HAMADRYAD.

whip-snakes began to ascend the tree by the aid of the small branches. The red bird, which proved to be a tree-frog, croaked in alarm, but seemed quite unable to use its powers of escape. Slowly the snake twined up the stem; the tree-frog, knowing its last moments were near, seemed perfectly petrified with fear, and in one of its feeble jumps the snake sprang at it and caught it by the hind leg; the snake hung half its length in the air with the croaking frog in its jaws. Having a natural antipathy to snakes, I shot it in this position, when it immediately relaxed its hold of the frog, which dropped to the ground. 'Just what I have been waiting for,' said the snake's spouse, as she seized the poor little wretch before it had reached many feet on the stem of a neighbouring tree. The little frog, however, stuck so firmly to the tree with its peculiar sucker-like toes, that I was able to approach near enough to administer a rap on the head to the snake, and so ended this unequal conflict. That snakes hunt in pairs there is little doubt. A friend in Sandakan showed me a snake 11 feet long that he had shot in the forest, whilst on its way to attack his dogs, which were at the moment busy worrying its companion."

The Tree Snakes are found in all the tropical regions. Those which are active by day are generally bright-coloured; the Nocturnal Tree Snakes, which constitute another family, are more soberly dressed.

The Sand Snakes, or Desert Snakes, range from Africa, where they are most abundant, to the Malay Archipelago. They live on the ground, and one species has been seen swallowing a viperine snake. It is not certain, however, whether this snake-eating is habitual or only occasional.

The Freshwater Snakes, rarely found on land, are natives of tropical and sub-tropical regions. They feed on fish and crustaceans. The Siamese Freshwater Snake is very strange in form. On the snout are two flexible projections, covered with scales, and probably serving as organs of touch, like the antennæ of a lobster or of an insect.

The Rachiodont, or Egg-swallowing Snake, is a native of South Africa. One received by the Zoological Society in 1894, lived for some time in their Reptile House, where it was fed on pigeons' eggs. The mouth is almost toothless, and the eggs are passed down into the gullet, where the shell is broken by what are called the "gular teeth" —really the tips of the spines of eight or nine of the first vertebræ. The fluid nourishment passes down into the stomach, and the shell is rejected in a pellet by the mouth. Dr. Andrew Smith, who kept several of these Snakes in confinement, observed that they always retained the egg stationary about 2 inches behind the head, and while in that position used great efforts to crush it.

The Ringed Snake (Plate VIII., No. 9) is pretty common in this country. It is much larger than the Viper, and has yellow markings behind the head, but lacks the lozenge markings on the back possessed by its venomous relative. These reptiles frequent warm, sunny places near water, to which they take readily, for they are capital swimmers. Mice, rats, frogs, lizards, birds, and their eggs form their chief food, but they are by no means disinclined for small fish. The female is larger than the male, and a Ringed Snake between 3 feet and 4 feet would be beyond the average size.

The Smooth Snake, somewhat smaller than the Ringed Snake, frequents dry, sandy places, and feeds principally on lizards. It will bite readily if irritated, and dislikes being handled. The first specimen recorded as British was taken at Bournemouth in 1859. Since then

BOA STRANGLING A JAGUAR.

many have been caught in that neighbourhood, and these Snakes seem to be confined in this country to Hampshire and Dorset.

To the same group belong the Indian Rat Snake, widely spread over Asia, where it is freely admitted into houses, and wages ceaseless war with the rats and mice; and the Black Snake, of America, which preys on the Rattlesnake.

The Boas and Pythons kill their prey by throwing their coils around it and squeezing it to death. Strange tales are told of the size which these reptiles attain, but probably 30 feet may be taken as the greatest length for which there is any evidence, and specimens of that size are very rare. They are natives of the tropics—the Boas in the western, and the Pythons in the Eastern hemisphere.

The Boa Constrictor (Plate VIII., No. 8) is well known, and may be seen in any zoological gardens or travelling menagerie. It is a common species in South America, and does not seem to be feared by the people there. When Bates was living in Pará, he was roused one night by a lamplighter, who had just captured a boa in the street and wished to show it to the English naturalist.

In October, 1894, a strange incident occurred at the Zoological Gardens, Regent's Park. When the house was shut up at night there were in one cage three Boas, two of which were feeding, each on a pigeon. When the keepers came in the morning, one of the reptiles, about 9 feet long, had disappeared. It needed but a look at the larger Boa, about 11 feet long, to see what had happened. The larger Boa (which died in April, 1895) had swallowed his pigeon, and then swallowed the second pigeon and the serpent to which it rightly belonged.

This seems extraordinary; but a still more extraordinary feat was recorded (in the *Field*) from Singapore. Five Pythons were put into a large enclosure together. One of these, a new-comer, was about 18 feet in length and very thick in proportion. The others had been kept together for some time, and had lived quite contentedly; but on the night after the large one was put with them, it swallowed three of the smaller snakes, each averaging 15 feet long. The remaining Snake, which was the second biggest, had a slight wound, as if there had been a fight between the two.

It was at one time supposed that the Hamadryad was the only Snake that devoured its kind, but the cannibal habit is now known to be much more common.

The Anaconda is the largest of the Boas. Bates, in his "Naturalist on the Amazons," says that these Snakes live to a great age, and attain an enormous size. He had measured skins over 20 feet, and had heard of specimens double that length; 30 feet, however, is probably the limit. There is no evidence, though there are many reports, of larger ones.

The Pythons do not differ in habit from their relations in the New World. They are found over Asia, as far as the Malay Archipelago, and the West and South of Africa. The largest nearly equals the Anaconda in size. Some of these Serpents incubate their eggs, piling them up and then coiling round them. This occurred in the Zoological Gardens, London, some years ago, and again in Leipsic in 1894.

Australia has some Serpents of this family—the Diamond Snake and the Carpet Snake, which are closely related.

AMPHIBIANS.

The Amphibians are even more closely allied to the Fishes than the Reptiles are to Birds. At one time Reptiles and Amphibians were grouped in one class, but there is this great difference between them : the former breathe by means of lungs all their life ; the latter always breathe by means of gills when young, and in some this method of breathing persists when they are adult, while others then breathe by lungs, like their higher relatives the Reptiles. There is usually a metamorphosis, and the larval form is called a Tadpole. Most of them swim well. Those in temperate climates hibernate in winter, generally at the bottom of ponds ; tropical forms bury themselves in the ground in the dry season.

The Frogs and Toads, in which the tail is lost when the tadpole stage ceases, are widely distributed over the globe. Insects, slugs, and worms form their chief food, but some of the largest will eat mice and small birds. In taking small prey, the tongue, which in most Frogs and Toads is fixed in front and free behind, is shot forth, and its free end, covered with a sticky secretion, touches the victim, which is then drawn into the mouth and swallowed whole.

HEAD OF FROG.
Showing the Tongue fixed at front of mouth.

The Common Frog, abundant in Britain, belongs to a group in which the digits are pointed, and its hinder toes are webbed. It needs no description, for everyone has seen it lurking in the shade by the side of a river or pond, into which it will spring on the least alarm. Most people, too, have heard its love-notes in the spring, though few would call them musical. From the quantity of noxious insects it destroys, this creature deserves to be reckoned as a friend to the farmer and gardener. The eggs of the Frog may be seen floating on ponds in the spring. The tiny black speck is the future tadpole, and the ball-like envelope in which each is set, is nothing more than its glassy covering enlarged by the absorption of water, for the eggs are always deposited below the surface.

The development of the tailed tadpole into the tailless Frog has been so often described that here it is only necessary to say that every boy may watch it for himself. One need only gather a lump of Frog-spawn, and put it into a vessel with some pond-water and water-weed. Light should be admitted from the top more freely than at the sides, and the vessel—a basin or pie-dish will do—kept in a cool place. In due time the tadpoles will make their appearance, and go through four well-marked stages, before becoming Frogs.

The Edible Frog (Plate VIII., No. 13), common on the Continent, in North Africa, and Asia, has been introduced into England. It is confined to a single county, and there called the "Cambridgeshire Nightingale," from its loud croak. Only the hind-quarters are eaten; in flavour they resemble spring chicken, and command a high price.

A large American species is called the Bull Frog, from its loud note, which may be heard half-a-mile off.

The Common Toad is much more stoutly built than the Frog, and has a warty skin. It frequents damp places, from which it comes out at night to feed on insects and slugs. An acrid secretion exudes from the skin, which would probably be injurious if introduced into a cut or wound. It is on this account that few dogs will mouth a Toad. But the old notion that Toads spit poison has no foundation in fact. The Natterjack Toad, another British species, with a yellow line running down the back, is more rare, and prefers dry places.

Strange stories are told of Toads being found in the heart of trees

TREE FROG.

and in the centre of blocks of stone, but none will bear investigation. These animals can live a long time without food, but experiments have shown that imprisonment in a hole in a block of stone, covered with glass, killed them in less than two years.

The Tree-Frogs are related to the Toads, and have on the end of each digit a sucker, which enables them to cling to the surface of leaves. Many of them are of a bright green, and some of the tropical forms are still more brilliantly coloured. The Common Tree-Frog (Plate VIII., No. 14) is widely distributed in Europe.

Some of the American Tree Frogs—Tree-toads, they call them—

have no tadpole stage, and "they pass the winter in a torpid state in the ground, or in stumps and hollow trees, instead of in the mud of ponds and marshes, like True Frogs."

The Ceratophrys is a large American Frog, from South America. Mr. Hudson considers them venomous, and in his "Naturalist in La Plata" gives an account of two horses that were killed by them. The small teeth in the upper jaw are not pierced or grooved, so that if the horses were really killed by the frogs, the secretion from the mouth was probably the cause of their death. Some of these animals have lived in the Zoological Gardens, where doubt exists as to their venomous qualities.

MIDWIFE TOAD.

The male of the Midwife Toad, a species common in some parts of France, fastens the string of eggs to his legs, and then burrows in damp ground. When the tadpoles are ready to come out, he betakes himself to a stagnant pond, and there the young burst through the jelly-like covering and swim away. Some of the tadpoles of this toad were born in the Zoological Gardens in 1893.

The female of the Pouched Frog, a native of Mexico, has a kind of pouch at the lower part of the back, into which the male puts the eggs as they are laid. Here the young tadpoles come out and undergo their metamorphosis, leaving the pouch as perfect frogs.

The Surinam Toad (Plate VIII., No. 15), when full grown, is nearly a foot long, and has neither tongue nor teeth. It is aquatic in habit and repulsive in aspect. The eggs develop in separate pouches, or pits, on the back of the female, but it is uncertain by what means they reach that situation. Some specimens were received at the Zoological Gardens in 1894; one female carried eggs for some time, but none of the tadpoles came out alive.

In the Newts and Salamanders the tail remains throughout life, and the general shape is lizard-like. In the Newts, which are aquatic, the tail is flattened to form a swimming organ.

We have three Newts in Britain—"evvets," country boys call them, and stone them without mercy in the erroneous belief that they are poisonous. In spring and summer the males may be distinguished by the crest which runs from the neck to the tail.

SURINAM TOAD.

The Great Water Newt is the largest and handsomest, and if well supplied with earthworms or bits of meat, may be kept, as may its smaller relatives, in a vessel of water in which some aquatic plants are growing. On the surface should float a piece of virgin cork, or some rockwork should allow the creature to leave the water and become for a time a dweller on land. The females generally deposit their eggs on the leaves of water-plants, wrapping a leaf with her hind-legs round each egg as it is deposited. The Newt tadpole is a very pretty creature, and its development may be watched as easily as that of the Frog tadpole, and under similar conditions. The fore-legs of the Newt and the hind-legs of the Frog are the first to appear.

The body of the Great Newt is warty; the smaller Smooth Newt has no warts, and the Palmated Smooth Newt has the hind feet webbed in the male.

The Salamanders live on land when adult, frequenting cool, moist places, but bring forth their young alive in the water. The Common Salamander (Plate VIII., No. 16) is found in Central and Southern Europe, and its yellow-and-black coat probably acts as a warning to birds that it is not good to eat. This creature was formerly believed to be able to live in the midst of flames.

The Axolotl is the tadpole of the Amblystome (Plate VIII., No. 18), which has its home in the lake near the city of Mexico, and an allied species is found in South Carolina. The larval and the mature form do well in an aquarium. The former will breed readily, and the tadpoles will develop rapidly if well supplied with food. The Axolotl will grow to a length of 9 or 10 inches, is stoutly built, of a dark slate colour spotted with black. When it loses its gills the tail becomes rounded, like that of a Salamander.

The Great Salamander, about 3 feet long, a native of China and Japan, is the largest living Amphibian. It is aquatic in habit, and a specimen may generally be seen in its tank in the Reptile House at the Zoological Gardens. Some American species are closely allied.

A few Amphibians from the Southern United States lose their gills when adult, but retain the gill slits. In others the gills are retained throughout life. The Proteus (Plate VIII., No. 17) is found in the underground lakes and streams of Carniola and Dalmatia. It is about a foot long, and as thick as one's finger. The eyes are rudimentary and hidden in the skin. The Siren, a larger animal, of similar shape, but of a dark colour, lives in the marshes and stagnant pools of South Carolina. It has no hind limbs.

The Limbless Amphibians are small burrowing creatures like large earthworms in appearance, peculiar to the tropics of both hemispheres. The young of one species are born alive.

TADPOLE OF NEWT.

CHAPTER XXIII.

FISHES.

FISHES are the lowest class of Vertebrates with jaws. They have cold red blood, and breathe the oxygen in the water by means of gills. Most of them have a swim-bladder, which lies under the back-bone. By compressing it, or allowing it to expand, they can sink to the bottom or rise to the surface without using their fins. In one order—the Double Breathers—the swim-bladder serves as a lung,

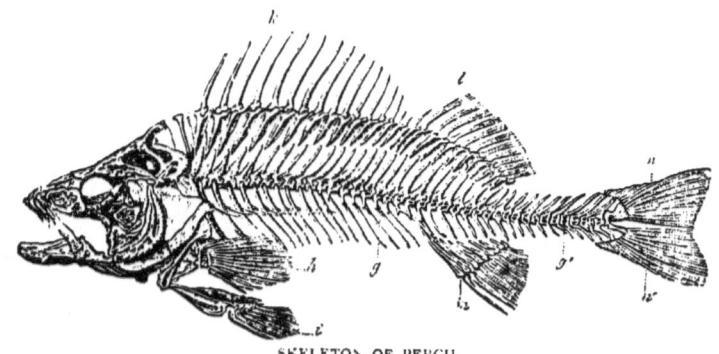

SKELETON OF PERCH.

g g, back-bone ; *h*, pectoral fin ; *i*, ventral fin ; *k l*, dorsal fins ; *m*, anal fin ; *n n'*, caudal fin.

and on that account they are sometimes classed with the Amphibians. The body is usually more or less spindle-shaped, and covered with scales. The pectoral fins of fishes correspond with our arms, and the ventral fins to our legs, though sometimes the ventrals are in front of the pectorals, as in the perch : these are called paired fins. The unpaired fins are the dorsal fin (or fins), on the back, the caudal or tail fin, and the anal fin, near the vent. The tail is the principal swimming organ, aided by the dorsal and anal fins, which from their position are called the vertical fins. The paired fins balance the fish in the water, and direct its course. Fishes propagate their kind by means of eggs, but these in some cases are hatched within the body of the parent.

The skeleton is more or less gristly in all except the Bony

Fishes, and even in these the internal framework is not so solid as it is in higher animals.

The Double-Breathers, found in Africa, South America and Australia, are known as Mud-fishes. The African Mud-fish, which may reach a length of about 6 feet, is found in the rivers of tropical Africa, and feeds on other fishes, frogs, and water-insects. Its shape is somewhat eel-like, but the body is flattened from side to side, and the limbs are mere filaments. During the dry season these

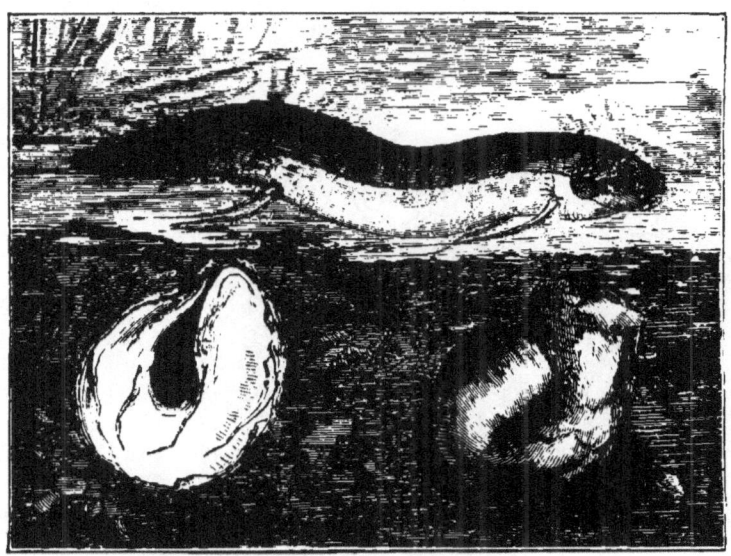

AFRICAN MUD-FISHES.

creatures bury themselves in clay, leaving a breathing-hole, which betrays their presence to the natives, who dig them out for food. Specimens brought to Europe are usually encased in clay. At one of the meetings of the British Association in 1894, Dr. Forbes exhibited some of these clay-cocoons, and then set the Mud-fishes free by immersing the clay in tepid water. The hard covering fell away, and the fish came out and swam round and round the jar. Some of these Mud-fishes are in the Insect House of the Zoological Gardens. The South American Mud-fish, from the Amazon, is closely allied. The Australian Mud-fish is covered with large scales, and has paddle-like limbs. Professor Spencer says that this fish lives only in the Burnett and Mary Rivers in Queensland. It does not form a cocoon

T

and probably never leaves the water. It comes continually to the surface, and passes out and takes in air, making a faint spouting noise. The lung is probably of the greatest service to the animal, not during the hot, but during the wet season, when the rivers are flooded, and the water thick with the sand brought down from the surrounding country. It appears to feed largely, if not entirely, on vegetable matter, such as the seeds of gum trees which fall into the water. The eggs, which are surrounded by jelly, are laid in strings.

The first division of the Bony Fishes have the bones of the throat separate, and in both divisions the anal, dorsal, and ventral fins are supported by spiny rays.

CHÆTODONT.

The Perches are widely distributed in the streams and round the coasts of temperate and tropical regions. The Common Perch (Plate IX., No. 1) is a good example of the family. It is found over the northern parts of both hemispheres, and in our own country is pretty plentiful, and generally swims in shoals. From its bright colouring it is a favourite fish for aquaria, and in confinement will soon become tame enough to take food from the hand. It is, however, very voracious, and will often make a meal of smaller fish. Some small American fish of this family build nests in company, and guard their young. The Bass, or Sea-Perch, is a common British fish.

The Chætodonts, chiefly from tropical seas, are brilliantly coloured, and ornamented with spots and bands. Here belongs the Archer-fish, from the East Indies, which owes its name to its habit of propelling water from its mouth at insects, so as to knock them into the water, where they are quickly snapped up. The Malays keep it in captivity and amuse themselves by watching its method of taking its prey. Some other fish of the same family, with tube-like snouts, have the same habit. Dr. Günther thinks that they use their snouts to capture small animals that have sought refuge in holes and crevices.

The Red Mullets are tropical sea fishes, with two long barbules, or fleshy appendages under the lower jaw, serving as organs of touch. One, the Mullet prized by the Romans above all other fish, is British. Like the Coryphæne, it undergoes a colour-change in death. The

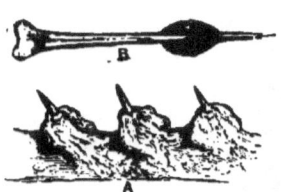

POISON ORGANS OF SCORPION
FISH.
A. Dorsal spines.
B. Spine and poison bag.

Romans used to have this fish brought living into the banqueting-room, so that the guests might enjoy the spectacle of its brilliant red colouring, which became brighter in the death struggle. The Striped Red Mullet is probably the female.

The Sea Breams, most of which are used for food, are plainly-coloured shore fishes, from tropical and temperate seas. A few occur on our shores. The Gilt-head, a native of the Mediterranean, sometimes found on the South Coast, is said to stir up the sand with its tail, so as to discover the shell-fish concealed in it.

The Scorpænoids are carnivorous sea fishes, having the head or fins, or both, armed with spines. Some live at the bottom of the sea, and these have appendages resembling seaweed, which may serve for concealment, or as bait for their prey. These species are also coloured so as to correspond with their surroundings. In one genus from the Indo-Pacific the dorsal spines are perforated and furnished with a poison-bag. These fish lie hidden in the sand near the shore, and when trodden on by natives wading, the pressure on the poison-bag as the spine enters the foot forces the venom into the wound, and death has frequently resulted therefrom.

The Maigres are found round the coasts of the Atlantic and Indian Oceans, in and near the tropics, especially near estuaries, up which many species travel, and some have become naturalised in fresh water. Almost all are used for food. The Common Maigre, which reaches a length of 6 feet, is plentiful in the Mediterranean, and sometimes strays to Britain. The Drum is an American species, and owes its name to the sounds it makes by bringing the large teeth sharply together, or by striking the hull of a vessel with its tail to free that organ from the parasites with which it is infested.

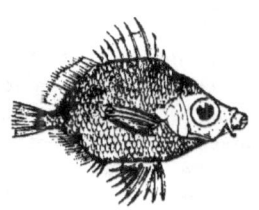

BOAR-FISH.

The Sword-fishes frequent the open seas of warm climates. The young at first differ very greatly from their parents, though it is not

T 2

long before the general form of the adult is assumed. The bones of the upper jaw grow together and, projecting far beyond the lower jaw, constitute a formidable sword-like weapon. In species from the Indian and Pacific Oceans the dorsal fin is very large, and it is said

SWORD-FISH.

that by erecting this fin they can sail before the wind like a boat. The Common Sword-fish is found in the Atlantic Ocean and in the Mediterranean Sea. It reaches a length of about 12 feet, and swords 3 feet long and about 9 inches round at the base are preserved. The Tunny-fishers take it in their nets, and the flesh, especially of young fish, is said to equal that of the Tunny. Sword-fish feed on smaller fish, and the reason of their attacking whales is not clearly made out. Very often they mistake ships and boats for the creatures to which they have such a dislike, and in many museums are ships' timbers which have been penetrated by the "sword," the weapon being broken off in the efforts of the fish to withdraw it.

Of the Horse Mackerel family, which contains numerous species living in temperate and tropical seas, the Scad, or Common Horse Mackerel, is the best known British form. It occurs in large schools on the Cornwall and Devonshire coasts, and its flesh is eaten, but is inferior to that of the Mackerel. The Boar-fish, with short, compressed body and peculiar snout, sometimes occurs on our southern coasts.

Perhaps the most remarkable of the family is the Pilot-fish, supposed to pilot Sharks to their prey. Dr. Meyen says:—" When we threw overboard a piece of bacon fastened on a great hook, the Shark was about twenty paces from the ship. With the quickness of lightning the Pilot came up, smelt at the dainty, and instantly swam back again to the

PILOT-FISH.

Shark, swimming many times round his snout and splashing as if to give him exact information as to the bacon. The Shark now began to put himself in motion, the Pilot showing him the way, and in a moment he was fast upon the hook." Here the Pilot, innocently enough, led the Shark to its death. The Pilot-fish probably keeps the Shark's company in hope of picking up the leavings, and for the sake of feeding on the parasites with which all large fish are infested. Professor Moseley said : " The Pilot-fish often mistakes a ship for a large Shark, and swims for days just before the bows, which it takes for the Shark's snout." And from the next sentence the reason for the attendance of the Pilot on the Shark may be gathered. ".After a time the fish becomes wiser, and departs, no doubt thinking it has got hold of a very stupid Shark, and hungrily wondering why its large companion does not seize some food and drop it some morsels."

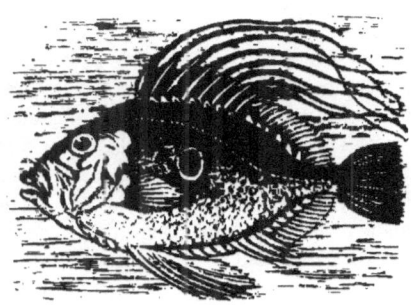

JOHN DORY.

The Dories inhabit temperate seas of the northern and southern hemispheres, and those of the genus to which our John Dory belongs are all highly valued for food. In Roman Catholic countries the black spot on the side of this fish is accounted for by the belief that it was from the mouth of the Dory St. Peter took the tribute money.

The Coryphænes are fishes of the open sea. The true Coryphænes, often miscalled "Dolphins," attain a length of 6 feet, are brilliantly coloured, and their flesh is valued for food. The Common Coryphæne, found in the Mediterranean, is the "dolphin" to which Byron refers when he compares its iridescent play of colour, when dying, to a beautiful sunset. All are noted for pursuing the flying-fish, on which they feed eagerly.

The Mackerel family comprises valuable food fishes, living in the open sea, and generally having a row of finlets behind the dorsal and anal fins. The blood is red, and the temperature is higher than that of other fishes. They are rapacious in habit, and approach the shore at certain seasons, probably in search of prey. The Common Mackerel is well known, and is one of the most beautiful of our British fishes. The upper surface is brilliant with green and blue marked with dark bands, and the silvery under-surface shows a play of colour like mother-of-pearl.

SUCKING-FISH.

On many parts of the coast Mackerel-fishing is an important industry.

The Tunny ranges from the Mediterranean northwards to the English Channel, and southwards as far as Tasmania. From 3 to 4 feet may be taken as the average length, though specimens of twice that size are recorded. The colour on the back is dark blue, and greyish on the under-surface. The flesh is highly valued for food, and is dried and prepared in oil for the market.

The Tunny fishery has been carried on in the Mediterranean from early times. The fish are netted, or surrounded and driven into an enclosure formed by nets, with a network bottom, which, when the fish are inside, is raised and fastened to the boats ranged all round. As the net is raised the fish are brought to the surface, when they are attacked with harpoons and boathooks. Quatrefages says: " Each hook that loses its hold is raised on high only to be buried still deeper in the quivering flesh, and soon the unfortunate animal is drawn to the side of the boat. In another moment it is seized by two men, who grasp the pectoral fins, lift it to the beam which is placed behind them, and throw it into the hold."

The Sucking-fishes, of which there are several species, have the spiny dorsal fin modified to form a sucking disk, which covers the back of the head. The Common Sucking-fish, which occurs in the Mediterranean, is about 8 inches long; some of the others attain a length of from 2 to 3 feet. These fish were well known to the ancients, and are mentioned by Aristotle as attaching themselves to the Dolphin. They probably attach themselves to Sharks for the same reason that Pilot-fish keep the company of those monsters (see p. 293), and when they fasten on to ships they do so imagining they are fastening on to a Shark. The back of the Sucking-fish is light-coloured and its under-surface chocolate. Professor Moseley says: "No doubt the object of this arrangement is to render the fish less conspicuous on the brown back of the Shark. Were its belly light-coloured as usual, the adherent fish would be visible from a great distance against the dark background. The result is that when the fish is seen alive it is difficult to persuade oneself at first that the sucker is not on the animal's belly, and that the dark

COMMON FISHING FROG.

exposed surface is not its back. The form of the fish, which has the back flattened and the belly raised and rounded, strengthens the illusion."

The Weevers are the type of a family of small carnivorous shore-fishes found in nearly all seas. Wounds from the grooved spines of the dorsal fin or the gill covers are exceedingly painful. The inflammation which follows is probably due to the mucus secreted by the skin of these fishes. The Greater Weever, about 1 foot long, and the Lesser Weever, about half that length, are British. The flesh of the former is excellent. In France and Spain this fish is sent to market, but the fishermen are bound to cut off the spines.

The Fishing Frogs have the head very large the pectoral fins developed somewhat like arms. Some feed on the bottom, while others are carried about on floating weed. They are strange-looking creatures, and in many the fin rays are elongated and the head is furnished with filaments. They lie half-buried at the bottom, and these waving filaments act as lures to smaller fish, which, when they come near, are swallowed at a gulp. The Common Fishing Frog, or Angler-fish, is British, and specimens 3 feet long are not uncommon. Dr. Gunther says that it attains a length of 5 feet. A tropical species of this family is often found clinging to floating gulf-weed with its arm-like pectoral fins. It forms a large round nest by cementing pieces of weed to-

BULLHEAD.

gether, and in the centre of this mass the eggs are deposited.

The Gurnards and Bullheads are small ground-fish found near the shores of all seas, only a few living in fresh water. The Common Bull-head, or Miller's Thumb, some 3 or 4 inches long, is plentiful in many British streams, and the Sea Scorpion is equally plentiful round the coast. The True Gurnards have three free pectoral rays, by means of which they can walk upon the bottom and climb over rocks. Some of them make a grunting noise when handled, and this is due to the escape of gas from the swim-bladder. They are valued for food.

The Mailed Gurnards have the body encased in bony plates. The Armed Bullhead, or Pogge, some 6 inches long, is found in estuaries in Britain. To this family belong the Flying Gurnards, from the Mediterranean, the tropical Atlantic, and the Indo-Pacific, with immense pectoral fins, which enable them to take long leaps out of the water. Professor Moseley believed that these fish move their "wings" during their flight, which he compares to that of many forms of grasshoppers, "which raise themselves from the ground with a spring, and eking out their momentum as much as they can by buzzing their wings, fall to the ground after a short flight."

To the family Labyrinthici, for which there is no English name, belong a number of small fish from tropical rivers, which, owing to the character of their gills, have the power of living for some time out of the water or in mud. The Climbing Perch, about 7 inches long, is a native of Asia. There is no doubt that it can travel for some distance over land by means of its fins and the spines on the gill cover. But although one was taken, some 5 feet above the water, in the act of ascending a palm tree, it has been suggested that this was exceptional, and not the habit of the fish. Sir E. Tennent could hear of no instance of a perch going up a tree ; but, on the other hand, the Malays call it by a name which means "Tree-climber." To this family belongs the Paradise Fish, which has the rays of the tail-fin greatly elongated, and which breeds readily in aquaria : the Gourami, which has been introduced into South America from the Eastern Archipelago, and the Fighting-fish of the Siamese.

FLYING GURNARD.

The Gobies are small shore fishes, mostly marine, and more abundant in tropical than in temperate regions. In many, as in our British species, the ventral fins are modified to form a sucker, by which these fish attach themselves to the rocks. The males of some species form nests for the eggs, and watch over the young when hatched. In one genus (Periophthalmus) from the Indo-Pacific, the fish leave their burrows at the bottom at low-tide and come upon the mud-flats in search of food. Mr. Hornaday says: "We shall probably never know the actual depth of mud on that bank at Selangore, but we sank into it to our knees at every step, and were fortunate enough to stop sinking at that point. There were probably a dozen fish in sight, hopping about or lying at rest on the mud ; but when we made for the nearest large specimens they developed surprising energy and speed, and made straight for their burrows. They moved by a rapid series of short jumps, bending the hinder part of the body and then straightening it suddenly, at the same instant lifting the front-half clear of the ground by means of the arm-like pec-

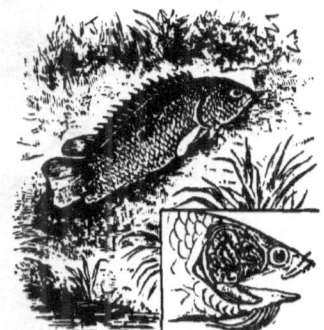

CLIMBING PERCH.

toral fins, which act like the front flippers of a sea lion." While Mr. Hornaday was trying to pass a string through the gills of the first fish he caught, it struggled out of his grasp, and as soon as it touched the mud started at top speed for the water. Instantly falling on his hands and knees he pursued the fish on all fours, while the Malays shouted with laughter to see a white man go over the mud like a crocodile.

The Blennies are shore fishes widely distributed. Many are found in brackish water, and some are freshwater fish. Among the British species, all of which are small, perhaps the most noteworthy is the Viviparous Blenny, the young of which are born alive. The Butterfly Blenny, pale brown in colour, marked with patches of a deeper hue, is found round the coast and in rockpools, and has on the

BUTTERFLY BLENNY.

dorsal fin an eye-like black spot ringed with white. The Cat-fish, or Wolf-fish, from the northern seas, is the largest of the family, sometimes reaching a length of 7 feet. It has large, molar-like teeth, which enable it to crush the shells of the molluscs on which it feeds. The natives of Iceland and Greenland use its flesh for food.

The Grey Mullets live in temperate and tropical seas, and some are common British fish. They are valuable food fishes, and Dr. Günther advocates their being kept and bred for the market in backwaters near the shore.

The Sticklebacks are small fishes. Most boys have netted them in streams or ponds, and very many have kept them in an aquarium. Most of them live in fresh or brackish water; but one, the Fifteen-spined Stickleback,

STICKLEBACKS AND NEST.

or Sea Adder, the giant of the family, some 7 inches long, is marine. The males of all the species build a kind of nest, in which the females deposit the eggs, the care of which is undertaken by the male fish, and he also watches over the young when they leave the egg, till they are fairly able to shift for themselves. The Common Three-spined Stickleback, which builds on the ground, or the Nine-spined Stickleback, which builds on and between weeds, will breed freely in the aquarium, provided there be plenty of water-plants, and small creatures to serve for food. Four females and one male may be kept in a fair-sized tank. No other fish should be introduced, for Sticklebacks are exceedingly voracious. Dr. Gunther tells of one that within five hours devoured seventy-four young dace, about ¼ inch long, and as thick as a horsehair. "Two days after it swallowed sixty-two more, and would probably have eaten as many every day could they have been procured."

The second order of Bony Fishes are sometimes described as having jaws in their throat, from the fact that the bones of the gullet, which bear teeth, are more or less united to form a single bone. Most of them are natives of warm seas, and are generally brilliantly coloured.

HEAD OF PARROT WRASSE.

Some of the True Wrasses, which spread from the Mediterranean round the coasts of Europe, are found on our shores, the commonest being the Ballan Wrasse, gorgeous in red and orange, spotted with bluish-green, and the Cook Wrasse, the male of which is so different in colour from his mate that she has been often called by another name. He is dark green striped with bluish-black, while she is red with dark spots on the tail. The Rainbow Wrasse sometimes visits us. In the whole family the lips are thick, and the teeth strong and well fitted for crushing the molluscs and coral which serve these fish for their chief food.

The Parrot Wrasses, from the tropical Atlantic and the Indo-Pacific, have the jaws produced into a strong beak. Mr. H. O. Forbes saw "shoals of these fish feeding in the surf on the living coral. They are furnished on the front of the head with soft pads, so as to be able to retain their position undisturbed among the breakers by squeezing hard up against the uneven wall, while they gnaw off the tips of the living polyps. One species requires to be prepared for the table with very great care, for if the gall bladder be ruptured and the contents escape into the body cavity, the flesh becomes quite poisonous." Several fatal cases occurred within his knowledge, and, what is more extraordinary, a frigate bird which swooped down on some entrails thrown into the water, "picked them up, and after rising some 30 feet in the air fell down dead." One family of this order is remarkable for the fact that the young are born alive; and another contains only freshwater fishes.

In the next order the vertical and ventral fins have no spiny rays, and the air-bladder, if present, does not open into the throat. The fish of this order are extremely important from a commercial point of view, two families furnishing a very large supply of food for man.

The Cod family comprises about half a hundred species, chiefly from temperate and northern seas. The home of the Cod Fish is in the North Atlantic Ocean. Iceland was in bygone times the seat of a great Cod fishery, but now the principal fishing grounds are in the North Sea, and on the Banks of Newfoundland. The appearance of the fish is well known: it reaches a length of from 2 to 4 feet, is greenish-olive in colour with darker spots. Large quantities are dried and salted in America, and exported to Catholic countries for consumption during Lent. Nearly every part of the fish is utilised, for the service of man or of the lower animals. The swim-bladder is made into isinglass, scarcely distinguishable from that of the sturgeon, or salted for food: medicinal oil is prepared from the liver; the hard roe is smoked and sold as a delicacy; and the Icelanders feed their domestic cattle on the bones. Norway is the centre for the preparation of cod-liver oil, and there the heads of the fish, mixed with seaweed, are given to cows, for the purpose of increasing their yield of milk. Cod are taken with long lines, and generally brought alive to the great fish markets in vessels furnished with wells to which the water has access.

The Haddock (Plate IX., No. 7) is common round our coasts; the largest are taken in the winter, when they approach the shore to spawn. Specimens of from 18 inches to 2 feet long are common. Like all the family, the Haddock is a voracious feeder. From the stomach of one 17 inches long were taken fourteen young whiting from 4 to 5 inches long, and a small shore crab measuring an inch across the back. The Haddock weighed 26 oz., and the contents of its stomach 6½ oz., which is as if a man should eat from 30 lbs. to 35 lbs. weight of food at a meal.

The Whiting, noted for its delicate white flesh, the Ling, the Hake, the Coal-fish, and the Burbot, from the rivers of the East Coast, all belong to this family, and are all used for food. The larger species are dried in the same way as the Cod.

To the family Ophidiidæ belong some strange forms. Some inhabit streams and ponds in caves in Cuba. One genus from Southern waters is known in New Zealand as "Ling," and "Cloudy Bay Cod." In the Mediterranean, Atlantic, and Indo-Pacific live the Fierasfers.

small eel-like fishes that take up their abode inside Jelly-fish, Star-fish, and Sea-cucumbers, doing no harm to the creature that gives them shelter, and feeding on the animalcules in the water that washes into

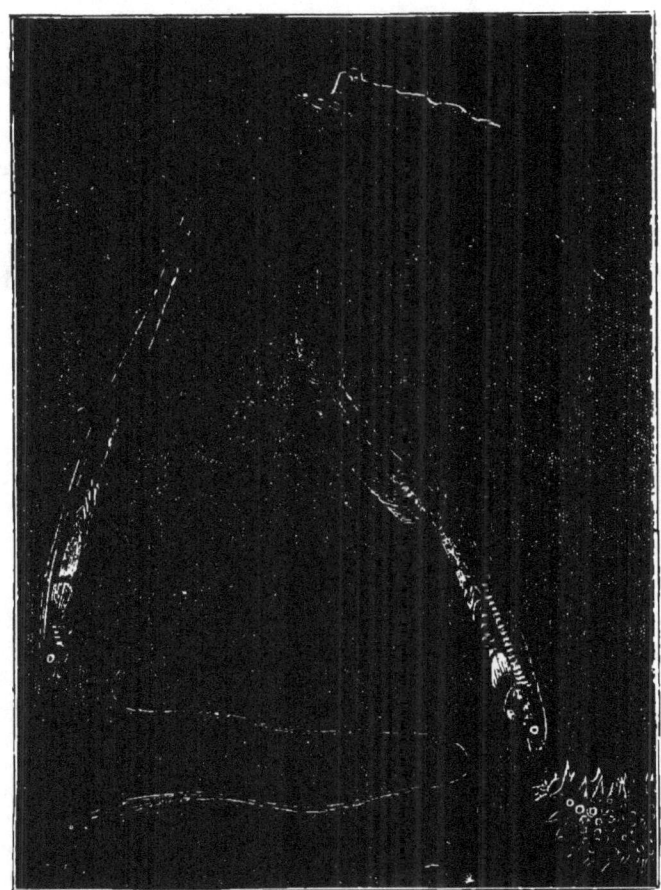

FIERASFERS AND SEA-CUCUMBER.

the cavity they dwell in. Here also belong the Launces, or Sand-eels, common on the sandy shores of Europe and North America. Porpoises prey upon them, and prevent them from seeking refuge in the sand by diving below the shoal. Fishermen use them for bait.

The Flat-fishes are as important a family as the Cod-fishes. When adult they live on the bottom, and swim with a wavy motion of the body. The young are generally met with in the open sea, and are formed and swim like other fish ; but as they grow the form of the body changes, and they no longer swim upright, but on the side. The upper side, which may be the right or the left, is coloured—on it are both eyes ; the lower side is whitish, and is called the "blind" side. All feed on other fishes, and some enter brackish and fresh water, though most are marine.

The Halibut, the largest of the family, is found on the northern coasts of Europe and North America. Fish from 4 to 5 feet are sent to the London market, and much larger specimens are recorded. The flesh is good eating, but inferior in flavour to that of the Turbot. The Indians of Vancouver Island fish for Halibut with a hook and line. Soon after the fish is hooked it rises to the surface, when a spear, to which a bladder is attached, is thrust into it. The fish then tries to descend, but the bladder, acting as a float, makes this difficult, and it once more comes to the surface, only to be speared again, and buoyed up with a second bladder. This process is repeated till the Indians manage to tow the fish to shore. Mr. Lord, from whose book our account is condensed, says : "The skill and tact of the uneducated men, pitted against a huge sea-monster of tenfold strength, was a sight a lover of sport would travel any distance to witness."

The Turbot (Plate IX., No. 8) is the king of the Flat-fishes, although it yields to the Halibut in point of size. The coloured side is brown, and is studded with little hard, round knobs. The Turbot fishery is carried on round our coasts, and on the shores of the North Sea from March to August, and the French and Dutch compete with our own fishermen in supplying English markets. The average weight of English Turbot is from 5 to 10 lbs., but much larger specimens are on record. One taken near Whitby is said to have scaled nearly 200 lbs. The Brill resembles the Turbot in appearance, but its flesh is not so highly valued. Both are taken with hook and line.

The Plaice, a common British sea-fish, may be readily known by the red or orange spots on the coloured sides. Its relation, the Flounder, has no spots, and enters fresh water freely. It has been taken in the Thames as high up as Teddington, and in the Avon within three miles of Bath. The Dabs are small fish of the same kind, common on the sandy parts of the coasts.

Of the smaller Flat-fishes the Sole is the most esteemed for the delicacy of its flesh. It lives on sandy bottoms, and feeds on molluscs and crustaceans. Like the Flounder, some species enter fresh water, and live there partially or entirely. All the smaller Flat-fishes are taken by trawling. Many of these fish have a chameleon-like power of colour-change in accordance with their surroundings. In the Jersey Aquarium some Plaice, put into a shallow tidal pond, where the colour of the bottom varied considerably and a portion was often in deep shadow, changed from a uniform grey to a dark blotched appearance within a few seconds.

In the next order the swim-bladder opens by a tube into the lower part of the throat.

The Cat-fishes form a very large family, exhibiting great diversity in the form and structure of their fins. The skin is naked or covered with bony plates instead of scales, and in many there are barbules, or fleshy filaments, round the mouth, which serve as organs of touch. Most of them are freshwater fish, the few that are marine keep near the coast. There is one European representative, the Wels, or Sheat-fish —next to the Sturgeon, the largest Continental river fish. It is found in rivers east of the Rhine, and Dr. Günther says that in favourable localities, such as the middle and lower course of the Danube, specimens are sometimes taken of from 400 lbs. to 500 lbs. weight. The flesh, especially of the smaller fish, is esteemed for food. Strange stories are told of the fierceness and voracity of these fish, but the evidence for a statement that the body of a child was once found inside a Sheat-fish is not convincing; and the tale of one "taken in Hungary," containing the body of a woman "with a marriage-ring on her finger, and a purse of money at her girdle," is an invention. About thirty years ago it was proposed to introduce this fish into British waters, but a great outcry was raised against the introduction of what was said to be no better than "a freshwater shark." Several gentlemen, however, procured some young fish from the Continent, and put them in their lakes and streams. Nothing was heard of them till August, 1894, when a strange fish, 4 feet 3 inches long, and weighing over 30 lbs., was taken in the Stour, at Stratford Mill, Suffolk. On examination this proved to be a Sheat-fish, probably one turned into a lake some few miles from the place where it was taken.

The Scopelids are fishes from the open sea and the depths of the ocean. Some of them are brilliantly phosphorescent. Dr. Günther says that in one genus dredged "at depths varying between 1,600 and 2,150 fathoms, off the coast of Brazil, near Tristan d'Acunha, and north of Celebes, the eye seems to have lost its function of vision, and taken on that of producing light."

To the Carp family belong many well-known British fish. The Common Carp (Plate IX., No. 2) was brought to Europe from the East, and introduced into England early in the seventeenth century. It frequents still waters, preferring those with muddy bottoms, and plenty of floating vegetation. It feeds greedily on small water-snails and aquatic larvæ, worms and vegetable substances, but can endure long fasts, and will live for some time out of the water. During the winter it lies buried in the mud. Carp are domesticated, and kept in fish-ponds, where they learn to come at a signal to receive their food.

TELESCOPE FISH.

The Gold and Silver Fish kept in ornamental waters and aquaria are varieties of the Carp, as is the Telescope-fish, with eyes on stalks and a large three-lobed tail. Other varieties, and some monstrosities, are known. The Japanese breed, with a double tail, is said to have been produced by shaking the eggs, and then breeding from the individuals in which the double tail appeared.

In the same group are the Barbels, so named from the barbules on the mouth of most of the species. They are freshwater fish, from the temperate and tropical rivers of the Old World. Our English Barbel, also common on the Continent, is usually about 15 inches long, but specimens more than double that size have been taken. The largest species is the Mahseer, with scales sometimes as large as the palm of one's hand, from Indian rivers. Sterndale records the capture of one that weighed 65 lbs., but says that fish from 15 to 25 lbs. give the best sport. They are chiefly animal feeders.

The Gudgeons are confined to the rivers of Europe, preferring

PLATE IX.

1. Perch. **2.** Carp. 3. Pike. 4. Flying Fish. 5. Herring.
6. Salmon. 7. Haddock. 8. Turbot. 9. Eel. 10. Sea-horse. 11.
Globe-fish. 12. Shark. 13. Ray. 14. Lamprey.

clear, running water, with gravelly bottom. They feed like the Barbels. The Common Gudgeon, a small olive-brown fish, with two barbules, is often kept in aquaria, where it soon becomes tame.

The following fish of the same family are British :—The Roach, common also on the Continent, easily distinguishable by the bright red of the fins on the under-surface ; the Dace, with the under fins tinged with red ; the Chub, which somewhat resembles the Dace, but has the hinder part of the anal fin rounded ; the Rudd, or Red-eye, not unlike the Roach in appearance; the Minnow, preyed upon by nearly every other freshwater fish ; and the Tench, noted for its slimy skin and its fondness for muddy water. The Golden Tench is an albino variety.

The Pikes are found in ponds and streams in the temperate parts of both hemispheres. The Common Pike (Plate IX., No. 3), or Jack, is a well-known British fish, which has the range of the family. It is extremely voracious, and causes great destruction in trout-streams. A writer in the *Field* (November 10th, 1894) records two instances of the voracity of Pike. One 23 inches long was found to have inside it a Trout that measured a foot ; and another that weighed 4½ lbs. was shot while attempting to swallow a Trout that scaled 3 lbs. The record Pike for Britain is said to have been taken in Loch Alva. Its length is given as 5 feet 4 inches, and its weight between 47 and 48 lbs. The Norfolk Broads afford good Pike-fishing.

The Cyprinodonts, or Toothed Carps, are small fish, widely distributed in fresh, brackish, and salt water. Most of them bring forth their young alive, and the fins of the males are generally more developed than in their mates, and sometimes brilliantly coloured. Some fish of this family in the Insect House in the Zoological Gardens were adorned with markings like those of the long feathers of a peacock's train. Closely allied to this family is the Blind-fish, from the underground rivers of the Central United States and the Mammoth Cave of Kentucky. It has no external eyes, but these organs, though of small size, are present in a nearly related species, the single known specimen of which was found in a rice-field in South Carolina.

To the Mackerel-Pikes belong the Garfish, of which there are many species, in which the body is long and thin, and the jaws form a long, slender beak. The Common Garfish is found around our coasts, and others live in tropical seas. Professor Moseley, in his "Notes of a Naturalist," says of a scene he witnessed in the Pacific : "I was interested in watching some Horse-Mackerel chase small shoals of young Garfish. The little fish, hotly pursued, dashed out of the water, and by violent lashing of their tails managed to keep themselves above the water in a nearly upright position for a distance of several yards, as

U

they moved swiftly from the danger. Their motion seemed a step towards that of the Flying-fish, their close allies."

To this family belong the Flying-fish, sometimes called Flying Herrings, to distinguish them from the Flying Gurnards. Nearly all the numerous species inhabit tropical seas, but the Common Flying Fish (Plate IX., No. 4) sometimes wanders to the English Channel. Mr. Murdoch, ("From Edinburgh to the Antarctic") says: "In colouring and shape they remind me of our blue dragon-flies : their bodies are deep blue with silver sides, and their gossamer wings shine with the colour of Venetian glass ; some are as large as herrings, and others we see taking very short flights are the size of minnows." Rapacious fish feed on them greedily, and the habit of taking "flying" leaps out of the water has probably been acquired by this species and by the Flying Gurnards as a means of escape from their pursuers.

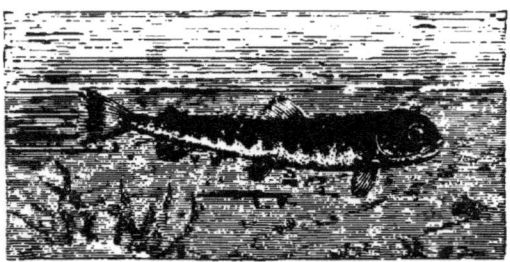

YOUNG SALMON, WITH PARR-MARKS.

The Salmon family are mostly found in the rivers of the northern hemispheres. Many of them descend to the sea after spawning, but some few live in salt water all the year round. All are food-fishes, and most of them are highly valued for the delicacy of their flesh. The young bear dark bands, called parr-marks.

The Salmon (Plate IX., No. 6) is found round the coasts on both sides of the North Atlantic, and in the rivers that fall into that ocean as far south as the Loire on the east and the Hudson on the west. The fish come into fresh water in the autumn to spawn, returning to the sea in the spring. The eggs take about three months or more to hatch out, but the time varies according to the temperature of the water, and few of the young fish go down to the sea in the first year. During the breeding season the lower jaw of the male is developed into a hook, which is used in fighting with rivals, and with which mortal injuries are often inflicted. In the course of their ascent Salmon encounter many obstacles. Rapids must be passed and cascades ascended before they get to the gravelly pools in the upper reaches of the stream. Fish ladders are placed in very many of the falls so as to aid the fish, by giving them platforms from which to leap. In spite of this assistance many Salmon are unable to make the ascent, and fall back into the broken water or on the rocks. On their return journey they are called "spent fish," and are then unfit for food. Fish of from 30 to 40 lbs. are fairly common, but specimens double that size are recorded. The Salmon fisheries of the United Kingdom are very valuable, and are said to yield about a million sterling every year. There are several American species, of which the most important is the King Salmon of the Columbia River. The annual take of this fish is said to average nearly 14,000 tons weight, of which a large quantity is tinned for the European markets.

The Common Trout, a finely marked fish, is common in most British rivers and lakes, and in many Irish waters. It affords good sport to anglers. A Trout from 1½ to 2 lbs. is reckoned a fine fish, but in July, 1894, one was landed from Lough Ennell that weighed 26 lbs. 2 ozs. There are several other species. Charr resemble Salmon and Trout in form and habit, but differ in the arrangement of the teeth. To this family also belong the Smelt and the Grayling, both British fishes, and the Candle-fish, from the North Pacific coast, which is so fat that it is "equally used for food and as candle."

The Paracarú is the largest of the freshwater Bony Fishes, and has been known to exceed 15 feet in length, with a weight of 500 lbs. It is found in the rivers of Brazil and the Guianas; its flesh is esteemed for food, and large quantities of it are salted at the fisheries inland and sent down to the seaports.

SALMON LEAP.
(*Kilmorack, Inverness.*)

The Herring family is extremely important in furnishing food for man. The Herring (Plate IX., No. 5) is exceedingly abundant in the German Ocean and the North Atlantic. It approaches the shore in vast shoals to spawn, and then it is that the Herring fishery commences, for the fish at this time are at their best. Vast quantities are taken and sent to market, while almost, if not quite, as many more are dried and cured as bloaters and red herrings.

The Pilchard greatly resembles the Herring, but is of smaller size, and is not so important commercially, the fishery being confined to

U 2

the south-western counties.   They are put in oil and sold as "Cornish Sardines."   The Sardines which are sent to England from Portugal and the Mediterranean are the young of the same fish.   To the same family belong the Sprat, a common British fish, large quantities of which are tinned like Sardines, at Deal, and the Anchovy—rare on our coasts.

The Electric Eel is the type of a family of South American fresh-

PIPE-FISH AND YOUNG.

water fishes, but none of the others has the power of giving electric shocks. The general colour is dusky above, and red below ; the body is scaleless, and the fish is said to attain a length of 6 feet. Two of these eels are in a tank in the Insect House at the Zoological Gardens, and visitors may receive an electric shock from them on payment of a small fee. Mr. Cornish, in his "Life at the Zoo," thus describes his own experiences :—" The first shock passed up the arm with a 'flicker' identical with that which a zigzag flash of lightning leaves upon the eye, and, as it seemed, with equal speed. A second and third felt like a blow on the funny-bone, and the hand and arm were involuntarily thrown back with a jerk which flung the water backwards on the pavement, and over the keeper, who was kindly assisting."

The Eels are distinguished by their long, snake-like bodies. There are two British species—the Sharp-nosed Eel (Plate IX., No. 9) and the Broad-nosed Eel, or Grig, the smaller. Although Eels are freshwater fish, they descend to the sea or to brackish water to breed, and the young—elvers they are called—make their way to inland waters up-stream, underground, and in some cases over the damp vegetation on the banks. The Conger-Eel is a British sea-fish ; its flesh is said to form the basis of a good deal of what is sold as "turtle" soup. It feeds on fish, cuttles, and lobsters, and when it has taken hold, it revolves rapidly on its axis, and so brings away the piece. Professor Morgan tells an amusing story of two young fellows who went out to fish for Conger. They did not return at the time they had fixed, and a coast-guardsman saw the boat drifting about apparently empty. Help was despatched, and when the boat was reached the two fishermen were in the water hanging on to the stern. They had caught a 40-lb. Conger, which drove them out of the boat as soon as he was hauled in. Here belong the Murænas, one of which was known to the Romans, and highly prized by them.

The Lophobranchs have the jaws produced into a snout, and the gills are composed of tuft-like masses. The males of most species carry the eggs in a pouch or sac till the young are born. The Pipe-fish are so named from their long, slender bodies. The Great Pipe-fish is British, and may often be seen in aquaria, and at the seaside. In many places fishermen dry their bodies and sell them to visitors as curiosities. In this group the tail is not prehensile, but carries a fin. The Sea-horses are small

SEA-HORSE.

fishes, with prehensile tail which carries no fin. They are natives of warm seas, and most in the head and fore part of the body bear a strange likeness to a tiny horse. The Common Sea-horse (Plate IX., No. 10), abundant in the Mediterranean, is rarely taken on our coast. It is a small fish, olive-brown in colour, spotted with bluish-white. It may often be seen in confinement, and presents a droll appearance upright in the water, grasping a piece of seaweed with its tail. The Australian Sea-horse is a very strange-looking creature, covered with spines and filaments. Its colour is nearly like that of the sea-weed among which it lives, and its waving filaments may easily be mistaken for pieces of weed.

The Plectognaths, mostly from tropical seas, have the bones of the upper jaw generally firmly united, and the skin covered with rough scales or bony plates, though in some it is naked. Here belong the

File-fishes, with teeth strong enough to break off pieces of coral, or to make a hole in the shell of a pearl oyster; the Coffer-fishes, with the body enclosed in a kind of box composed of six-sided plates; and the Globe-fishes (Plate IX., No. 11), so called from their power of filling themselves with air, and thus assuming a nearly round shape, and then the spines on the skin stick out, whence they are also called Sea-hedgehogs. A Globe-fish has often been found alive and distended in the stomach of a Shark, and has been known to eat its way not only through the coats of the stomach, but through the side of the Shark, thus avenging itself on its swallower.

COFFER FISH.

The Sun-fishes have a short body flattened from side to side. The Common Sun-fish, which may be 7 or 8 feet long, is often seen basking on the surface in fine weather on our southern coast and off the coast of Ireland.

In the Ganoid Fishes the scales are large, or the skin bears bony plates arranged in rows. In the Sturgeons, confined to the rivers of the North Temperate Zone, the snout is produced, and carries four barbules on the under-side, and the bony plates are disposed in five rows from head to tail. Sturgeon are ground-feeders, and live in fresh water, or, like Salmon, come from the sea into fresh water to spawn. Caviare is prepared from the hard roe, and isinglass from the inner coats of the swim-bladder. There are several species, and the Common Sturgeon is often taken round our shores. Its flesh is said to resemble veal in flavour, and it is often salted for future use; but it is more valued on the Continent than in England. Russia is the chief seat of the Sturgeon fishery. Closely related is the Paddle-fish, or Spoon-bill, of the Mississippi, with a flat, shovel-shaped snout. To this order also belong the Bichir, from the Nile, and other African rivers, having the dorsal fin divided into a number of finlets, the Gar-pikes or Bony-pikes, and the Bow-fin from America. The Bow-fin has swim-bladders that act in some sort like lungs.

The Cartilaginous Fishes are the Sharks and Rays, and the Chimæra or Sea-cat. The skeleton is gristly, the mouth is on the under-surface, the gills open by slits, and there is no gill-cover. Few eggs are produced, and in many cases they are developed within the body of the female. Sometimes they are deposited in oblong capsules, with filaments at each corner serving to attach them to marine plants. These egg-cases are very common on the shore, and are called skate barrows, butcher's trays, and mermaid's purses. Nearly all the fishes of this order are marine, and all are animal feeders.

Sharks are fierce and rapacious. The mouth is armed with rows of sharp, jagged teeth, but fortunately these fish are obliged to turn on the side to seize their prey, and this often allows the intended victim to escape. The White Shark (Plate IX., No. 12), ranging over the tropical seas, is the dreaded man-eater, so hated by sailors, though other species attack bathers and divers. It grows to a large size, and can bite off the limb or even sever the body of a man at a single bite. Sharks often follow vessels for days for the sake of the offal thrown overboard, and fishing for Sharks is a favourite amusement on long voyages, though not always rewarded with success. The late Professor Moseley, after describing the capture of a Shark by the men of the *Challenger*, says that "the sailor has

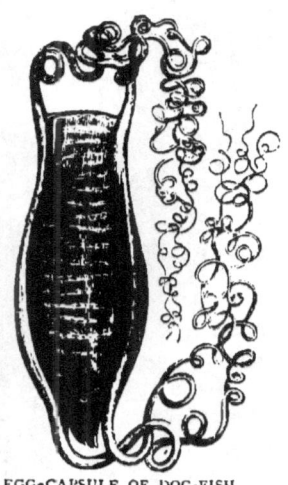

EGG-CAPSULE OF DOG-FISH.

absolutely no pity upon a Shark. I have heard one of our men say to a Shark which he had just hauled on to the forecastle with a line, 'Ah! thou beggar, thou'd hurt I if I was in the water, and now I'll hurt thee,' whereupon he caught it a vicious kick, and proceeded to gouge it. When a big Shark like the present one is landed it is regarded as a general enemy, against whom everybody has an old score to pay off. Mr. Cox (the beatswain) shoves the boat-hook about five feet into its mouth and down its throat. The others job the beast in the eyes with sticks and knives, and make a deep slash across the tail to prevent its lashing out, and proceed to open the belly, where the usual miscellaneous collection is found—lots of ship's beef bones, a two-pound lead sinker of a fishing line, with chop-stick and hooks complete."

The Hammer-headed Shark owes its name to its T-shaped head. It has been taken on our shores. The Topes are small British Sharks, detested by fishermen because they take the bait off the lines, and drive away the fish. The Fox-Shark, or Thresher, sometimes occurs on the British coasts, and feeds on small fish. It is said to attack Whales, but Dr. Günther believes this to be an error. The Basking-Shark, which grows to a length of 30 feet, and is valued for its oil, belongs to the same family. The Dog-fish live on the bottom near the shores, feeding on dead fish, molluscs, etc. They are as obnoxious to fishermen as the Topes. Some of the same family frequent the open sea, as the Tiger-Shark, from the Indian Ocean, which attains a length of from 10 feet to 15 feet. The Greenland-Shark, which rarely strays to Britain from its northern home, is the foe of the Whale, biting pieces out of it while living, and feeding on the carcase when dead.

CATCHING A SHARK.

The Angel-fish, or Monk-fish, some 5 feet long, is common in temperate and tropical seas. It owes its name to its large pectoral fins, which spread out something like wings.

The Australian Saw-fishes may be distinguished from the True Saw-fishes by their smaller size, the gill-slits on the side, and the tentacles at the lower side of the jaw.

The Rays have the body (usually) flattened, the tail long and slender, the pectoral fins wide, and the gill-slits on the under-surface.

Most of them are ground fishes, swimming by means of their pectoral fins, the tail acting as a rudder.

The Saw-fishes have the upper jaw produced into a long, flat plate, armed with teeth on each side. All use this weapon for tearing pieces out of the bodies of other animals, for their true teeth are far too small to seize or hold prey.

There are six or seven species of Electric Rays, one of which, the Common Torpedo, sometimes reaches our southern coasts. The discharge from a large specimen will disable a full-grown man, and these fish may, therefore, be a source of danger to bathers.

The True Rays, in which the back is studded with spines, are more widely distributed in the northern than in the southern hemispheres, and we have several species round our coasts. The Thornback (Plate IX., No. 13) is a common British fish and often attains a large size. Its flesh is used for food, but is less highly valued than that of the Skate, though neither commands a high price. The Sting Ray, or Fire

SAW-FISH.
a. Saw, on larger scale.

Flaire, is so called from the toothed spine in its tail, with which it can inflict a severe wound, often aggravated by the secretion from the skin of the fish. The Eagle-Rays, chiefly from the tropical seas, are of enormous size. Some of the largest are generally seen in pairs. Professor Moseley tells us how a pair of these huge fish, "as big as an ordinary dining-room table, came round the *Challenger* at Cape Verde. One, supposed to be the male, was struck with a harpoon, but after some time managed to draw it out by its struggles. It twisted up the harpoon, and was said even to have moved the ship in its throes."

The Chimæras are strange-looking fishes with Shark-like body and the long whip-like tail of the Rays. One species, the Northern Chimæra, or Sea-Ape, is sometimes called the King of the Herrings, from its following shoals of these fish in their migrations.

CYCLOSTOMES, OR ROUND MOUTHS.

To this group belong the Lampreys and Hags—creatures with eel-like bodies but with no jaws, the mouth being leech-like. There are no paired fins, and the gills are enclosed in little pouches. The young of the Lampreys pass through a metamorphosis.

There are three British Lampreys. The Sea-Lamprey (Plate IX., No. 14) is common round our coasts, and ascends the Severn and Thames.

It is often called "Nine-Eyes," though, of course, it has but two. A glance at our illustration will show the origin of the name—for one can see the single nostril, one eye, and the seven gill-openings on one side, and these counted together make nine. This species grows to a length of from 2 to 3 feet, and, like the smaller River Lamprey and Pride, is esteemed for food. All feed on fish, to which they attach themselves with their leech-like mouth and rasp off the flesh with the teeth on their lips and tongue.

Less is known of the Hags—or Slime-fishes, as the Germans call them, from the fact that they secrete a great quantity of mucus from the skin. This was known to the ancients, who said that the Common Hag "could turn water into glue." One species is found round our coast, where it bores into living fish, a bad habit shared by other species. Professor Lloyd Morgan says :—"Sometimes at Scarborough a haddock may be drawn up on one of the long lines. From the external view, that is to say, it is haddock; but within it is all Hag-fish. For these curious eel-like creatures pass through the gills of recently dead or dying fishes, and devour the whole of the soft materials inside, leaving nothing but bones and skin."

" NINE-EYES."

CHAPTER XXIV.

THE BORDERLAND.

BEFORE we get to the various sub-kingdoms of Backboneless Animals, there is a Borderland to be traversed, wherein the creatures found are "links" between the higher and lower groups. Or perhaps it would be better to say they represent stages in the path up which the higher animals travelled. For of this nothing is known, though much may be inferred with probability. But whether the path has been upward or downward, there can be no doubt that these dwellers in the Borderland show many striking affinities with their higher relations and neighbours.

The Lancelets, of which there are several species or varieties, are found round the coasts of most warm and temperate seas. The

LANCELET.

Common Lancelet is taken in the British Channel, with the dredge, and on the coast of Jersey it has been dug up from the sand. Its length is about 2 inches, whitish in colour, almost transparent, and pointed at both ends. Lancelets are dull and sluggish in habit, lying buried in the sand, with the mouth, fringed with tentacles, protruding, and they feed on the minute organisms in the water, which is sucked in at the mouth, and passes out at a pore a little in front of the vent. They are said to swim about in the evening, and the young are more lively than the adults.

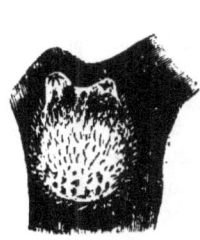
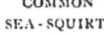
COMMON SEA-SQUIRT.

BOTRYLLUS.

The Sea-squirts were known to Aristotle, who dissected and described some of them. It was not, however, till about the middle of the nineteenth century that their true place was ascertained. Their name, Tunicates, refers to the fact that they are clothed in a leathery test or tunic, which encloses the body, while the popular name, Sea-squirt, is due to the fact that they can, when disturbed, emit a tiny jet of water from either of the two apertures. The body may be compared to a flask with two mouths; through one the water enters, bearing with it the minute organisms on which the Sea-squirts feed, and passes out by the other. The motion of the water is due to the waving to and fro of the cilia, or fine hairs, with which the passage is fringed. The young are like tiny tadpoles in shape, and possess an eye, a mouth, a brain, a notochord in the tail, and gill-slits. Up to a certain point the development is extremely like that of the frog; and though in some few cases the high standard of the larval stage is maintained through life, they generally settle down, the tail with its notochord is absorbed, the eye and mouth are lost, and they become sad examples of degeneration. Many are common round our shores.

There are three distinct groups. In the lowest of these, which are of small size, the larval condition is retained through life. Next come the Common Sea-squirts, some of which multiply by budding from a stolon. Some are compound, like Botryllus, which may be found on the fronds of the bladderwrack on all our coasts. Each individual of a colony has an orifice at which the water enters, but there is a single opening in the centre of the colony for its exit. In the Fire-flame, which is brilliantly phosphorescent, the union is very close, and the whole colony swims like one creature. The last group consists of barrel-like forms, the water flowing in at one end and out at the other. These swim in the open sea, and it was in one form (*Salpa*, which unite in chains, phosphorescent at night) that the poet-naturalist Chamisso observed what is called the Alternation of Generations—a sexless individual, or "nurse," giving rise to male and female buds, the progeny from which develop into "nurses," and so on.

ACORN WORM.

The Acorn Worms live in sand or mud in the English Channel, the Mediterranean, and on the eastern coast of North America. They vary in length from 1 inch to 6 inches. They live at a depth of about 8 inches below the surface of the sand, and the body is generally coiled in a corkscrew-like spiral. The proboscis and the hinder part are usually held upright; the latter moving up and down in a shaft in the sand opening to the surface, through which casts like those of the earthworm are thrown up.

CHAPTER XXV.

MOLLUSCS.

HE classification of backboneless animals is far more difficult than was that of the Vertebrates, which possess so many characters in common that it is easy to compare the creatures of one class with those of the classes below them. But when we pass the limits of the Borderland, nothing of this kind is possible. Our knowledge is far too limited to enable us to set out, with anything like certainty, the relationship between the various sub-kingdoms (Molluscs, Arthropods, Echinoderms, "Worms," Stinging Animals, Sponges, and One-celled Animals) which are grouped together as Invertebrates, or animals without a backbone. Below the "Worms," we meet with creatures of a different grade. From Man down to "Worms" every living creature has a body-cavity between the food canal and the body walls, and this may be compared to one tube inserted in a larger one. In the Stinging Animals and Sponges there is but a single cavity, that of the food canal. The lowest animals of all consist of a single cell, or a collection of single cells, each independent of its neighbour.

We commence with the first of the above groups or sub-kingdoms.

Molluscs are soft-bodied animals, not divided into segments. The body may be naked, or covered with a shell, secreted by a fold of the skin called the mantle. The higher forms have an internal hard shell or cartilage. They fall into three groups, of which the Octopus and Cuttle-fish, the Whelk, Snail, and Slug, and the Oyster, Scallop, and Mussel are familiar examples; and these three groups are called respectively Cephalopods, Gastropods, and Bivalves.

The Cephalopods are so named because their "feet" (better called arms or tentacles) surround the head, which is armed with a pair of jaws, not much unlike the beak of a parrot. These arms are long, and, except in the Pearly Nautilus, bear on one side suckers, by means of which these Molluscs can hold securely anything they may seize. With these arms they walk, fix themselves to rocks, and capture prey. They can only swim backwards, and this is effected by forcibly ejecting the water from the gill-cavity through the funnel. In all the brain-mass is protected by cartilage. The gills lie on each side: in most forms there are two; the Pearly Nautilus alone has four. Of those with two gills, some have eight arms, and others ten. Only the Paper Nautilus and the Pearly Nautilus have external shells, the others are called Naked Cephalopods. All but the Pearly Nautilus secrete an inky fluid, which can be expelled at will through the funnel, so as to cloud the water and baffle an enemy. To the first group belong the Paper Nautilus and the Common Octopus.

The Paper Nautilus, or Argonaut (of which there are four species), is found in all open seas in the warmer parts of the globe. Aristotle

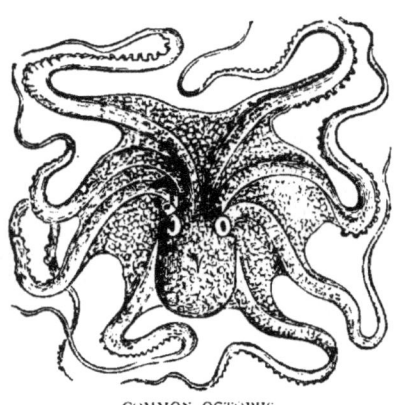

described the common species, and gave currency to the story that the creature floated on the surface in fine weather, spreading out its two sail-shaped arms to catch the breeze. This was repeated, in prose and verse, down to very recent times. It is now known that only the female has a "shell," and that this is nothing but an egg-case, secreted by the large arms, which also hold it in position, for its occupant is not attached to it in any way. The male is about an inch long.

COMMON OCTOPUS.

The Common Octopus is the type of a family in which the arms are united at the base by a web, and the internal shell is quite rudimentary. The Octopus and its fellows of the same genus are found on rocky coasts, and in all tropical and temperate regions. They are active and voracious, and feed on living prey, and some are large enough to be dangerous to man. Mr. Lord, when in Vancouver Island, measured the arm of an Octopus that was 5 feet long, and as thick at the base as a man's wrist, and says : "Were it by any chance to wind its sucker-dotted cable arms round a luckless bather, fatal would be the embrace, and horrible to imagine the being dragged down and drowned by the eight-armed monster — a worse death than being crushed by coiling serpents, like ill-fated Laocoön."

In the next group, in addition to the eight arms, there are two longer ones, generally wider at the ends. In most cases they can be wholly or partially retracted into pouches. In the Squids the internal shell is called a "pen," from its resemblance to an untrimmed quill. The sides of the body are produced so as to form fins.

The Common Squid is very plentiful round the British coasts, and is used as a bait for ground-fish. Fishermen call it the "pen-and-ink" fish, from the readiness with which it discharges its ink-bag. To the Squids belong some of the largest Cephalopods, and their appearance in mid-ocean may have given rise to some of the stories of the "Sea-serpent." Some of the larger Squids have been known to attack pearl-fishers. The Flying Squids, or Sea-arrows, often leap out of the water, and sometimes fall on the decks of ships.

The Common Cuttle-fish (Plate XIII., No. 4) is well known round our coasts, and most people have seen their eggs, which fishermen call "sea-grapes." It is beautiful and very active, but seldom lives long in confinement, though specimens may generally be seen at the various English aquariums. It is exceedingly voracious, and some localities on our coast are sometimes so infested by Cuttle-fish that drift netting has to be abandoned in consequence of their devouring the fish, or mangling them with their horny beaks as they hang in the meshes.

The Spirula, from the Indian and Pacific Oceans, is best known by its shell, which is not altogether internal. These shells are plentiful enough, but of the habits of the creature nothing is known.

The Pearly Nautilus, from the Indian Ocean, has a chambered shell (Plate XIII., No. 5), in the last and largest chamber of which it lives. There are four gills on each side, and the arms are replaced by retractile tentacles, which also act as feet, enabling the animal to crawl on the bottom in search of small crabs and molluscs, on which it feeds.

SEA-BUTTERFLIES.

These creatures form a distinct class, and have a pair of fin-like organs developed from the sides of the mouth, which, when flapped backwards and forwards, propel their owners through the water. Some have a glassy shell, others are shell-less. They are extremely abundant in all open seas. One Northern species forms a considerable item in the food of the Greenland Whale.

GASTROPODS.

The "belly-footed" Molluscs crawl or glide by alternately expanding and contracting the broad under-surface of the body, which is called the "foot." One may watch this mode of progression by putting a Periwinkle or Limpet into a vessel of seawater, or by fixing a garden snail or slug to the outside of a window-pane. Some few, however, are free-swimmers. The head is more or less developed, and there is a rasping tongue on the floor of the mouth. When a shell is present it consists of a single valve. They may be grouped in four divisions : (1) The Prosobranchs, all of which have a shell, with the gills between

the head and the heart; (2) The Opisthobranchs, with the gills behind
the heart, and the body naked, or only the gills protected by a shell;
(3) The Nucleobranchs, with the gills at the hinder part of the back,
sometimes protected by a shell; and (4) the Pulmonates, or Air-breathers,
in which gills are replaced by a breathing-cavity in the mantle.

The Prosobranchs fall into two groups, which may be distinguished
as flesh-feeders and vegetable-feeders. In the first, of which the Whelk
is a good type, the siphon which carries the water to the gills is long,

COMMON WHELK.

and is lodged in a
canal between the
outer lip of the
shell and the last
or body whorl. In
the second, of which
the Periwinkle is a
representative, the
siphon is short, and
there is no canal.

The Wing Shells
are the largest
of the Gastropods.
There are numerous
species, from warm
seas. They act as
scavengers, and
move with a kind

of jumping motion. The Fountain Shell, from the West Indies, is a
common ornament for garden fountains and rockwork. It is sometimes used for cameo-carving, and largely for the manufacture of
porcelain. The popular name refers to the wide wing-like lip. Here
belong the Scorpion Shells, or Spider-claws, in which the outer lip
is produced into spines.

The Rock Shells are noted for the spines with which many of the
species are furnished. They feed chiefly on other molluscs, boring
through the shell and devouring the victim piecemeal. The Sting-
winkle, noted for this bad habit, is very common on the shores of the
English Channel. The celebrated Tyrian purple was obtained from
Mediterranean species. Here belong the Mitre Shells, so called from
their shape. They live in tropical and subtropical seas near the shore.
The Bishop's Mitre (Plate XIII., No. 7) is common in all collections of
shells. The Spindle Shell, or Red Whelk, the Buckie of the Scotch
fishermen, is not unlike a Whelk, but has a much longer canal.

The Whelk family has several British representatives. The Common Whelk is known to everyone; its shells are scattered on almost every shore. It is largely used for food, and by fishermen for bait. The Whelk in the aquarium is exceedingly pretty, as with its large white foot extended it crawls up the sides of its glassy prison, and its white is rendered more brilliant by contrast with the dark operculum, or plate with which the mouth is closed, when the animal retires into its shell. The Dog-whelk is a much smaller species, common on the shores of the Channel, and has the bad habits of the Sting-winkle. The Common Purple Shell is abundant round our shores, and is very destructive to mussel-beds. Its eggs, like tiny stalked cups, may be found on the under-surfaces of overhanging rocks. Magilus lives as a parasite in coral; and though the young have spiral shells, as the animals grow the shell is produced into a tube. The ribbed Harp-shells, and the Olives, so called from their shape, belong to this family. Both are from tropical seas.

The Helmet Shells are valued for cameo-cutting, as they consist of different coloured layers. The Trumpet-shells, or Tritons, are used by the natives of the Pacific Islands as musical instruments. A hole in the top of the shell serves as a mouthpiece, and the player inserts his hand in the aperture of the shell to modulate the sound.

The Cones, so called from their shape, are tropical. CONE-SHELL.
They are exceedingly beautiful, brilliantly coloured, and variously marked. On this account they are valued by collectors, and some of the rarer species fetch very high prices. The Auger shells are greatly elongated, and are not carried by their owners, but dragged along after them on the bottom.

The Volutes, also from tropical seas, have a short spire. Like the Cones, they are valued by collectors, but do not fetch such high prices. The young of the Boat Shells and Melons are born alive.

The Cowries when adult have a shell in which the last whorl wholly or partly conceals the rest. Nearly all are brilliantly coloured, and the Tiger-Cowry is a common mantelpiece ornament. The Money Cowry passes as coin in Africa; and one very small species, the European Cowry, is found on our own shores. To this family belongs the Weaver's Shuttle, with a long canal at each end. It feeds on the barked corals, and some of its smaller relations resemble the Corals which form their food, and among which they live.

The Periwinkles are found near the shore, and in brackish water,

V

and feed on seaweed, especially on the smaller kinds that cover the rocks. Vast quantities of the Common Periwinkle are consumed year by year, and it is said that in London alone the annual value of those brought to the markets is about £15,000. If a few of these animals be brought home from the sea-side, and kept with some seaweed in a glass jar of sea-water, their mode of walking, and their feeding with

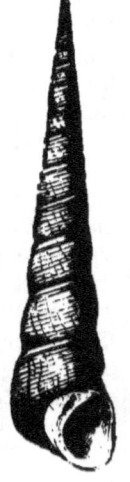

their rasp-like tongue on the vegetable growth that will soon cover the inside of the jar, may be watched at leisure. To this family belong the Carrier Shells, which cement to the outside of their dwellings stones, coral, and pieces of other shells, probably to serve as a protection, as they live on rough bottoms usually strewn with dead and broken shells.

WENTLETRAP.   The Naticas, some of which are British, have a roundish shell with few whorls. The animals of the type genus are blind, and burrow in the sand for small bivalves.

The Staircase Shells, from tropical seas, owe their name to the fact that the spiral edges of the whorls seen on the under-surface, look something like a tiny winding staircase.

The Wentletraps are widely distributed, and have the whorls ribbed. The Precious Wentletrap was at one time very rare, and £40 has been given for a single specimen, which could be bought now for about as many pence.

Some of the Horn Shells are found in brackish water, but most are marine. A species from Borneo attaches itself to water-plants above the surface by glutinous threads. The shells of another species are burnt for lime.

The Tower Shells are so called from the long, tapering shape. They are widely distributed, but most abundant in tropical seas. Of the same family are the Worm Shells. These are free and of a spiral form when young, but afterwards lose their regular shape, and are found massed together or attached to rocks and stones.

Our Common River Snail is the type of a family found in fresh-water all over the world. The young are brought forth alive, and it is easy to watch the development and growth of the young Snails for oneself in a fresh-water aquarium. Specimens may be obtained from any pond or marshy stream, and a few sprays of water-weed will supply ample food. The Apple Snail is a tropical species, which buries itself in the mud in the dry season.

TOWER SHELL.

The Top Shells are marine. The opercula of some of the tropical species are very beautiful and are often mounted as jewellery. Several British species are almost as beautiful as the foreign ones.

The Ear Shells adhere firmly to rocks with their large foot. The interior of the shell is pearly, and is used for inlaying. We have none on our own shores, but in the Channel Islands the Ormer is abundant, and used for food, after being well beaten to make it tender.

The Violet Snails congregate in the open sea, and are sometimes washed to our western shores by the Gulf Stream. They secrete a float or raft, to the under-side of which the eggs are attached, and then raft and eggs are set adrift.

The Limpets need no description. Everyone has seen their conical shells studding the rocks on the shore. The Common Limpet is largely used for bait, and in some parts it is eaten. Like the Periwinkle, it does well in the aquarium, and will help to keep the inside of the glass free from vegetation. If one has a Periwinkle and a Limpet it will be interesting to notice the difference in marks left by their rasping tongues. A few species may be found on the stems and in the root-like disks of the larger seaweeds, and their shells are thinner than that of the common species. The Mexican Limpet has a shell a foot across at the base. The Keyhole Limpets have a hole on the top of the shell.

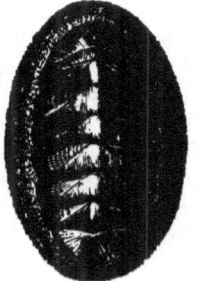

CHITON.

The Chitons, or Coat of Mail Shells, have the back covered with eight shelly plates, that overlap like the slates on the roof of a house. They can roll themselves into a ball, like the Pill-bugs or Wood-lice. In some tropical forms the back is studded with eyes, and as many as ten thousand have been counted in one animal.

The Tooth Shells are in appearance not unlike an elephant's tusk in miniature. They have no head, eyes, or tentacles; and burrow in the sand for minute animals. Some are British, and one North American species was used by the Indians for money.

To the Opisthobranchs belong the Bubble Shells, in which the shell is well developed, and the Sea Hares, in which it is rudimentary and serves to protect the gills. The Common Sea Hare is found round our shores, and is common in the Channel Islands. It may be 2 or 3 inches long, and as it crawls along, with its upper tentacles erect, it has some resemblance to a tiny hare or rabbit. The Sea Lemons have the gills naked. These and their allies are popularly called Sea-slugs, a name also given to the Sea-cucumbers. About a hundred, mostly small, are found round our shores and in rock-pools near low-tide mark. They are exceedingly beautiful, but from their flesh-eating habits are dangerous inmates of an aquarium.

The Nucleobranchs comprise a few free-swimming forms with glassy shells. That of the Glassy Nautilus at one time was so rare that a specimen was worth £100. Prices have fallen since then, and these shells may now be bought for a few shillings.

The Pulmonates, or Air-breathing Molluscs, comprise the Snails, Slugs, and Pond Snails, etc. Few Land Snails have an operculum. Most of them bury themselves in the winter, and then secrete a covering for the mouth of the shell. In this condition they have remained for years, and revived when exposed to warmth. The Common Snail needs no description. Everybody knows it, and specimens are always to hand—too plentifully gardeners would say. The Roman, or Edible Snail (Plate XIII., No. 6), though common on the Continent, is rare with us, and is said to have been introduced by the Romans, who were as fond of them as many people are now. This, the largest British snail, is small in comparison with some tropical forms, for the African Agate Shell is from 6 to 8 inches long. The foot of the Slugs extends the whole length of the animal, and there is no true external shell. They frequent damp places, and do great damage to growing crops. Testacella, rare in this country, has a small ear-shaped shell on the hinder part of the body. In the genus Oncidium, the animals have eyes on the back like the Chitons, though not nearly as many.

The Pond Snails are abundant in streams and standing water all over the country. They make good inmates of the aquarium.

THE BIVALVES.

The shell of these Molluscs, which have no head, true eyes, or rasping tongue, consists of two valves. The point of union is called the hinge, and at this point the valves are connected by an elastic horny ligament, and projections or teeth which interlock may also be present. Powerful muscles close the shell. Many are attached to rocks or live on the bottom; some burrow into sand, coral, stone, and wood, and a few swim by opening and shutting the valves. All are aquatic, and most of them marine. In the first group there is no siphon for respiration.

The Common Oyster (Plate XIII., No. 9) is the type of a family. In the illustration the shell is open and the animal exposed; the ligament may be seen and the radiating lines round the oyster mark the gills, fringed with tiny hairs, which waft to the mouth the organic particles forming the food of all Bivalves. Oysters are widely distributed, and most of the species are eaten. The British Oyster takes high rank as a delicacy, and oyster culture is an important industry. Oysters are in season from September 1st to April 30th. The close season prevents their being taken while spawning. Dog-whelks do great injury to oyster-beds.

Scallop-shells may be seen in plenty on the shore, and in the winter the Scallop, with its brilliant white and orange, ornaments every fishmonger's slab. The shell of an allied species was the badge of a pilgrim to the Holy Land in the Middle Ages. The Squin, a small form also valued for food, is common round our shores.

The Pearl Oysters, or Wing Shells, are tropical. The Common Pearl Oyster (Plate XIII., No. 10), from which are obtained most of the pearls, so highly valued as ornaments, are fished in the Persian Gulf, and off the coasts of Ceylon and West Australia. The shell is lined with thick layers of mother-of-pearl; and pearls are outgrowths from this layer, or are found loose among the fleshy parts of the animals. A fish and a small crab covered with mother-of-pearl, and so made into pearls, may be seen in the Natural History Museum. Here belong the Hammer Oysters and the Fan Mussels, from the byssus, or mooring-threads of which, mixed with silk, gloves and stockings have been manufactured.

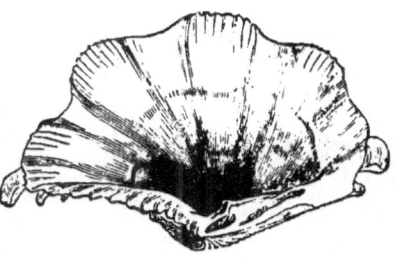

GIANT CLAM SHELL.

The Sea-Mussels have wedge-shaped shells, and live attached to rocks, piles, and the piers of bridges by their mooring-threads. Large quantities of Mussels are sent to market; but at some seasons they are unwholesome, though the cause is not known. The Common Mussel may be kept in a small aquarium, where it will moor itself to the sides; and if one situation does not suit the creature, it will shift from place to place. The Date Shell bores into hard limestone, and the holes made by it in the columns of the Temple of Serapis at Puteoli, show that the level of the coast there has changed within modern times. The Zebra Mussel has been introduced into British docks and rivers by foreign timber ships. To the River Mussels belong the Pearl-bearing Mussel of our northern streams, whence Scotch pearls were once obtained, and the Swan-Mussel, found all over Europe.

Of the Bivalves with a siphon, the Common Cockle, found on all sandy shores, is probably the best known. The foot is bent and used for leaping. Kingsley compares the foot of one of the larger British Cockles to a great red capsicum, and tells how a lady, when she saw it, exclaimed, "Oh, dear! I always heard that my pretty red coral came out of a fish, and here it is all alive!"

The Giant Clams of the Indian Ocean are the largest known bivalves. The shells may often be seen in the windows of oyster-shops. One in the Natural History Museum weighs 510 lbs. and measures 3 ft. the longest way.

, The Venus Shells, noted for their beauty of form and colour, generally live buried. One species was used by the North American Indians for money.

The Razor Shells are to be met with on all sandy coasts. They burrow end-on in the sand, and their presence is shown by a small hole. On the Devonshire coast people bore for them with iron rods armed with a screw, and they are sold for food.

PIDDOCK IN ROCK.

The Stone-Borer may be found in holes in rock, and in the disks of limy seaweed in rock-pools. There is but one species, which varies so greatly that it has received a number of names. The Watering-pot Shell, of the same family, are tropical, and burrow in mud or sand. The shape is that of a tube, closed at one end by a perforated disk, and furnished with limy frills or ruffles at the other.

The Piddocks bore into rocks, wood, and sand. The submarine forest near Hunstanton abounds with them, and the Common Piddock may be met with all round our coast. In Devonshire it is used largely for bait, and some foreign species are eaten. The foot is probably the boring tool, the ridged shell being used, by turning from side to side, to enlarge the hole, while rubbish is removed by the siphon.

The Ship-Worm (Plate XIII., No. 11), with a tiny shell at one end, and the body enclosed in a tube, is from 6 inches to a foot long. The two tubes at the other end are the siphons; the shorter conveys water to the gills, the longer carries off the water charged with the woody pulp excavated by the foot. These Molluscs are very destructive to ships, piles, and piers, and the only defence against them is sheathing the wood with metal, or studding it with flat-headed iron nails.

CHAPTER XXVI.

ARTHROPODS. INSECTS.

THE Arthropods have the covering of the body divided into a number of rings or segments, hardened by the deposition of a horny substance, as in the Beetles, or of carbonate of lime as in the Crabs, and united by soft skin, in which is little or no hard matter. Here belong the Insects, the Spiders, Scorpions, and Mites, the Centipedes and Millipedes, and the Lobsters, Crabs, and their near relations.

In INSECTS, three distinct parts may be made out—the head, thorax, and abdomen. The head bears the eyes, mouth organs, and antennæ, or feelers. The eyes are usually very large, and composed of a great number—sometimes many thousands—of small facets; and in addition to these compound eyes there may be simple eyes, of a single facet, on the top of the head between the compound eyes. The antennæ, or feelers, vary very much in size and shape, and probably serve partly as organs of touch, in some cases for smell, and it may be also for hearing. The thorax bears the wings and legs: of the former, there are usually two pairs, but there may be only one pair, or none at all; and of the latter there are always three pairs in a perfect insect, varying in shape according as they are used for walking, burrowing, or swimming. The abdomen contains the vital organs. Insects breathe by means of tubes which open to the atmosphere by breathing-pores, or stigmata. There is generally a metamorphosis, more or less complete. From the egg is hatched the active larva, which, after a period of growth, passes into the third stage as a pupa, there undergoing transformation, and finally appearing as a perfect insect.

The Insects with a complete metamorphosis may be divided into two groups, according as the mouth-organs are formed for biting or sucking. To the first belong the Beetles, the Bees, Wasps, and Ants, and the Ant-Lions and Scorpion-Flies. To the second belong the Butterflies and Moths, the Flies, and the Fleas. In the Bugs, with sucking mouths, and the Straight-winged Insects and the Spring-tails, both with biting mouths, the metamorphosis is incomplete, or the young are miniature copies of their parents.

BEETLES.

In this Order the fore pair of wings are changed into horny coverings for the hinder pair, which are used for flight. It is easy to lift up the wing-cases, when the wings will be seen lying neatly folded above the top of the abdomen.

The Tiger Beetles are well represented in Britain. The Green Tiger Beetle (Plate X., No. 1), a very common species, is so active and rapacious, that it is no easy matter for any insect of small or moderate size to escape its attack. There are about a thousand species of the family to which this Beetle belongs; most of them frequent sunny places, and in the East they are so numerous that they rise like swarms of flies.

The Ground Beetles are a much larger family, and no part of the world is without some of them. The Large Gold Beetle (Plate X., No. 3) is very rare in this country, as is the Gold Beetle (Plate X., No. 2). The latter is very common on the Continent, and the few taken here were probably brought over with foreign vegetables.

The Bombardier Beetle (Plate XI., No. 4), a common British Beetle, is the type of a group with a curious means of defence. When attacked, they discharge from the hinder part of the body an acrid fluid, that explodes on contact with the air, with a sound like that of a popgun, whence they are often called Artillery Beetles. The explosion is accompanied with a whitish vapour. To the Ground Beetles belongs the Fiddle Beetle of the Japanese, while another strange form from the Malayan Peninsula might be compared to a miniature double-bass.

The Margined Water Beetle, found in almost every pond, is extremely voracious. The hind-legs, flattened and fringed with hairs, are capital swimming-organs, and the male has a peculiar pad, or cushion, on the first pair. It may be kept in an aquarium by itself, and fed with tiny pieces of meat or fish; but where this beetle is, there should be no other living creature that one hopes to keep, for it is fierce, and strong, and has an insatiable appetite. It rises to the surface from time to time to take in a fresh supply of air, which flows under the wing-cases to the openings of the air-tubes. The Whirligigs, which spin about on the top of the water, are near relations. The Great Water Beetle, which does not paddle, but swims like a dog, is to some extent a vegetable feeder. The female encloses her eggs in a kind of cocoon.

In the Rove Beetles the wing-cases are very short. Of this group the best known is the Devil's Coach Horse, a large black beetle, common on roads, and emitting a strong, fetid smell. When alarmed, it raises its abdomen in threatening fashion, as if it were a scorpion

PLATE X.

1. Green Tiger Beetle. 2. Gold Beetle. 3. Large Gold Beetle.
4. Bombardier Beetle. 5. Red Rove Beetle. 6. Bacon Beetle. 7.
Museum Beetle. 8. Stag Beetle. 9. Sacred Beetle. 10. Rhinoceros
Beetle. 11. Hercules Beetle. 12. Rose-chafer. 13. Click Beetle.
14. Cucujo. 15. Glowworm. 16. *Trichodes apiarius*. 17. Blue
Oil Beetle. 18. Blistering Beetle. 19. Apple Weevil. 20. Musk
Beetle. 21. Poplar Beetle. 22. Lily Beetle. 23. Common Ladybird.

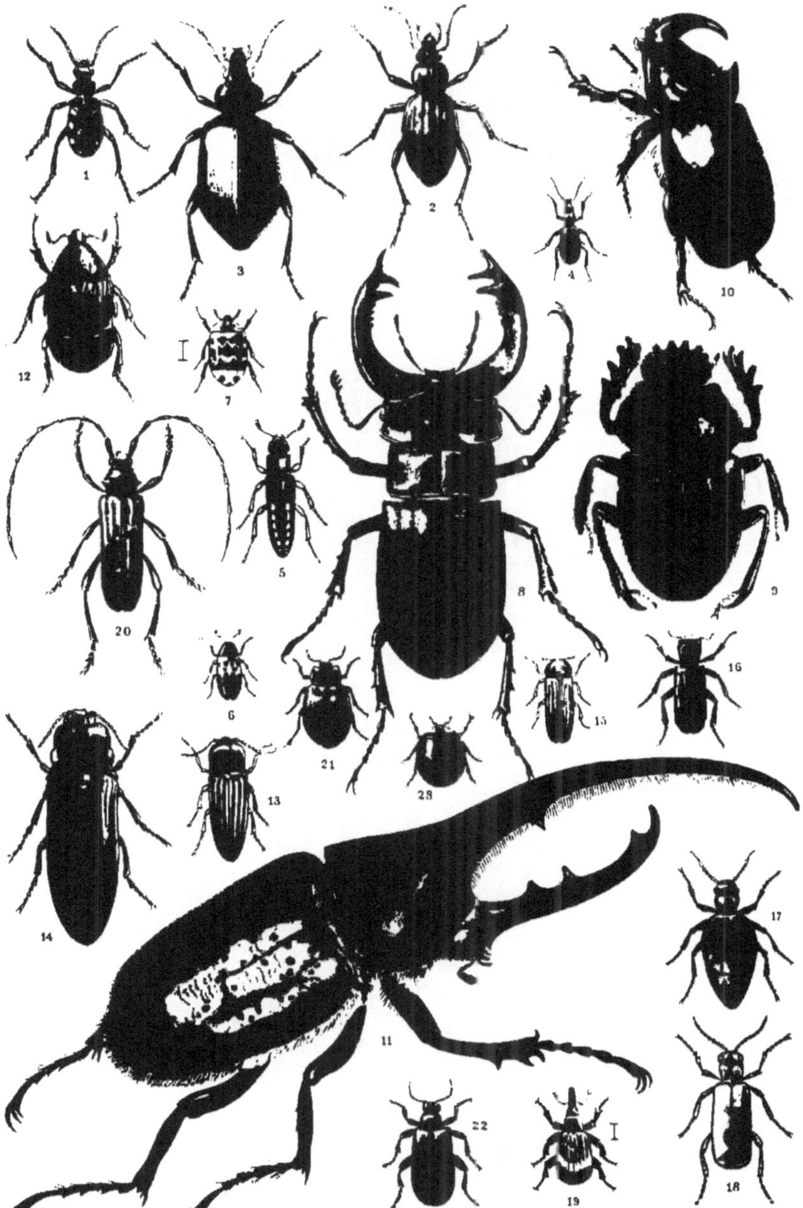
1
2
3
4
5
6
7
8
9
10
11
12
13
14
15
16
17
18
19
20
21
22
23

about to sting. The Red Rove Beetle (Plate X., No. 5) is rare in Britain. The rare Humble-Bee Beetle, found in dung-heaps, owes its name to its likeness to the Humble Bee.

Many of the Burying Beetles feed on, and some of them deposit their eggs in, carrion, which serves as a food-supply for the larvæ when they leave the egg. The Common Burying Beetle, plentiful enough in this country, is black, with two orange bands across the wing-cases. Four of these creatures have been known to bury, in the space of seven weeks, two moles, four frogs, three little birds, a couple of grass-hoppers, the entrails of a fish, and two pieces of meat. Near relations are the Bacon Beetle (Plate X., No. 6), which feeds principally on cured meat, and the Museum Beetle (Plate X., No. 7), which feeds on dry animal matter, destroying stuffed specimens, and when it invades insect collections, often leaving nothing but the pins on which the specimens were mounted.

The Stag Beetles feed on the juices flowing from wounds in trees; the larval stage is passed in tree-trunks, and at this period the diet is wood. The Common Stag Beetle (Plate X., No. 8) is found in the southern counties of England, and reaches a length of 3 inches. The large antler-like mandibles of the males are used as weapons in their combats; the smaller ones of the females are employed in making holes in the trunks of oaks and willows to receive the eggs. The larva was highly prized as a delicacy by the Romans.

The Sacred Beetle of the Egyptians (Plate X., No. 9) is the type of a group which form pellets of dung, with their long hind-legs, to form a store of food for themselves or their larvæ, though the fact of the Sacred Beetle depositing its eggs in dung-pellets has been disputed. Closely allied is the Dor Beetle, about an inch long, and blue-black in colour, whose loud humming flight is well known.

The Cockchafer is extremely common, especially on the Continent. It has large fan-shaped antennæ.

The Rhinoceros Beetle (Plate X., No. 10), so called from its pro-jecting horn, is a native of the East, and very destructive to the cocoa-palm. Closely allied is the Hercules Beetle (Plate X., No. 11), from Guiana and the West Indies, in which the head and thorax are pro-duced into long processes, often mistaken for jaws.

Of the Rose Chafers, the common species (Plate X., No. 12) is abundant in the South of England. It has numerous allies, and to it is related the Goliath Beetle, a specimen of which lived for a short time in the Zoological Gardens in 1893.

The Click Beetles, or Skip-Jacks, owe their name to their power of springing up and regaining their feet when turned over. A spine on the thorax fits into a notch on the abdomen. When the body is raised the spine is released; it returns to the notch when the body is straightened, and this check is enough to throw the insect two or three inches into the air. The larvæ of some species are called wire-worms, and feed on roots. The Common Click Beetle (Plate X., No. 13) is abundant in this country. Some foreign species are luminous, as is the Cucujo (Plate X., No. 14), from Mexico and South America.

' The Glow-worms, despite their name, are true Beetles. The Common Glow-worm (Plate X., No. 15) is British, and the wingless female emits a brighter light than her winged mate. *Trichodes apiarius* (Plate X., No. 16), a near relation, is found on flowers; the larva is destructive to bees. The Death Watch, a small brown Beetle, less than ½ inch long, lives in old furniture and wainscoting. The male makes a ticking noise by knocking its head against the wood, and this is supposed to presage the approaching death of some member of the household.

The Blue Oil Beetle (Plate X., No. 17), is a vegetable feeder, though its larva, like that of Trichodes, preys on bees and honey. To the same family belongs the Blistering Beetle (Plate X., No. 18), or Spanish-Fly, found in this country; but those used for making blisters are imported from Spain.

The Weevils may be readily distinguished by the beak, into which the front of the head is produced. They form a numerous group, feeding on plants, and boring into fruit, seeds and wood to deposit their eggs. The harm done by the Apple Weevil (Plate X., No. 19), the Nut Weevil, and the Pea Weevil is well-known, and everybody has met with the maggot-like larvæ of these Beetles.

The Musk Beetle (Plate X., No. 20), which has a strong smell of musk, is common on old willow trees. The group to which it belongs has very long antennæ, and most of these Beetles can make a chirping noise by rubbing them against the wing-cases.

COLORADO
BEETLE.   THE LARVA.

Of the Plant-eating Beetles we can only mention the Colorado Beetle, which caused such destruction to potato crops in the United States between 1860 and 1870; the Poplar Beetle (Plate X., No. 21), and the Lily Beetle (Plate X., No. 22). The larva of the Lily Beetle covers itself with its droppings, probably for protection.

The Lady-Birds are found on plants and feed on plant-lice. When touched a yellowish liquid exudes from them. The Common, or Seven-Spot, Lady-Bird (Plate X., No. 23) is abundant in this country.

ANTS, BEES, WASPS, ETC.

These insects have two pairs of membranous wings, the hinder and smaller pair furnished with hooks by which during flight they are fastened to the hinder margin of the front pair. The females have an appendage at the end of the abdomen for depositing their eggs in leaves, wood, or the bodies of other insects. In some, this organ is modified into a sting.

The Ants form a very numerous group, widely distributed in temperate and tropical regions, and are exceedingly interesting from their social habits and great intelligence. They have been known

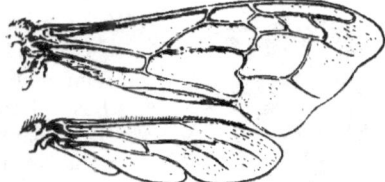

WINGS OF BEE, SHOWING HOOKS.

from early times, and their industry and thrift are proverbial. Some few kinds are solitary; but the large majority build dwellings, more or less complex, often furnished with chambers connected by galleries and passages. All sorts of material are utilised: our Common Yellow Ant makes an underground dwelling; others bore into wood, or mould clay, or weave a kind of tissue; and one South American Ant thatches its nest with little circles cut out of leaves. One European Ant lays up a store of grain for winter use, as do others in the East; but the habit is unknown to our English Ants, which become torpid in cold weather. The Agricultural Ant of Texas not only stores grain, but in some fashion prevents its germinating; and if the seed should begin to sprout puts an end to its growth by biting off the rootlet. Some ants enslave other kinds and live on the fruit of their labours. Many keep plant-lice in their nests for the sake of the sweet liquid they secrete; but the highest pitch of development is reached by the Leaf-cutting Ants of South America and the West Indies, who make gardens and plant a kind of fungus, on which they live and on which they feed their larvæ. In all nests males, females, and neuters, or workers, are found; and in some cases there are separate castes of workers. The various forms are probably determined by the kind of food given to the larvæ.

The Honey Bee, domesticated for the sake of the honey it collects and the wax it secretes, must serve as a type of Social Bees. A community consists of one fertile female or queen, the workers, and the males, or drones, who have no other duty than to perpetuate the race. These last, when winter comes, are driven out of the hive to die, or are stung to death. The nest, which is made of wax, is called the comb, and consists of six-sided cells, arranged back to back. In some of these cells are placed, singly, the eggs laid by the queen bee ; whilst others are used as storehouses to contain honey. This is collected from flowers by means of the proboscis, passed into the honey-bag, and thence disgorged into the cells when the bees return to the hive. Pollen is also stored for food ; and is brought to the hive in the pollen baskets on the last pair of legs.   Swarming is

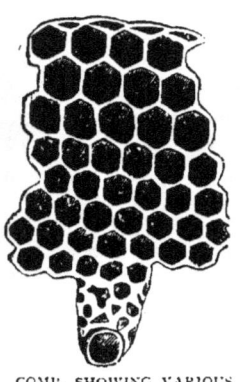

the name given to the flight of the young from the hive to found a new colony.

Our Humble Bees are also social, but their communities are very small compared with those of the Hive Bee. One of the Humble Bees coats its nest with moss, and the workers stand in a line from the nest to the place where the moss grows. The first bee pulls out a piece, makes it into a ball, which it passes under its body to the bee behind, and so the moss travels on until it reaches the nest, where a band of workers receive it, and use it as a kind of thatch.

Of the Solitary Bees (which are male and

COMB, SHOWING VARIOUS
SIZES OF CELLS.

female) there are many British species. Some make their nests or burrows near each other, showing how social life may have begun. The Carpenter Bee burrows into wood : the Mason Bee makes a nest of sand-grains cemented together with a secretion from its mouth ; the Leaf-cutting Bees line their nests with leaves, and one need only examine a rose-bush to find traces of their work in the leaves, from which semi-circular pieces have been cut. The Upholsterer Bee seems to like bright colours, for it lines its nest with the petals of the poppy. The Cuckoo Bees lay their eggs in the nests of other species.

Wasps, like Bees, may be social or solitary ; and the nests of the social forms are found in holes in banks or trees, or suspended from branches. The material is a kind of coarse paper made from the woody matter of plants, worked into a solid mass by the jaws of these insects. There is a single row of cells in each comb, and the combs are ranged one above another, not side by side, as are those of the Bees.

The food consists in part of honey, some of which is got honestly, and some by plundering Bees. The gardener knows that Wasps are fond of fruit, and they plague the grocer for the sake of his sugar and other sweets. Nor do they disdain animal food. The Solitary Wasps provide for their larvæ a store of insects, which in some cases they paralyse by stinging.

The Hornets resemble the Solitary Wasps in habit, but exceed them in size, and the sting of a Hornet is more painful than that of a Wasp. There are several British species. To this family belongs the Common Sand Wasp (Plate XII., No. 3), about an inch long, clothed in red and black. The female stores the cells of her nest with caterpillars as food for the young when they leave the egg.

The Golden Wasps, of which some are British, lay their eggs in the nests of Bees or Wasps. The Common Gold Wasp, or Ruby Tail (Plate XII., No. 1); is common in this country.

Most of the Gall Flies lay their eggs in leaves, buds, and other parts of plants and trees, thus giving rise to galls and blights. Examples of both may be seen on the oak-tree, in the shape of oak-apples and the fungus-like disks on the leaves. The Rose Gall Fly (Plate XII., No. 2) punctures the stem of the wild-rose, and deposits several eggs. The gall resembles a ball of moss, and each larva is in a separate chamber.

The Ichneumon Flies deposit their eggs in the bodies of other insects, chiefly in the larvæ of Butterflies and Beetles, the bodies of which serve to nourish the young when they come into the world.

The Tailed Wasps are often mistaken for Hornets, and the long ovipositor of the female of the Great Tailed Wasp is supposed to be some terrible kind of sting. It is, however, used for boring into timber, in which the eggs are deposited. Here the larvæ are developed, and they feed on the wood. Some of the smaller species attack grain crops.

The Saw Flies, of which many are British, have the ovipositor notched like a saw and hidden in a sheath. The Gooseberry Saw Fly devours the leaves and buds of gooseberry-bushes.

LACE-WINGED FLIES.

These insects possess two pairs of transparent membranous wings, usually of equal size, the nervures forming a network.

The largest and most brilliant are the Dragon Flies, of which about fifty different kinds are found in this country. All feed on other insects, and on a bright sunny day they may be seen hawking over ponds for insects, like so many swallows. The Demoiselle, or Damsel Fly (Plate XII., No. 5), is found near running water. Another common British Dragon-Fly is popularly called the Horse-Stinger; but none of the Lace-winged Flies has a sting. The largest Dragon Flies are American. The females drop their eggs into the water, where the larval and pupal stages are passed, and the young are as rapacious as their parents.

The May Flies live but a few hours as perfect insects. They have no mouth, and can take no food. The larva and pupa are aquatic, and bear on each side of the body fringed gills, which serve also as swimming-organs. Both the larval and pupal forms of the common May

Fly are abundant in ponds all the year round, and if kept in an aquarium will probably develop. Some May Flies are so common on the Continent that their bodies are swept up for manure.

The Aphis Lions owe their popular name to the fact that Aphides or Plant-lice form the chief food of their larvæ. The Golden Eye (Plate XII., No. 6), very common in Britain, emits a most disagreeable smell.

Some of the Scorpion Flies justify their name from their abdomen being armed with forceps. This is so with the Common Scorpion Fly, which often raises its tail as if about to sting.

The Ant Lions, in their perfect condition, are not unlike small Dragon Flies. But their chief interest lies in the fact that the larvæ excavate

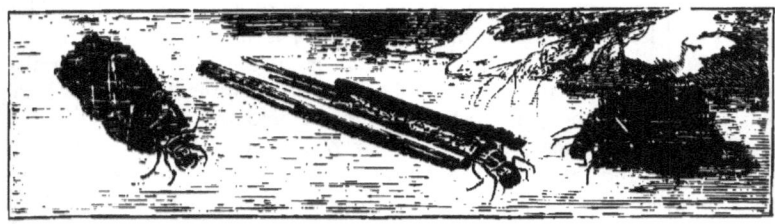

LARVÆ OF CADDIS FLIES.

little pitfalls to entrap unwary insects, and throw up showers of sand and dust to hasten the descent of their prey.

The Caddis Flies have the wings clothed with fine hairs. The larvæ live in streams and ponds, and make for themselves little cases of sand, bits of vegetable matter, stones, and shells, bound together by silky threads. When the larva is about to pass into the pupal stage, it closes each end of its tube with silk of its own manufacture, so as to keep out intruders, while allowing the water to pass through freely. These larvæ, or Caddis Worms, as anglers call them, are used for bait.

The small, wingless Bristle Tails and Spring-Tails are covered with scales, sometimes mixed with hairs. The Silver-fish, about $\frac{1}{3}$ inch long, is silvery-white, and common in decaying wood. Some near relations are found in damp ground and under stones. The Aquatic Podura, which lives on the surface of salt and fresh pools, may be easily recognised from its blue colour.

BUTTERFLIES AND MOTHS.

These insects pass through the four stages of egg, larva or caterpillar, pupa or chrysalis, and perfect insect. The body of the larva forms twelve segments, and the three immediately behind the head bear on each a pair of true legs. Some of the hinder segments bear false feet (usually four pairs), and on the last segment is a pair of claspers. Many larvæ are covered with hairs or spines, which in some cases sting like nettles. In all there is a pair of powerful jaws ; and many are very destructive to vegetation. Sometimes they occur in unusual numbers, and become quite a scourge. In 1894 there was a plague of caterpillars at Hong Kong. These attacked the pine-trees, and the Government took active measures to check the pests. Hand-picking was resorted to, and the caterpillars gathered in about two months weighed nearly 36 tons. One need not, however, go so far from home, for the ravages of the caterpillars of the White Butterflies are known to everybody who has a garden. The chrysalids of Butterflies are suspended head downwards, or fastened by a thread to some support. Many of the Moths pass the pupa state in a cocoon spun by the larva ; and the larvæ of the Hawk Moths make a cell in the ground, whilst others form tunnels in trees for the same purpose. The mouth of the perfect insect has a proboscis, with which the nectar of flowers is drawn out for food. There are two pairs of wings, more or less thickly covered with scales, which, when the wings are touched with the hand, come off in the form of fine dust. The antennæ of Butterflies are more or less thickened at the top. Most Butterflies are day-fliers, but some tropical species come out at dusk, and a few are probably night-fliers.

In the Swallow Tails the hinder pair of wings generally end in a point, and the upper pair are triangular. To this group belong most of the large gorgeously-coloured tropical species, chief among which are the Bird-winged Butterflies from Africa and the Eastern Archipelago. Mr. Wallace thus describes his feelings when he captured one of the finest :—" I could hardly believe I had succeeded. in my stroke till I had taken it out of the net, and was gazing, lost in admiration, at the velvet-black and brilliant green of its wings, 7 inches across, its golden body and crimson breast. It is true I had seen similar insects in cabinets at home ; but it is quite another thing to capture such one's self—to feel it struggling between one's fingers, and to gaze upon its fresh and living beauty, a bright gem shining out amid the silent gloom of a dark and tangled forest. The village of Dobbo held that evening at least one contented man." We have one British species— the Swallow-tail Butterfly (Plate XI., No. 3*a* [larva], 3*b*), now confined to the Fen district.

The Whites owe their popular name to their colour, and bear no hook or tail on the wings. Here belong the Cabbage Butterflies, whose caterpillars do so much damage in gardens, and would do much more did not the birds keep down their numbers. The Black-veined

White (Plate XI., No. 4) is now rarely found in Britain. When the perfect insect comes out of the chrysalis it discharges a few drops of a red fluid, and this has given rise to stories about showers of blood. The Orange Tip (Plate XI., No. 5) may be found in meadows in April and May.

To the Fritillaries, in which the first pair of legs are not fully developed, belong several large brilliantly-coloured British Butterflies, such as the Red Admiral (Plate XI., No. 1c; 1a larva, 1b chrysalis) and the Peacock Butterfly (Plate XI., No. 2.) Here belongs the Leaf Butterfly of the East, remarkable for the resemblance of the under-surface of the wings to a dead leaf. One of our own Whites—the Small Cabbage Butterfly—is scarcely distinguishable from pea-blossom, when at rest.

HUMMING-BIRD MOTH FEEDING.

The Meadow-Browns, with the first pair of legs very small, have eye-marks, or rings, on the wings. Many of the European Butterflies belong to this group. The Blues, Coppers, and Hairstreaks have many representatives in this country. The Blues, named from their colour, are common in the South on chalk lands. The Large Copper is nearly extinct in England, but the Small Copper is fairly plentiful.

The Skippers, so called from their jerky flight, are moth-like in form and are generally of a tawny hue marked with white spots. The caterpillars pass the pupal stage in a cocoon.

The Moths have no club at the end of the antennæ, which are often feathered in the males. The majority are evening or night-fliers, and when at rest the wings usually droop, being rarely carried erect like those of the Butterflies.

To the Hawk Moths belong our largest British forms. The proboscis is long, and these Moths take their food on the wing, hovering above flowers and extracting the honey. The Death's Head Moth (Plate XI., No. 6), the largest British insect, owes its name to the skull-like mark on the back of the thorax. When seized or disturbed this insect can produce a sound, somewhat like the squeak of a mouse. It is a foe to bee-keepers, for it enters hives and steals the honey, the stings of the bees being powerless against its thick skin. The

PLATE XI.

1. Red Admiral Butterfly (*a*, caterpillar ; *b*, chrysalis ; *c*, perfect insect). 2. Peacock Butterfly. 3. Swallowtail Butterfly (*a*, caterpillar; *b*, perfect insect). 4. Black-veined White. 5. Orange-tip Butterfly. 6. Death's-head Moth. 7. Six-spot Burnet. S. Hornet Clearwing. 9 Silkworm (*a*, caterpillar ; *b*, cocoon ; *c*, chrysalis ; *d*, perfect insect). 10. Lackey Moth (*a*, caterpillar ; *b*, perfect insect). 11. Black Arches (*a*, caterpillar ; *b*, perfect insect). 12. Clifden Nonpareil (*a*, caterpillar ; *b*, perfect insect). 13. Winter Moth (*a*, caterpillar ; *b*, perfect insect).

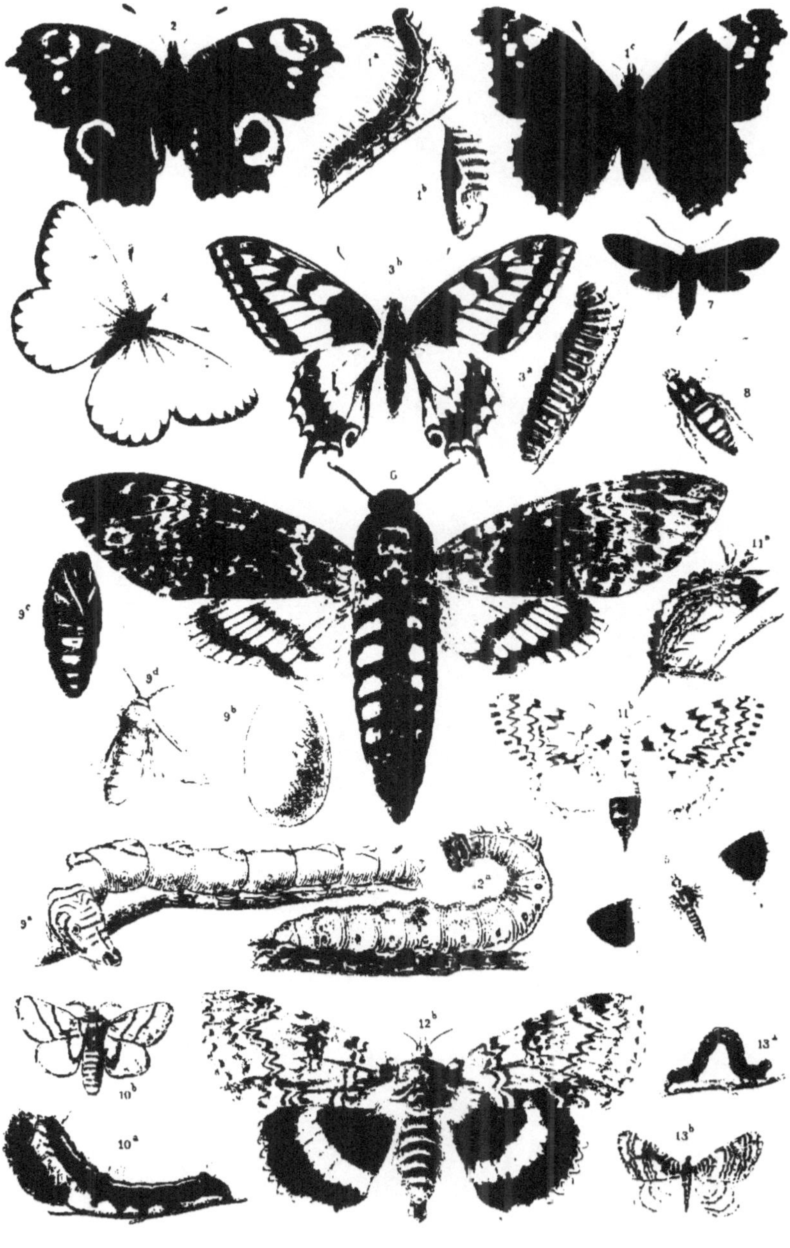
2
1a
1c
1b
4
3b
7
3a
8
6
11a
9c
9d
9b
11b
12a
5
9a
12b
13a
10b
10a
13b

Humming-bird Hawk Moth is common in Britain in some seasons. The illustration shows how it feeds. The flight of this Moth and of some of its allies is so like that of Humming Birds, that naturalists have shot Moths in mistake for the Birds. The Six-spot Burnet (Plate XI., No. 7) is common in meadows, and has some relations with five spots, differently arranged in each. The Hornet Clear-Wing (Plate XI., No. 8) is the type of a group, with transparent wings, which fly by day. The Currant Clear-Wing closely resembles a gnat.

Of the Silk-spinners, which are natives of the East, but one, the Common Silkworm (Plate XI., No. 9*d*) is reared in Europe, where it was introduced from China in the reign of Justinian, for commercial purposes. It bears our climate well, and even here spins silk of good quality, but the thread is too short to be of value. The domesticated Moth has lost the power of flight. The caterpillar of the Silkworm is figured on Plate XI., No. 9*a*, the chrysalis at 9*c*, and the cocoon secreted by the caterpillar, from which the silk is obtained, at 9*b*. There are many other silk-producing Moths in the

CATERPILLAR OF ROLLER MOTH IN TUBE.

East, and closely related is the Atlas Moth, from China, the largest species known. The Lackey Moth (Plate XI., No. 10*b*) is a common British species. The female lays her eggs in bands round twigs, and the caterpillars (10*a*) live in company under a kind of web. The Black Arches (Plate XI., No. 11*b*, caterpillar 11*a*) is also British.

The Clifden Nonpareil (Plate XI., No. 12*b*, caterpillar 12*a*) is rare in Britain, but some allied species are fairly common.

The Loopers, or Geometers, are so called from the way in which the caterpillars move, by stretching out the body and fixing the front legs, then arching the body and bringing the hind-legs up to the front ones. The Winter Moth (Plate XI., No. 13*b*, caterpillar 13*a*) is very common, and destructive to fruit trees.

The caterpillars of the Roller Moths make tubes of leaves in which they pass their pupal stage.

The Clothes Moths belong to a group which prey on woollen stuffs, furs, feathers, and the like, or feed on vegetable substances. The Plume Moths have the wings cleft into feathery segments.

W

THE FLIES AND FLEAS.

The insects of this order have but one pair of wings, the hind pair being represented by the balancers or poisers, or true wings may be altogether wanting. The mouth is suctorial, and in the Gnats there are also bristles or lancets. The larvæ are maggot-like, and the pupal stage is generally passed in an oval case.

The Wheat Midge and Hessian Fly belong to a group which does great damage to wheat crops. The larvæ of the Wheat Midge feed on the pollen, and prevent the fertilisation of the seed; those of the Hessian Fly live in and feed on the stalks.

METAMORPHOSES OF GNAT (*magnified*).

The Gnats may be recognised by their long legs and proboscis, or trunk. The females feed on blood, and inject a poisonous liquid in the wound. The larger species from tropical countries are called Mosquitoes. The females lay their eggs in a raft-like mass in the water, where the larval and pupal stages are passed. When the perfect insect is about to emerge, the pupa floats at the top of the water. When the insect is partly free, the pupa-case is very liable to be upset, and the gnat often perishes by drowning. It is easy to procure gnat larvæ and watch their development in a small aquarium. In our illustration two pupæ are shown on the left, and two larvæ on the right; above the pupæ a perfect insect is coming out of the pupa-case, and a little farther away a female is depositing eggs. To this group belongs the Crane Fly, or Daddy

Long-legs, which lives on the juices of plants. Its larvæ burrow and feed on roots of plants.

The Flies have short antennæ, as one may see in the House Fly, the Bluebottle, which lays its eggs on meat, or the Flesh Fly (Plate XII., No. 4), the larvæ of which feed on carrion. The Drone Fly owes its name to its likeness to a drone; its larva is a rat-tailed maggot, which develops in stagnant water and cesspools. The females of Breeze Flies, or Gad Flies, are bloodsuckers; and the Bot Flies deposit their eggs in the skin of cattle. The larvæ burrow under the skin, and damage the hides.

The Tsetse Fly is a native of tropical Africa, and its bite is certain death to horse, ox, and dog. Wild animals are not affected by it, nor is the mule, ass, or goat; while in man its bite is followed by no worse consequences than is the sting of a wasp. Gordon-Cumming says that the puncture of the skin by a Tsetse Fly "will cause one when seated to spring up as if pricked by a needle. As they are possessed of a long probe, a thick flannel shirt offers no protection against these most abominable of all created insects—direct descendants, no doubt, of the flies that plagued Egypt."

The Fleas need no description. The Common Flea confines its attentions to Man; but the Cat, the Dog, and other animals have their special flea-parasites. The Sand Flea, or Jigger, of America, burrows in the skin of the feet of Man and the lower animals, and there the female produces her eggs, but the larvæ are not parasitic.

THE BEAKED INSECTS.

In these the metamorphosis is incomplete, and the mouth is formed for sucking. In the first group the fore-wings are harder than the hinder pair; in the second both pairs are of similar texture.

The Air Bugs live on the ground, on plants, and on the surface of water, feeding chiefly on the juices of other insects. The Fly Bug preys upon its near relation the Bed Bug (Plate XII., No. 16), an insect that must be dismissed with mere mention. The Water-Measurer (Plate XII., No. 10), about ½ inch long, is common in Britain, and may be seen walking on aquatic vegetation or skimming over the surface of stagnant or slowly-running water. The bugs of the genus Gerris are met with in similar situations, and a near relation in the Atlantic and Pacific. One may find several of the Water Bugs in almost any pond. The largest and most rapacious is the Water Scorpion, probably so called from the tail-like organs, which form a breathing tube, being mistaken for a sting. The Pond Skater has a much more elongated body; and some American species reach a length of 3 or 4 inches. The Water

W 2

Boatman, which swims on its back, and its allies have the hind-legs thickly fringed with hairs, thus forming admirable swimming-organs.

The Insects in the next group feed on the juices of plants, and their fore-wings are not thickened. First come the Cicadas, noted for their "song." The Common Cicada of Southern Europe is sometimes kept in small cages, as was the fashion in classic times. Only the males have the power of producing sounds. The Cicada feeds on the ash, and from the punctures it makes in the bark flows a secretion which hardens into manna, though most of that substance is obtained by making cuts in the trees.

The Lantern Flies are natives of the tropics, and have on the head an empty projection called the "lantern," and said to be luminous. The evidence is not very convincing, and it is probable that these insects are luminous only at certain seasons, or it may be that, as in the Glow-worm, the power of emitting light is confined to one sex. Most of these insects secrete a waxy substance, which in some cases is collected and sold.

Here belong the Cuckoo Spits, or Froghoppers, the young of which exude froth as a protection; the Aphides, or Plant-lice, some of which are destructive to roses and hops in this country, while another species does great harm to vines on the Continent; and the Scale Insects, detested by gardeners. Closely allied to the last mentioned are the Cochineal Insect and the Lac Insect, both commercially valuable. The dried females of the Cochineal Insect yield a red dye; and the West Indian Lac Insect produces shellac, as the Cicada produces manna.

The Lice constitute a group parasitic on Man and the lower animals.

THE STRAIGHT WINGED INSECTS.

These insects have four wings—the first pair narrow and parchment-like, and the hind pair, used for flight like those of the Beetles, broad and membranous. The young are more or less like the adults; the mouth is formed for biting; and in many the females bear at the end of the abdomen an organ for depositing their eggs. Many make various noises by rubbing one part of the body over another.

The House Cricket, oftener heard than seen, is common in Britain, especially in bakehouses. Its "chirp," which proceeds only from the males, is produced by rubbing one of the wing-covers over the other, causing them to vibrate. The Field Cricket, a rather larger insect, is rare in Britain. The Mole Cricket, which burrows, has the fore-legs flattened for digging. It damages vegetation by destroying the roots of plants; and worms and insects constitute its chief food. The Crickets jump, but not so well as the Grasshoppers or Locusts.

The Great Green Grasshopper (Plate XII., No. 9) is common in Europe, but less so in our own country. The body is over 1 inch long, and its colour harmonises so well with the leaves on which it settles, that it is no easy matter to find one of these insects, though one may know that it cannot be far off, having just seen it jump. The Common Grasshopper, a much smaller and more soberly coloured insect, "chirps" by rubbing the last pair of legs against the wing-cases. It is a relation of the Locusts, which do so much damage to vegetation in some other parts of the world. No true Locust lives in Britain, though one European species, the Migratory Locust, sometimes wanders hither. The ravages of Locusts are well known, and some idea of their immense numbers may be gained from the fact that in one year in Cyprus alone 150,000,000,000 eggs were destroyed by persons employed by the British Government.

LOCUST DEPOSITING EGGS.

In the next tribe the hind-legs are used for walking or running, but not for leaping. The Praying Mantis (Plate XII., No. 8) should rather be called the "Preying" Mantis, for it is a rapacious little creature which kills and devours other insects, and in confinement will eat its fellow-prisoners. When on the look-out for food, it assumes the attitude in which it is represented, resting on the two hinder pairs of legs, and with the front pair and the head raised, as if engaged in an act of devotion. As soon as any hapless insect comes within reach, the fore leg is extended and the prey seized, the spines on the leg serving to hold it fast, in spite of its struggles. A story appeared in a scientific magazine of an American species that seized and killed a humming-bird. These insects are confined to the warmer parts of the world, as are the Leaf Insects and the Stick Insects, so called from their resemblance to leaves and twigs. Some of the Praying Insects and Stick Insects occur in Europe.

Cockroaches are found all over the world, but are most abundant in the tropics. Our household pest the Common Cockroach was introduced into this country from Asia. Wingless forms are met with, and these are pupæ or larvæ. In the females the wings are small. The American Cockroach is a larger insect of redder hue.

The Common Earwig (Plate XII., No. 7) may be found in any garden.

These insects do great damage to vegetation, but there is no truth in the story that they creep into the human ear. Even if they did, they could not get far enough to do harm ; but their popular name in most European languages has reference to this supposed habit.

The Termites are mostly tropical, but a few species have been introduced into some parts of Europe, though none is found in Britain. They are often miscalled White Ants, but the chief point in which they resemble Ants is their habit of living in communities. They attack furniture, woodwork, and merchandise of every kind ; and it is said that every article of furniture in a house may be damaged by them without their presence being suspected, as they form their burrows underground, and make their places of exit immediately beneath the legs of tables, and chairs, of which they eat away the interior, leaving only a thin outer shell, which crumbles to dust when moved. The nests vary in form : some species form large conical structures covered with clay ; many tunnel in trees, and the nests of some are attached to tree-trunks.

MYRIAPODS.

To this class belong the Centipedes and the Millipedes—long, worm-like animals divided into a number of similar segments. In the Centipedes each segment carries one pair, and in the Millipedes two pairs, of many-jointed legs, armed with claws. The head bears a pair of antennæ, or feelers. The former have the body flattened, and feed on insects and worms, and their bite is poisonous, for the mouth organs are pierced like the fangs of venomous snakes, and pour a poisonous secretion into the wound. The latter are round ; they feed chiefly on plants and root-fibres, and cannot bite. Myriapods are found in all parts of the world, but the largest occur in tropical regions. All of them love dark places, and they live in holes in buildings, or under stones or the loosened bark of trees. The Shield-bearing Centipedes are distinguished by the great length of their limbs, which often exceeds that of the whole body ; and the plates of the segments overlap on the back. Several small Centipedes occur in Britain—one, the Lithobius, about 1 inch and reddish brown, is very common. If handled it attempts to bite, though it is too small to do any harm. The Ground-loving Centipedes are also British. They are very worm-like in appearance, and live in the ground. They emit a faint phosphorescent light, and this has given rise to the stories of luminous Earthworms, which appear from time to time in the newspapers. Some of the Scolopendras from India and South America are 9 or 10 inches long. Hornaday was bitten by one in Borneo, and he compares the pain to that caused by three or four hot needles plunged deep into the flesh.

Tincture of arnica was his remedy, and the pain ceased in about a couple of hours. The Common Millipede is a native of Britain, and may be met with among the clods of a ploughed field, under stones, or the bark of trees. It can roll itself up in a spiral form as shown in the illustration. There are many forms, some of which occur in Europe, and others in hotter parts of both hemispheres. All these creatures breathe by means of air-tubes regularly placed in the skin uniting the segments. Relationship to Insects is shown in the difference between the young and adult forms.

COMMON MILLIPEDE.

Another order has been formed for Peripatus, a worm-like creature with unsegmented body and numerous legs, found in dead wood and the stumps of trees in the island of St. Vincent early in this century. Since then about a dozen other closely allied forms have been met with. The skin is richly coloured and of velvety texture. They live under stones, in rotting wood, in moist places, are nocturnal in habit, and feed on insects, which they ensnare by ejecting slime from tube-like projections at each side of the mouth. The air-tubes are scattered all over the skin; and the young are born alive, the eggs being hatched within the body of the parent.

PERIPATUS.

CHAPTER XXVII.

ARACHNIDS AND CRUSTACEANS.

RACHNIDS include the Spiders, Mites, and Scorpions. In these animals the head and thorax are fused together, and the result of this union (the cephalothorax) and the abdomen may or may not be divided into segments. (This will be understood if we compare the Scorpion with the Spiders in Plate XII.) Breathing is carried on by air-tubes or by "lung-books," or both may be present.

Spiders have four pairs of walking legs, a pair of palps for seizing prey, and a pair of falces, or claws, with poison-glands. They feed on living prey, which they kill with their falces, and with the same organs they suck out the juices of their victim. All are spinners, and most of them make webs as snares.

TRAP-DOOR SPIDER.

The Bird-Catching Spider (Plate XII., No. 13) is South American, and there seems to be no doubt that birds are sometimes taken in its web. This need not surprise us, for a specimen in the Zoological Gardens has been known to kill a mouse, though its ordinary diet was Cockroaches and Mealworms. There are several species, and it is said that their bite is attended with serious consequences to Man. But in a case recorded in *Nature*, where one of these spiders darted from a heap of bush, and bit a labourer on the heel, "fright was the only ill-effect."

To the same family belong the Trap-door Spiders, which burrow in the ground, lining their dwellings with silk, and generally closing them by means of a door. These spiders make no web, but come out at night in search of prey. There is one British species, fairly common in some places on the South Coast, but its dwelling has no door.

The Jumping Spiders which make no web, are cat-like in the mode of hunting, stealing up, and their pouncing on their prey with a sudden spring. The Wolf-Spiders, like the creature from which they are named, run down their prey. Some of them may be found on the surface of aquatic plants, and they sometimes go under water, carrying down

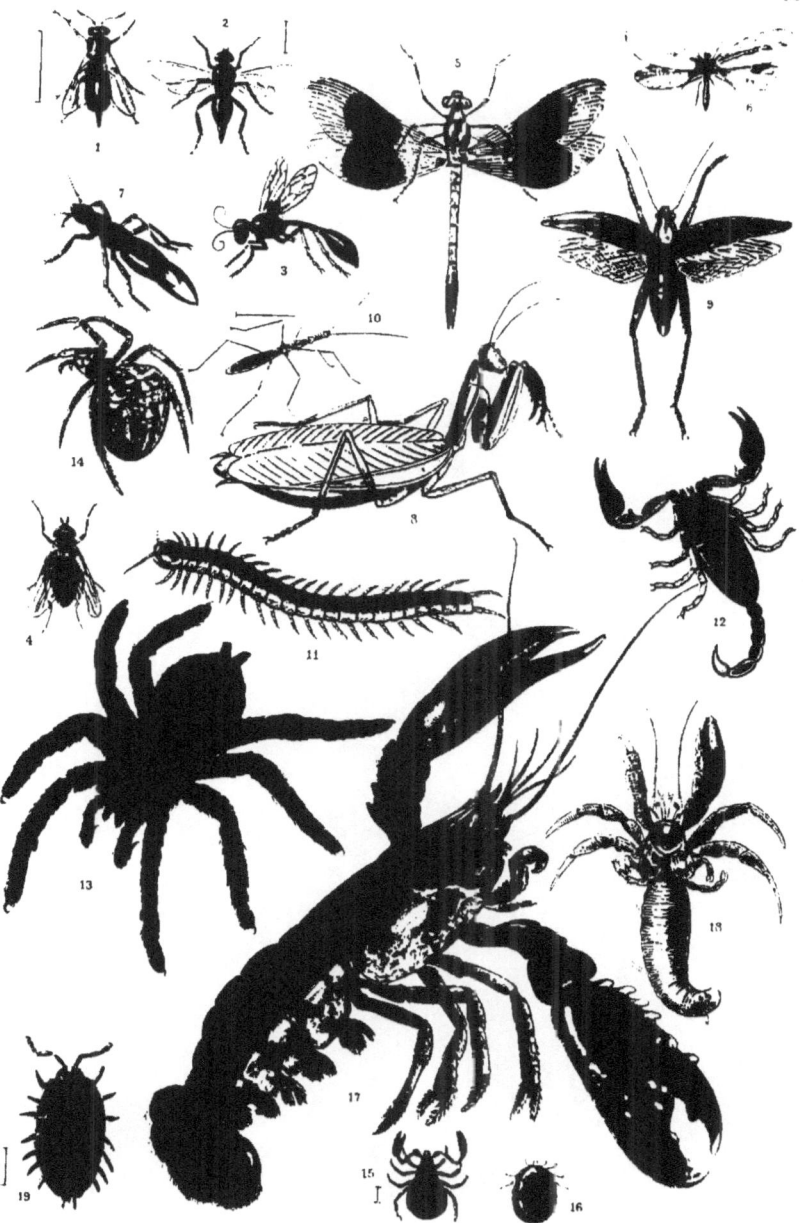

sufficient air, entangled in the hairs with which the body is covered, to serve for respiration for a short time. Here belongs the Tarantula, the bite of which was formerly said to produce a dancing madness.

The Common House Spider and the Diving Spider belong to a family the members of which make a web, with a tube-like part, in which the owner lies in wait for its prey. The Diving Spider is fairly common in ponds, passing most of its life below the surface, where it builds a nest and deposits its eggs. It is no difficult matter to keep it in a small aquarium, in which are small crustaceans and insect larvæ, or it may be fed with flies.

The Common Garden Spider (Plate XII., No. 14) is known to everyone. We may, however, learn something about its habit by dropping one into a wide-mouthed bottle, and covering the top. Then the operation of web-making may be watched, and the anchorages, as shown in the illustration, will be dotted all over the side of the glass. It should be fed with flies, and if a Bluebottle be given it, the Spider will hold the Fly behind it, and turning the victim round and round with its third pair of legs, pour forth a supply of silk that swathes the Fly like a winding-sheet.

ANCHORAGE OF WEB.

The Mites are small creatures, widely dispersed, and some of them aquatic. Many of them are parasitic, some are rapacious, and others are scavengers feeding on decaying matter. Here belong the Cheese Mite, the Itch Mite (the cause of a loathsome disease), the Harvest Mite, Water Mites (from salt water as well as fresh), and the Ticks which infest dogs, sheep, and other animals. The Water Bears are microscopic animals, with four pairs of clawed limbs.

The King Crabs, or Horseshoe Crabs, from the shores of North America and the Moluccas, are strange looking creatures, whose chief interest lies in the fact that they are survivals from a long-distant past.

The Sea Spiders have four pairs of long, slender legs and a small body. Some live in the open sea, but a few may be found on weeds in rock-pools, and under stones on the shore.

There are no Scorpions in Britain, but a few species occur in Europe, though the largest are found in tropical countries. They have a pair of large, pincer-like claws, four pairs of walking legs, and the last joint of the abdomen carries a sting—fatal to the insects and spiders on which these creatures chiefly live, and in some cases productive of serious consequences to Man. Scorpions are light-shunning animals, and pass the day in holes, and under stones, coming forth at dusk to feed. The European Scorpion (Plate XII., No. 12) ranges as far north as France and Germany. Closely allied are the Harvestmen, easily recognisable by their long, slender limbs, like those of a Crane Fly, or Daddy Long-legs —a name given to them in America—and the Book Scorpions, some of which are found among dusty books and papers, while others live as parasites, under the wing-cases of beetles.

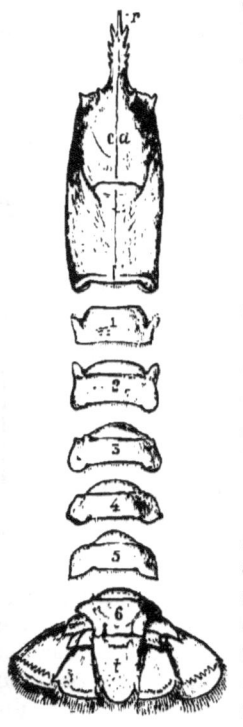

SEGMENTS OF LOBSTER.

CRUSTACEANS.

Some of the members of this class are familiar objects, even to persons who have never seen the sea, for Lobsters, Crabs, Prawns, and Shrimps are common in every fishmonger's shop. A search under stones in the garden will yield other specimens, for there we shall be pretty certain to find Wood-lice or Pill-bugs; and in almost any pond or stream the Freshwater Shrimp may be met with.

The body is segmented, and the normal number of segments is twenty or twenty-one, according as the telson (t) is reckoned as an appendage of the last segment or as a distinct segment. But all the segments cannot be made out in the same animal. Our illustration shows the segments of a Lobster, in which the carapace (ca) consists of the united segments of the head and thorax (fourteen segments), while those that follow belong to the abdomen. In the Lobsters and Crabs there are five pairs of walking legs, and the first pair of these are usually modified into large claws. The eyes are fixed on footstalks, and in one deep-sea form, dredged by the *Challenger*, these are longer than their owner's body. Most undergo a metamorphosis.

The Common Lobster (Plate XII., No. 17) is abundant round our shores, and is highly valued for food. Vast numbers are sent to the London markets from British fishing-stations and from Norway. The Spiny Lobster, or Crawfish, has no claws and its flesh is less delicate than that of the Lobster. Its larva was formerly known as the Glass Crab. To this group belongs the River Crayfish, a small form found in many of our streams. It is used for food and for garnishing dishes.

The Shrimps and Prawns can only be mentioned, but the latter may be recognised in a moment by the notched rostrum, or beak, similar to that of the Lobsters (*r* in illustration on p. 346). Prawns may be met with in almost any rock-pool, and one species is noted for changing its colour in accordance with its surroundings.

It will be interesting to compare the form of the Common Crab with that of the Lobster, piece by piece. This will be easy if we bear in mind the fact that the broad shell of the Crab corresponds to the part marked *ca* in the illustration on p. 346 ; and that by laying the Crab on its back we shall be able to turn up the abdomen from the groove into which it fits. If we succeed in doing this, we shall have taken the first step,

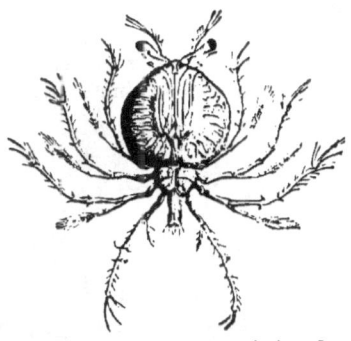

LARVA OF SPINY LOBSTER (*enlarged*).

and that no small one, towards gaining some practical knowledge of the external structure of these animals.

The Edible Crab needs no description, and the Shore Crab may be easily distinguished by the five teeth on each side of the carapace. The Swimming Crabs have the last pair of legs flat and fringed with hair. The Masked Crab lies buried in the sand at the bottom, with the tips of the antennæ projecting. To Mr. David Robertson is due the discovery that the antennæ thus form a breathing tube, and he saw a specimen which he kept in an aquarium discharge its eggs into the

PEA CRAB.

water by this passage. The tiny Pea Crabs live as parasites in bivalve shells. The species figured lives in the Mussel.

The Land Crabs of the tropics rarely visit the sea, except to deposit their eggs. Many of them have lost their aquatic habits, and while some are becoming, others have actually become, air-breathers.

The Spider Crabs have long, slender legs, and the carapace narrowed in front. Some of them are used for food, especially on the South Coast and in the Channel Islands. The carapace is generally covered with bristles, and many of these crabs deck themselves with weeds, sponges, hydroids, and sea-squirts, for purposes of concealment. Some that were kept in the Aquarium at Jersey, when deprived of weed, were seen by Mr. Hornell to cover themselves with pebbles. A small one taken by the writer, in 1893, was covered with red weed when put into a small aquarium, in which there was only green weed. During the night the crab had decked itself so completely that it was difficult to distinguish it from the weed in the jar. The experiment is easy enough to repeat, and every reader may see crab-decking performed under his very eyes.

Of the Hermit Crabs, most of which live in borrowed shells, the best known in books is the Cocoa-nut Crab, from the Pacific Islands, which lives on cocoa-nuts, though it may be doubted whether it climbs

TUBE-BUILDING SHRIMP.

trees to get its favourite food. This crab makes a burrow in the earth, and does not utilise a shell for a dwelling. Of our Common Hermit Crabs we may find plenty on the shore. From Plate XII., No. 18, we may get a good idea of the formation of the body, and notice in what respects it differs from, and in what it resembles, that of a lobster. We shall see that one claw is much larger than the other, and this serves the same purpose as the operculum, or shell-cover of a mollusc. When danger threatens, the crab retires into its house, and—so to speak—shuts the door by laying its large claw across the mouth of the shell.

The Mantis Shrimp, from the Mediterranean, is interesting from the armature of the first pair of legs which resembles that of the Praying Insects. The Opossum Shrimp, which forms a large part of the food of the Greenland Whale, has a brood pouch, in which the young pass their earlier stages. Both these forms have gills on the legs.

The next group have no eye-stalks. There are two divisions—the Amphipods (so named because their feet spread in all directions, while some are used for swimming and others for jumping) and the Isopods (in which the legs may be said to be equal). Most of them are aquatic, and the species that live on land prefer damp places.

The Sandhoppers are abundant everywhere on the sea-coast, but most numerous on sandy shores. Their office is that of scavengers, and they are charged with cannibalism, but Mr. David Robertson could find no evidence of this bad habit, though he kept some under circumstances that would encourage it. The Freshwater Shrimp—which, however, is not a True Shrimp—belongs to this group, as do very many marine forms. Some of these latter build tube-like dwellings, and on the South and West Coasts one may find pieces of weed on which every branch bears a little conical mud tube, from the mouth of which project the antennæ of one of these creatures. Some, again, make tubes of vegetable *débris* and sand, cemented with a glutinous secretion. In a four-ounce bottle on the table at which this is written is such a tube, constructed

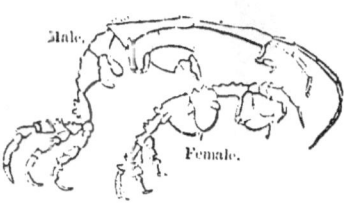

SKELETON SHRIMP (*magnified*).

against the side. It is fixed to the bottom for a little more than an inch, and then bends sharply upwards for about half that length. The builder has lived there for two months, and given me considerable pleasure in watching its movements, and anyone who will hunt a rock-pool for some of these little animals, may have the same enjoyment.

The Skeleton Shrimp clearly has a right to its name. It is very common in rock-pools, and will live for some time in an aquarium. The Whale Louse, parasitic on the whale, is a near relation ; and the Rev. T. R. R. Stebbing (in *Good Words*) has shown how, through change of habit and abundance of food, the Whale Lice have "become comfortably broad in shape and demurely flat in posture."

Of the Isopods the Common Woodlouse (Plate XII., No. 19) is

WHALE LOUSE (*enlarged*).

a familiar example, and it has some near relations which can roll themselves into a ball when disturbed, and hence are called Pill Bugs. Several of these creatures live on rocks and in pools near the shore, and one—the Gribble—does great damage to timber under water.

The Entomostraca are small aquatic animals, most of them of microscopic size. The Phyllopods have gills, or breathing-plates, on at least four pairs of feet. Here belongs Apus—"a giant among Entomostraca"—over 1 inch long. It is found in ponds on the Continent. To this order belong the Common Water Fleas, so abundant in ponds in this country. Estheria, which looks like a little bivalve mollusc, has representatives all over the world, the eggs of an Eastern form, brought to this country from Palestine in dried mud, have repeatedly developed when put into water.

The Ostracods are all enclosed in a small bivalve shell. Cypris is a freshwater form, and some of its relations live in rock-pools and in the sea.

ESTHERIA.

The Copepods have no shell. The freshwater Cyclops has many marine representatives, and these, like most of the other Entomostraca, form an important part of the food of fishes. Many are parasitic, and are called fish-lice; in these the mouth is fitted for

APUS.

blood-sucking. Nicothoë, parasitic on the gills of the Lobster, is readily distinguishable from its red colour. Our illustration shows a female carrying eggs, and should be compared with the free-swimming Cyclops.

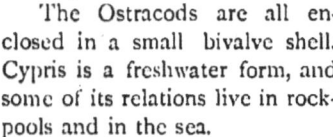

NICOTHOË WITH EGGS.

The Barnacles form a separate group. They are free-swimming in early life, but soon settle down, secrete a shell of many valves, and pass their lives, in Professor Huxley's words, "fixed by their head, and kicking the food into their mouth with their legs." This one may see by dropping into a glass jar of sea-water some of the Acorn Barnacles that cover rocks and stones and shells on the shore. The Ship Barnacle, which fixes itself to floating timber and to ships' bottoms, has a long stalk, at the end of which is the shell containing the body of the animal, which feeds in the same way as the Acorn Barnacle. The small Pyrgoma lives on the Cup Coral.

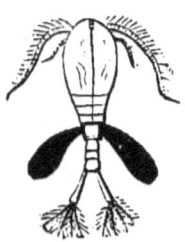

CYCLOPS, WITH EGGS.

CHAPTER XXVIII

ECHINODERMS AND "WORMS."

TO the Echinoderms belong the Feather-stars, Brittle-stars, Starfishes, Sea-urchins, and Sea-cucumbers. The body is produced into more or less definite rays, generally five in number. There is usually a system of tube-feet, or suckers, worked by water-vessels. These creatures undergo a meta-morphosis, and have the power of renewing parts of the body when injured or cut off. All are marine, and most of them feed on organic *débris* in mud and sand, while some take living prey.

Of the Feather-stars, most of which are stalked when young and some during life, there is one British representative—the Rosy Feather-star, by no means rare. The free-swimming adult is generally taken in the dredge, but it may be seen swimming in rock-pools. The larval forms may be found attached to the stems of the common coralline and looking like very short white threads. Dr. Hickson speaks of the beauty of the tropical forms, and says that he saw one, " lively and happy, apparently, climbing

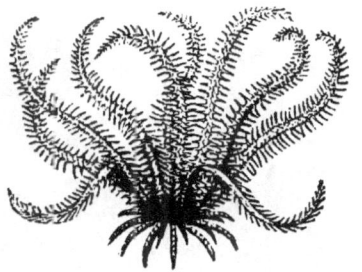

ROSY FEATHER-STAR.

on a pile of Talisse pier, a couple of feet above the water." In these creatures the mouth is on the upper surface, and the arms, used also for walking and swimming, waft minute particles of food to the mouth. The forms that are stalked throughout life live in the deep sea, and recent expeditions have made known many new ones.

The Brittle-stars have a small central disk, with five long, snakelike arms, and tube-feet too small to be of much service. They wriggle along the ground, or clamber over rocks and weed. One of the strangest of the Brittle-stars is the Basket-fish, or Gorgon's Head, from the North Atlantic. The arms divide and sub-divide, and twine round the disk like a mass of basket-work, in which it is said small marine animals are entangled and captured.

Of the True Starfish the Common Five-fingers is a good type, and anyone living at the seaside need not wander far to find a specimen which will repay examination. The mouth is on the under-surface, and on the top, between two of the arms, is a porous plate, which admits water to supply the tube-feet. It is easy to see the tube-feet at work, by dropping the creature into a glass vessel full of sea-water. It will not, however, live long, unless the water be changed frequently, or growing vegetation be added. One will not have long to wait before the captive begins to explore his prison. If dropped on its back it will right itself, and crawl up the glass by means of the suckers at the end of the tube-feet. Professor Romanes puts the pace of a Starfish at 2 inches a minute, while a Brittle-star will scramble over

STARFISH.

6 feet in the same time. Scattered over the body are pincer-like organs mounted on stalks, and called pedicellaria, or "little lice-like things," for when they were named they were thought to be parasites. These are used for grasping, and so help the animal to climb.

"Comet" Starfish is a name used when one arm that has been cast off, sprouts out, and develops four new arms and a central disk. In confinement Starfish have been seen to break up, each ray moving independently when separated from the disk.

The Sun-star (Plate XIII., No. 14) is more deeply coloured, and has from eleven to thirteen rays. It is to be met with round our coasts, though not so frequently as the common form. The small Gibbous Starlet, about an inch across, is common in rock-pools. There is one British Starfish like a big pincushion, and another is as flat as a pancake.

Sea-urchins may be compared to Starfish with the rays bent upwards to form a covering for the body. This covering, called a shell or test, is studded with spines, and pierced with holes for the tube-feet. They live off rocky coasts, and some make holes for a dwelling-place. The Common Sea-urchin (Plate XIII., No. 13) is abundant on British coasts, and the shell, stripped of the spines, is to be bought for a penny or twopence at almost any seaside resort. But a search among the weeds

PLATE XIII.

1. Earthworm. 2. Leech. 3. Segments of Tapeworm. 4. Cuttle-fish. 5. Nautilus Shell. 6. Roman Snail. 7. Mitre Shell. 8. Purple Shell. 9. Oyster. 10. Pearl Mussel. 11. Ship-worm. 12. Sea Cucumber. 13. Sea Urchin. 14. Sun Star. 15. Jellyfish. 16. Sea Anemone. 17. Red Coral. 18. Sea Pen.

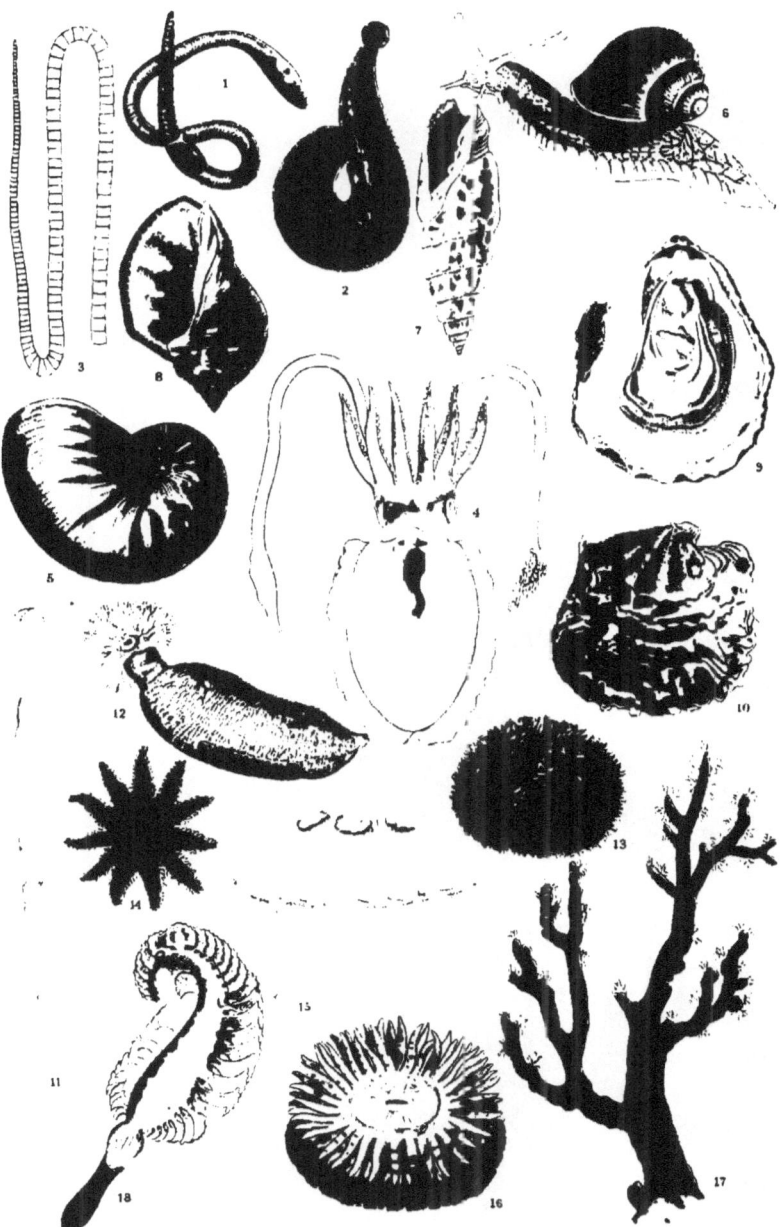

in a rock-pool, half-way between tide-marks, would probably result in finding some small ones.

The Sea-cucumbers are marine, and have an elongated body. The mouth, fringed with tentacles, is at one end. Some have, and others have not, tube-feet like the Starfish. These animals grow to a large size in warmer seas, and when dried form the *trepang*, or *bêche-de-mer*, so highly prized by the Chinese. Regular fisheries are carried on for the capture of these creatures, and the annual value of the export from Queensland alone is about £30,000. Small specimens may be found round our coasts, and larger forms are brought up in the dredge. That figured in Plate XIII., No. 12, is found in the North Sea.

"WORMS."

This is a useful name—but nothing more—for a number of creatures, whose relationships to each other, and to higher and lower groups, are not clearly made out. They are of similar shape, and move head foremost, and the higher forms are divided into rings or segments.

The Bristle-footed Worms fall into two sets, of which the Earthworm and the Sea-worms are the respective types. The former has few bristles; in the latter they are abundant, and there are also outgrowths from the sides, which serve as oars and feet.

The Common Earthworm (Plate XIII., No. 1) is of great importance to man, for by its burrowing in the soil the earth is loosed and broken up. It does the work of the plough, for by its castings—that is, earth which has passed through its body—it covers the surface with soil brought from beneath. Darwin calculated that in an acre of land there were 53,000 Earthworms, that ten tons of soil passed through their bodies year by year, and that their castings would cover the surface with earth to a depth of 3 inches. If gardeners would think of this they would not be so ready to kill worms, but would rather encourage them as helpers. There are many related forms, and one giant, from New Zealand, is about 6 feet long.

Several of this group live in the mud, in fresh water, some of them forming tubes, while others are free swimmers. The tiny Naïs, to be met with in every pond, is remarkable for the way in which it buds off new individuals. The number of its segments increases, and the hinder part of the worm is cast adrift, producing a head, proboscis, and eyes for itself, while the original grows a new tail. The tail that belonged to A, now belongs to B, and may be passed on to C, D, and E, all the way down to Z, unless the worm dies for want of food, or is snapped up by some water-beetle.

Some of the marine Bristle-footed Worms lead wandering lives, while others are quiet stay-at-homes, dwelling in tubes.

x

To the first group, in which there are horny jaws and a proboscis, belong a number of British forms. Among the best known are the Nereid Worms—long, centipede-like creatures, which may be found in rock-pools and under stones, or by turning up the sand. One of these may often be found dwelling in a whelk-shell with the Hermit-crab, and the Devonshire fishermen say there is no better bait. The Sea-mouse is a lovely creature, some 4 or 5 inches long, common on the southern coast. The gills on the back are covered with scaly plates, one pair to each body-ring, and from the sides project bundles of spines, which reflect the light in rainbow hues, and which Cuvier compared to the plumage of the humming-bird. Another scale-backed worm is Polynoë, common on the shore. The Syllids reproduce by budding, to a much greater extent than Naïs.

GROUP OF SERPULÆ.

Of the Tube-dwellers, which have no jaws but feed on minute organisms, drawn into the mouth by the play of the tentacles and gills, the best known are the Serpulæ, whose limy shells on the shells of molluscs, on rocks, stones, and any object that has lain long under water, are familiar to every-one. A Serpula, with its gills extended, is a charming spectacle, which every one who visits the seaside may view for himself by putting some of these tubes into a jar of sea-water with some growing green weed. Before long the branched fans will expand, aërating the blood of the tiny worm and wafting its food towards its mouth. There will be seen a broad trumpet-shaped stopper, which closes the tube when the gills are drawn in, and through which the worm breathes when concealed in its tube.

These little worms have formed a natural breakwater off Pernambuco, which owes its Spanish name Recife, or the Reef, to that fact. They have in vast numbers settled on, and so rendered solid, a sandbar, under the shelter of which the harbour was formed. Darwin says : "Without their protective aid the bar of sandstone would inevitably have been long ago worn away, and without the bar there would have been no harbour."

The tubes of the tiny Spirorbis on the fronds of bladder-wrack are quite as common, and in this worm the stopper serves also as a brood pouch. The tubes of the Shellbinder, made up of grains of sand and tiny pieces of shell, may be readily found in pools, and those of the Honeycomb Worm form large honeycombed masses.

The Lobworm burrows, and is very common on sandy shores, where its castings may be seen as the tide goes out. It forms no tube. It is about 8 inches long, and bears brilliant crimson gills on its central segments. Fishermen dig it out of its burrows, for it is largely used for bait. Closely allied is the Palolo Worm from the coral reefs of the South Seas, valued as a delicacy by the Fijians.

The Gephyrean Worms are not divided into segments; they have a protrusible proboscis. The Siphon Worm, which lives in sand, is British, but is not often seen alive.

LOBWORM.

The Leeches are distinguished by a suctorial mouth, and a sucker at the tail. Most are aquatic, and swim well; they also move by a looping motion, fixing themselves by the mouth, then drawing up the tail and holding by the sucker thereon, while the body is thrust forward, and so on. The Medicinal Leech (Plate XIII., No. 2), though rare in Britain, is common on the Continent; and the usual method of taking them is for the collector to wade into a stream, and allow them to fasten on his legs. It is not so much used for blood-letting as formerly. The Horse Leech is probably so named from its size, for its teeth are not strong enough to penetrate the human skin, much less a horse's hide. Several leeches are parasitic on fish; and some that are common in our ponds carry the young about attached to the under-surface.

The Land Leeches of the tropics live among vegetation near water. Of a species common in Ceylon, Sir E. Tennent says that "they are about 1 inch long, and as fine as a common knitting-needle; but capable of distension till they equal a quill in thickness, and attain a length of nearly 2 inches . . . The bare legs of the palankin-bearers and coolies are a favourite resort; and, their hands being too much engaged to be spared to pull them off, the leeches hang like bunches of grapes round their ankles; and I have seen the blood literally flowing over the edge of a European's shoe from their innumerable bites."

X 2

The Rotifers, or Wheel Animalcules, are so called because the motion of the minute hairs on the disk or lobes at the front end present the appearance of a wheel in rapid motion. The lashing of these minute hairs produces currents in the water which bring food particles to the mouth, and in many forms they serve as swimming organs. The cup-like body, sometimes enclosed in a crystal case or lorica, generally ends in a tail, or "foot," often forked, and functioning as a rudder to guide and as an anchor to moor. All these creatures are microscopic. Some are fixed : the Floscules and the Crown Animalcule secrete tubes, while the Common Melicerta builds a tower-like dwelling of pellets. Lacinularia is a tube-dweller, living in clusters, and Conochilus greatly resembles it, but the colony is free-swimming. Pedalion, a remarkable form, not unlike the larva of a crustacean, has foreshadowings of limbs. If dried slowly, some Rotifers secrete a gelatinous sheath in which they can live for some time, and revive when put into water.

COMMON ROTIFER (*magnified*).

To the Round Worms belong the Hair Worms—formerly, and perhaps in some places still, believed to be vivified horsehairs—the Paste and Vinegar Eels, and some forms parasitic in Man, of which the most dreadful is the Trichina, for the attacks of which there is no known remedy. The larval state of all is passed in a different host from that which they infest when adult. Some attack plants.

The Nemerteans, or Ribbon Worms, are mostly marine, and some may be found under stones near low-tide mark. They are flesh-feeders, and mostly small ; but one, Lineus, is known to attain a length of 12 feet. In Kingsley's

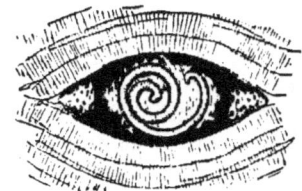

TRICHINA IN MUSCLE (*magnified*).

"Glaucus" is a vivid description of an incident that took place in his aquarium—the capture of a small fish by one of these worms, and how when he had swallowed his prey "the black murderer slowly contracts again into a knotted heap, and lies, like a boa with a stag inside him, motionless and blest." There are, however, few boas large enough to swallow a stag.

Many of the Flat Worms are parasitic. Those that affect us the most are the Flukes and the Tapeworm. The Liver Fluke passes its larval state in a small pond-snail, which when eaten by sheep, passes to its liver and causes "rot." The loss to the flocks of the United Kingdom from this disease in one year has been put at no less than 3,000,000 sheep. The Tapeworm finds its way into man's body from his eating measly pork. The head bears suckers and hooks, and there are a number of joints (Plate XIII., No. 3), each of which contains embryos ; but these will not develop in the human body. Some of the lower animals are infested with other Tapeworms.

The Bryozoa, or Polyzoa, are aquatic animals which—with the exception of Loxosoma, parasitic on the tail of the Siphon Worm—form colonies. The mouth is surrounded by tentacles, arranged in a crescent or circle. The larvæ are free-swimming, and these animals also reproduce by budding, and the freshwater forms give rise to statoblasts, or resting-buds. Each of the colonists lives in a tiny cell, which may be leathery, jelly-like, or limy. Most of the freshwater forms are branching, as are some of the marine, but the latter generally spread over weed, stones, and shells. The Common Sea-mat, often mistaken for a seaweed, may be found abundantly on almost any coast. Along the shores of the Channel it may be gathered by bushels, and though a microscope is necessary to make out the details of the animal, a cheap pocket lens will show the cells pretty clearly. A calculation of how many of these cells are packed into a square inch, and a rough guess at how many square inches lie on half a mile of the beach, will give the observer a better idea of the marvellous abundance of life in the sea—whence all life came—than all the books one could read, or all the lectures one could hear on the subject. It would probably make him think And to think is better than to read.

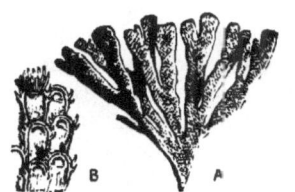

SEA-MAT.
A Natural size.   B Portion magnified.

CHAPTER XXIX

"STINGING ANIMALS" AND SPONGES.

THE creatures called "Stinging Animals" differ greatly in form, but nearly all of them possess stinging-cells, with the power of ejecting therefrom a long poisoned thread by contact with which their prey is paralysed. Nearly all are tubular, as are sea-anemones, and have the mouth fringed with tentacles, or they may be bell or umbrella-like, as are the jelly-fish. Some in early life more or less resemble jelly-fish, and then settle down into forms like the Common Hydra.

In the Comb-bearers, or Ctenophores, the stinging-cells are usually so changed as to have an adhesive power. The sexes are combined, and the young resemble their parents. They live in the open sea, feeding on small living creatures, and moving by means of bands of very fine hairs on the body, which is generally globular. Some of them are called Sea-gooseberries, and of these one is very common on our shores. It resembles a tiny crystal globe gleaming with all the hues of the rainbow; and, like its near relations, is one of the sources of the phosphorescence of the sea. In this form there are two long tentacles, fringed on one side with spirally-twisted filaments, and capable of being drawn back into pouches on each side. Venus's Girdle, a ribbon-like form, a yard long, from the Mediterranean is closely allied. Beroë, larger than the Sea-gooseberry, has no tentacles.

SEA-GOOSEBERRY.

"Coral" is a name used somewhat loosely for any of the Stinging Animals that form hard skeletons, which may be internal and act as a support, as in the Red Coral, or external, and serve as a dwelling, as in the Reef-Builders; and of these two types there are modifications. All are colonial. The young is for a time free-swimming, then settles down and becomes the starting-point of a new colony, which increases by budding in plant-like fashion. In the first group there are eight feathered tentacles.

The Sea-Pen (Plate XIII., No. 18) is a feather-like form, with a central axis, on each side of which the polyps, or hydra-like animals, are borne. There are allied forms from shallow water and the deep sea.

The Fan-Corals are so called from their flat-spreading shape. They live on the floor of the sea; and though the dark skeleton is an unattractive object, must present a charming picture when the polyps come out from the brilliantly-coloured flesh-mass, and expand their tentacles in search of food.

Everybody knows the Red Coral (Plate XIII., No. 17), so highly prized as an ornament. Colonies of it live on the rocks at the bottom, in most parts of the Mediterranean, whence it is dragged by a rude kind of dredging-apparatus worked by a capstan from the deck of the fishing boat. The investing flesh is stripped off, and the Coral—that is, the supporting axis—is sent to market in one of the Italian ports, and sold to the jewellers, who mount it in various ways for ornament.

The Organ-pipe Coral, from tropical seas, is deep red in colour, and consists of a number of tubes, ranged row above row, each separated from that above it by a kind of platform. The greenish tentacles contrast well with the red of the tubes.

The simplest of this group of Corals is the Dead Men's Fingers, or Cow's Paps, in which the skeleton is represented by tiny spicules scattered throughout its substance. It is washed up on our shores, especially after stormy weather, and one may often see it on Scallop-shells, an uninviting flesh-coloured, leathery mass. If placed in a jar of sea-water, the mass will expand, little pits or cells appear on its surface, and from each of these issues forth a living flower—for such it seems in shape—which gradually expanding, till it has attained its full development, reveals itself to be a Hydra fishing for surrounding prey, by means of petal-like tentacles placed around its mouth. Even if the colony be dead it will repay examination, for one may see the star-shaped orifices through which the polyps come forth.

The next group of the Stinging Animals contains the Stone Corals, or Reef-builders, and the Sea-Anemones, between which there is very close relationship. If we could take a slice out of a Sea-Anemone and another out of some of the Stone Corals, and place them side by side, we should see how closely the hard skeleton of the one follows the lines of structure of the other. The "Corals" of this group are the shell, or skeleton secreted by and supporting a number of small anemone-like animals, called the polyps. Nearly all form colonies.

The branching Madrepores, in which there is a separate cup for each polyp, are among the most active of the reef-builders. Their skeletons are common objects in our museums, and look like branching shrubs turned to snow-white stone; but when living they are of a bright olive brown, with growing points and polyps of emerald-green or violet. The Stone Corals form large solid masses, and the orifices whence the polyps protrude their tentacles resemble those of the Dead Men's Fingers. The Brain Coral is so called because the skeleton is marked with ridges like the winding curves in the human brain. The Mushroom Coral, the skeleton of which is not unlike a gigantic mushroom resting on its top and deprived of its stalk, is a solitary form. All these forms are tropical.

The Small Cup Coral—miscalled a Madrepore—is to be found on the shores of Devon and Cornwall. It is a capital object for the aquarium, for as Gosse says, "scarcely any species is more hardy, more beautiful, or more changeable in its aspects. It is free in expanding in

captivity . . . and especially at night the animal expands to the full, and rears its lovely form far above the level of its stony walls. This condition may, however, at any time be induced by a proffer of food : an atom of raw flesh cautiously laid on the half-exposed disk is a great temptation, too great to be resisted."

It is usual to class all the Corals as flesh-feeders ; and in the main they doubtless are so ; but, like creatures higher in the scale of life, if they cannot get what they like best, they get what they can. Dr. Hickson says that there is some reason for believing

SMALL CUP CORAL.

that many of them may be, partially at least, vegetable feeders, for in the neighbourhood of mangrove swamps, whence much vegetable matter must be carried to the sea, the coral banks are often much more vigorous than those near steeper shores.

The Sea-Anemones have a short, column-like body, fixed by a broad base, and the disk at the top, in which is the slit-like mouth, is surrounded with tentacles. The Common Sea-Anemone (Plate XIII., No. 16) is abundant on all our coasts between tide-marks. It is an extremely hardy creature, and in the aquarium will thrive under conditions that would be fatal to most, if not all, of the other inmates. When any prey is stopped by one of the tentacles the others bend over and fasten on to the victim, gradually enfolding and bringing it nearer and nearer to the mouth, and holding it there until it is engulfed. This species varies greatly in colour, crimson and green being most common—

mixtures of these two often occur—and they are sometimes striped and spotted. Gosse called this Anemone the Beadlet, from the row of blue bead-like bodies generally, but not always, present at the bases of the outermost tentacles. These were supposed to be sense-organs, but are really batteries of stinging cells. Our coasts and the seas near shore are fairly rich in Anemones, of which we can only mention the Plumose Anemone—the noblest of them all—from a depth of about 20 fathoms; the Daisy, that lives embedded in crannies of the rock, and only to be obtained with much patience; the Parasite Anemone, that lives on the house of the Hermit Crab; and the Gem, which is perhaps as beautiful as the Plumose Anemone.

The Common Medusa, or Jelly Fish (Plate XIII., No. 15), is very abundant round our shores in summer and autumn, and numbers of them may be seen stranded on the shore as the tide goes out. It swims by alternate contractions and expansions of the umbrella, and its stinging cells are sufficiently powerful to inflict considerable pain on the unwary bather. Mr. George Milner, in his "Studies of Nature on the Coast of Arran," says:—" Having been once stung myself, I know the sensation, and remember it well, though it is more than thirty years ago. I shudder as I recall the . . . instantaneous injection of the poison, the slimy touch as I caught the creature in my hand, and flung it over the sea; and then the rubbing of my skin with dry salt, and

PLUMOSE ANEMONE.

the incessant pacing up and down my chamber for seven long hours. Only in that way could the virus be got rid of. In my case, to sit down for a moment was to sleep, and then to fall into great torture." But all persons are not affected by them to this extent. This creature begins life as a tiny sexless free-swimming animal, which settles down and becomes a polyp, from which the free-swimming Jelly Fish is budded off. Some Jelly Fish are free-swimming throughout life.

Some Medusas are found in fresh water. The best known is Limnocodium, which made its appearance some years ago in the Victoria Lily tank at Kew, and at the Botanic Gardens. Its development resembles that of the Common Medusa. One occurs in Lake Tanganyika, in Africa, and another in a lake in Persia.

The Portuguese Man-of-War is a free-swimming colonial form, from tropical seas, and has been compared to "a child's mimic boat shining in all the gaudy painting in which it left the toy-shop." It consists of a bladder-like float, on the under-side of which are the polyps. Some of these procure food for the colony, while others serve for reproduction, and there are long tentacles armed with stinging cells. The poison of this creature is very virulent; and Mr. Saville Kent, says of an allied form recently discovered near the Great Barrier Reef, that it can sting a man to death.

The Hydroids constitute the group that were formerly called Zoophytes. At one time they were considered to be plants—they are really animals of plant-like growth, though one may see at Hastings and Ramsgate, and a hundred other places, the branching dwellings of these creatures exposed for sale, and boldly ticketed "Seaweed." Of the Hydroids there are two main divisions; in one the polyp is enclosed in a cup, which is a continuation of the covering of the colony, and in the other the polyp is free. Some Hydroids in each group pass through a medusa-stage, and swim about for a time.

The Plumularians and Sertularians (or Sea-Firs) are branched. In the former the polyp-cups

PORTUGUESE MAN-OF-WAR. are on the upper side of the branches; in the latter the branches are set with these cups or cells above and below. The Campanularians are so called from the beautiful crystal bell-shaped cups that enclose the polyps.

Of the Hydroids in which the polyp has no protective cup, the most beautiful is Tubularia, the Oaten Straw Coralline, masses of which may be picked up on almost any sea-beach. The living colony looks like a number of straws closely packed together, and from the top of each protrudes a lovely crimson flower-like polyp. Coryne is the most abundant British form : and there is scarcely a place that has a rock-pool, or a shore on which weed is thrown up by the sea, where half-an-hour's diligent search would

CLAVA.

be fruitless.   Clava is almost as common.   Cordylophora, closely allied to the Common Hydra, is the only fresh-water branched form.

The Common Hydra is such an interesting little creature, and so easily procurable in almost any numbers, that one should examine it for oneself. It is sufficient to procure from a pond a gathering of duck-weed, and put into a vessel of water, which should be set in the sunlight. Before long there will be seen, hanging from the stems and under-sides of the leaves, some small creatures from ¼ inch to ¾ inch in length, green, or brown, or orange of hue, and having the mouth fringed with long tentacles. If, as will pro- bably be the case, some small crustaceans have been brought home with the weed, the Hydras will soon show how the tentacles are used to capture prey. These are the lowest of the Stinging Animals, and are inter- esting from the fact that they may be cut in two, lengthways and across, and each part will develop into a new animal. It was at one time believed that if the Hydra was turned inside out, the outside would do duty as a stomach, while the stomach would take over the duties of the outside and act as a skin. But this belief was founded on a mistake. It is not very difficult, if one has a steady hand, to turn a Hydra inside out; but the creature will soon die unless it can turn back again, and this it generally succeeds in doing.

GREEN HYDRA.
A. Natural size.
B. Magnified.

SPONGES.

Of this group, most people know but the horny skeleton of the Bath Sponge; yet British Sponges are exceedingly numerous, and may be found covering rocks, boring into shells, or lying on the beach washed up from their home in the sea. Some of them are brightly coloured. Mr. Hornell calls the Jersey shore, between tide marks, "Spongeland," and tells us how scarlet, brick-red, orange, yellowish-green, yellow, white, grey, and black patches relieve the sombre brown of the weed-covered rocks. Even in Britain we can show something like this, and sponge-hunting is a pleasant occupation for a seaside holiday.

Professor Huxley compared a sponge to a city under the water, where the people are arranged about the streets and roads in such a manner that each can easily appropriate his food from the water, as it passes along. If we examine the Bath Sponge, we shall see very many small openings and some large ones; and if we cut it through, we shall discover that the small holes are the ends of tubes that lead to cavities in the interior. In life the skeleton is covered with jelly-like flesh, and the tubes and cavities are lined with cells armed with whip-like lashes. The motion of these lashes draws in at the pores water bearing food - particles, and washes out waste products at the larger holes, or oscules, as shown by the course of the arrows in the diagram; and thus the people in the city—that is, the separate cells —have their food brought to their doors. In the Bath Sponges the

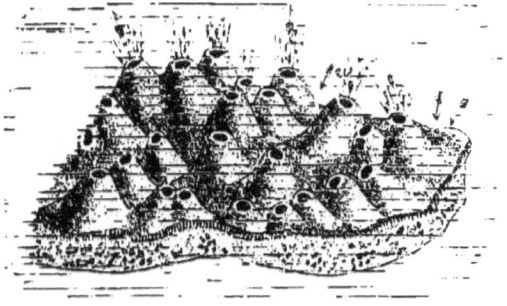

framework is horny; and others are distinguished as Flinty or Limy, from the nature of the spicules in the sponge flesh.

Sponges multiply by eggs, and by budding, and the plant-like mode of increase is taken advantage of, cuttings being successfully bedded out. The best

LIFE OF A SPONGE.

Bath Sponges are brought from the Mediterranean, and the West Indies produce an inferior kind. Sponge-fishing in the Mediterranean is carried on by divers from small half-decked boats, and each diver will go down from fifteen to twenty times a day.

Some of the Flinty Sponges are extremely beautiful, and the skeleton of one known as Venus's Flower-Basket is often mounted under a shade as an ornament. At one time these were very rare, and consequently fetched very high prices; but in 1875, when the *Challenger* visited Cebu, baskets-full were brought off to the ship, and the sponges were sold at twopence apiece. The skeleton of the Japanese Glass Rope Sponge is also a charming object. A near ally is found off the coast of Portugal, where the fishermen took it to be the nest of some marine animal.

We have on our shores many of the Flinty Sponges, of which the largest is the Mermaid's Gloves, a branching, plant-like form that is often thrown up by the sea. In its fibrous skeleton it approaches the Bath Sponges, but scattered through this are needle-like spicules.

VENUS'S FLOWER BASKET.

The Crumb-of-Bread Sponge, ranging from greyish-yellow to various shades of green, covers the surfaces of rocks between tide-marks.

Recently, on a piece of rock in one of the tanks at the Brighton Aquarium, there was a large patch of this Sponge. The young, free-swimming embryo had, doubtless, been introduced with the sea-water, and had settled down. The constant introduction of water from the sea supplied plenty of food, and the colony consequently throve.

Clione, the Boring Sponge, lives in oyster-shells, and burrows into limestone. Suberites covers the whelk-shell, and eats away its substance, sometimes to the discomfort of the hermit-crab who may tenant it. But the hermit-crab has been known to turn the destruction of his shelly house by the Sponge to his own profit. For he has declined to shift, and hollowed out a dwelling in the Sponge itself. The Common Freshwater Sponges, which may be spreading or plant-like in form, have flinty spicules.

The Limy Sponges, generally of cup-like shapes, are scattered all over the world, ranging from between tide-marks to a depth of about 400 fathoms. Several of them may be found in our own rock-pools, where they live attached to the fronds of seaweed.

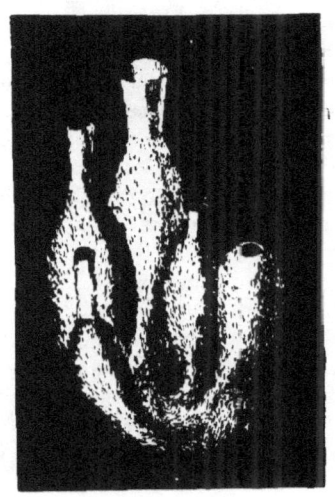

COMMON BRITISH LIMY SPONGE.

CHAPTER XXX.

THE OLDEST AND SIMPLEST ANIMALS.

THE vast majority of the creatures of this grade are exceedingly minute, and a large number of those that can be discerned by the unassisted eye, look like moving points in the water. At one time they were all called Infusory Animalcules, because those first known and studied were discovered in infusions of vegetable substances. But they have other and far different dwelling-places. Some are found in the sea, others in sand and on weed, some in moss and damp earth, and a few are parasitic in higher animals. Most of them consist of a single cell, and the lowest resemble a minute speck of jelly. Some

consist of a mass of cells, all doing the same work; these show how the gulf that separates the Simplest from the Many-celled Animals (which range from the Sponges up to Man) may have been bridged. But all animals of the higher grade *begin* life as simple cells.

EUGLENA
(*magnified*).

Of the Simplest Animals we need notice only the Infusorians which are furnished with cilia or with whip-lash processes, and those that resemble Amœba in many respects; in these last the protoplasm, or living matter, flows out into finger- or thread-like processes. Both are found in ponds and streams, and the latter occur abundantly in the sea.

Euglena is a tiny speck of a bright green colour, having at its front end a lash-like process, or flagellum, by means of which it swims. One may see how small it is by remembering that 350 of them, placed end to end, would cover a space as long as our picture. It is so common in many ponds as to tinge the water with a greenish hue. If we imagine a number of creatures somewhat like these united by a common jelly, we can see how forms like Volvox may have arisen.

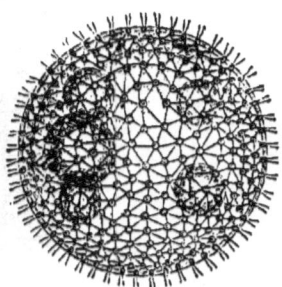

VOLVOX.

Such a form may be represented thus and if this were pushed in, as one would push in a hollow indiarubber ball with one's finger, the result would be somewhat like this 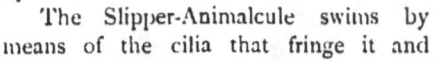 which is not far from the shape of the simplest sponge. To see the beauty of Volvox one needs a microscope. Prof. Rymer Jones describes it " as a microscopic globe,

rotating on its own axis—a tiny world rolling majestically through the little drop of water that forms its space." And more than fifty years ago he discovered that " every speck that dots its surface is a perfectly formed animal," and knew that the motion of the colony was due to the lashing of the two flagella possessed by each colonist. The diameter of the Volvox in our picture is about equal to the space occupied by seventy-five of the living animals placed side by side.

BELL ANIMALCULES (*magnified*).

The Slipper-Animalcule swims by means of the cilia that fringe it and drive food-particles into its "mouth"—a mere opening in the body mass.

We may compare the Bell-Animalcules to tiny cups, set on stalks ; and these stalks may contract and expand, or be straight and rigid. These are free-swimming in their youth, and when they settle down only the edge of the cup is fringed with cilia, which procure food.

The Stentors or Trumpet Animalcules are not unlike a post-horn in shape, but as they are generally somewhat curved, the cornucopia, or "horn of plenty," may afford a better comparison. These little

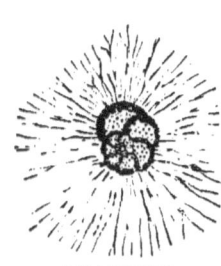

FORAMINIFER.

creatures seem to have gone some way towards forming colonies, for though but one Stentor is generally found in the sheath it secretes, occasionally a group will be met with, all having their bases fixed in one jelly-like mass, into which they shrink down when disturbed.

The Marine Night-light is one of the sources of the phosphorescence of the sea. It occurs in countless myriads round the British coast, and to these tiny creatures is due much of the sparkling that one sees when the tide breaks on the shore on a calm summer evening.

Of the creatures that resemble Amœba, the Sun Animalcules may be mentioned first. They are minute specks of jelly, set round more or less regularly with spine-like processes, with which the food is drawn into the body. About 400 of these little creatures would, if ranged in a line, measure an inch. They are plentiful in ponds, and are found also in the sea.

The Foraminifera, most of which live in the sea, have a shell or test, in which, besides an opening—like the mouth of a univalve shell —there may be numerous openings all over the surface. Through these the protoplasm streams in thread-like masses in search of food. The shell may be of lime, or flint, or grains of sand may be gathered. In one case sponge-spicules are used, and for a long time the builder was supposed to be a sponge. The Radiolarians are marine, and the flinty or horn-like skeleton, of beautiful geometric forms, is embedded in the living matter, which spreads out on all sides.

Still simpler than the Foraminifera is Amœba. It has no shell, and the processes are thick and finger-like, and it can only be described

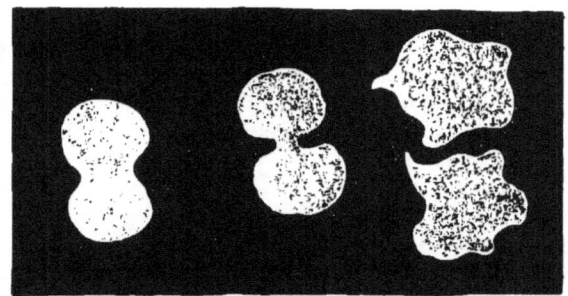

AMŒBA DIVIDING.

as a tiny blob of living matter without definite shape. It was formerly said to be without structure; but the microscope shows structure in the shape of a network of living matter. These creatures feed by flowing round some particle of animal or vegetable matter, the nutritive parts of which are absorbed in the body-mass, which flows away from the indigestible parts.

All these simple animals increase by dividing into two, so that birth and death may be said to be unknown among them. The late Prof. Marshall Milnes put the case very clearly, and our picture will illustrate his remarks. Tom is the Amœba on the left-hand: he gets a good food-supply, and grows till he is too large for a single individual, and contracts in the middle. The contraction goes on with the result shown on the right-hand, and there are two Amœbæ (Dick and Harry) where but one was before. "Tom has disappeared without having died, while Dick and Harry have come into existence without having been born. Where is Tom at the end of the process?"

INDEX.

The asterisk (*) denotes an illustration in the text.

www.ingramcontent.com/pod-product-compliance
Lightning Source LLC
Chambersburg PA
CBHW021321110726
47900CB00005B/1301